Mycotoxins and phycotoxins

IUPAC

AOAC INTERNATIONAL

Mycotoxins and phycotoxins

Advances in determination, toxicology and exposure management

Proceedings of the XIth International IUPAC Symposium on Mycotoxins and Phycotoxins, May 17-21, 2004, Bethesda, Maryland, USA

edited by:
Henry Njapau
Socrates Trujillo
Hans P. van Egmond
Douglas L. Park

Wageningen Academic
Publishers

ISBN-10: 90-8686-007-9
ISBN-13: 978-90-8686-007-4

Cover photos
Aspergillus flavus
Rob Samson, Fungal Biodiversity Centre (CBS), The Netherlands
Pyrodinium bahamense
Ben Tall, Sherwood Hall and Jennifer Rios, Food and Drug Administration, USA

First published, 2006

Wageningen Academic Publishers
The Netherlands, 2006

Preface

Over the last several decades research has highlighted the potential deleterious effects to the health of man and animals some mycotoxins and phycotoxins may cause if consumed in quantities sufficient to elicit harm. The hazards may be more significant in economically challenged regions of the world where the means to detect and segregate contaminated commodities, and alternatives to consuming food that could be adulterated, may be lacking.

Although recorded history is replete with evidence of mycotoxin-associated food poisoning incidents or outbreaks such as those arising from ergot poisoning during the middle ages in Europe and food spoilage and shortages during World War II in Russia and Japan, the aflatoxicosis outbreak affecting ducklings and Turkey poults that occurred in England in the 1960s, commonly called Turkey-X disease, brought to the forefront the potential for mycotoxin occurrence in animal feed with no apparent visual evidence of fungal contamination. In the Turkey-X incident, a feed ingredient, peanut meal, produced in Brazil and subsequently included in the feed rations in England, was shown to be the source of the toxic agent, later identified as aflatoxin, that had been naturally produced by *Aspergillus* species. This aflatoxicosis incident demonstrated that there could be significant animal and/or human health problems from consuming products with no apparent visual presence of moulds. This fact prompted the scientific and regulatory communities to conduct in-depth research on factors leading to mycotoxin contamination of foods and feeds including the presence of other non-aflatoxin mycotoxins.

Similarly, phycotoxin-associated toxic events could have occurred in history but may have been associated with other causes. Although highly disputed by others, some scholars believe that the biblical story in Exodus 7: 20-21, where Moses turned Egyptian waters into blood could have been the phenomenon called red tides. The significance of red tides and harmful algal blooms (HAB) has become apparent with time. Currently incidents of human poisoning due to the consumption of phycotoxin-tainted seafood have been reported worldwide and the events are often named after the causative agent, i.e. ciguatera fish poisoning (CFP), amnesic shellfish poisoning (ASP), diarrheic shellfish poisoning (DSP), neurotoxic shellfish poisoning (NSP) or paralytic shellfish poisoning (PSP). The importance of these hazards to local communities and globally is perhaps best illustrated by the increase in regulations targeted at minimizing exposure to the phycotoxins. In addition to these marine biotoxins, toxins from blue-green algae, found in fresh and brackish waters are a hazard to health and therefore need continuous attention.

The XIth International IUPAC Symposium on Mycotoxins and Phycotoxins was part of an ongoing series of symposia initiated by the International Union of Pure and Applied Chemistry (IUPAC) Division of Chemistry and the Environment, Subcommittee on Food Chemistry (formerly Food Chemistry Commission); the first, covering only mycotoxins, was held in Kungalv, Sweden in 1972. Since then nine symposia have been held in Pulawy, Poland; Paris, France; Lausanne, Switzerland; Vienna, Austria; Pretoria, South Africa; Tokyo, Japan; Mexico City, Mexico; Rome, Italy and Guarujá, Brazil. The symposia series has grown in scope, scientific novelty and value. It has become the principal international interdisciplinary meeting on mycotoxins and phycotoxins focusing initially on analytical methodology necessary to detect and control contamination in human food (and animal feed). Later symposia topics included the toxicological potentials of the mycotoxins and phycotoxins, occurrence and management of the risks they pose to the consumer globally. The XIth symposium assessed progress or advances made in relation to the overall focus of the series and the evolving global perspective of food safety.

The meeting attracted about 240 participants from 40 countries underscoring the international character of the symposium series. A significant number of attendees were from developing countries; many participating under the auspices of the U.S. Food and Drug Administration (FDA)-led international outreach program on mycotoxins. The outreach program has, since its inception in 2002, also received substantial financial and material support from the U.S. Department of Agriculture, Foreign Agricultural Service (USDA/FAS); the Joint Institute for Food Safety and Applied Nutrition (FDA/ University of Maryland); the United Nations' Food and Agriculture Organization (FAO) and World Health Organization (WHO), and several US-based food and agribusiness industries. In addition to participating in the main symposia, the mycotoxins outreach program held a satellite session at which 8 country reports were presented.

The quality and novelty of scientific research and discovery that was shared at the XIth Symposium on Mycotoxins and Phycotoxins demonstrated that the scientific community is ready to meet the evolving challange of ensuring global food safety. This book is not a compendium of all the 225 oral and poster presentations made at the sympsoium: it contains only selected peer-reviewed papers. Submitted manuscripts were reviewed by experts in the specific topic areas and the editors selected recommended manuscripts that exceeded set quality standards in order to uphold the established pre-eminence of the symposia series. Nevertheless, the contents of this book encompass the current state of understanding of mycotoxins and phycotoxins, critical ecological parameters and interactions, advances in methodology, toxicology, and strategic management interventions that could minimize the negative effects of mycotoxins and phycotoxins on the well being of consumers. The symposia series has been, and will continue to be, of prime importance to public health managers, educators and researchers worldwide. Preparations for the XIIth International IUPAC Symposium on Mycotoxins and Phycotoxins are progressing well. This conference will be held in Istanbul, Turkey, May 21-25, 2007.

The editors

Henry Njapau
Centre for Food Safety and Applied Nutrition
Food and Drug Administration, College Park,
Maryland, USA
Email: henry.njapau@fda.hhs.gov

Socrates Trujillo
Centre for Food Safety and Applied Nutrition
Food and Drug Administration, College Park,
Maryland, USA
Email: socrates.trujillo@fda.hhs.gov

Hans P. van Egmond
Laboratory for Food and Residue Analysis
National Institute for Public Health and the Environment,
Bilthoven, The Netherlands
Email: hp.van.egmond@rivm.nl

Douglas L. Park
Centre for Food Safety and Applied Nutrition
Food and Drug Administration, College Park,
Maryland, USA
Current Address: 20 Gettysburg North, Cabot,
Arkansas, USA
Email: dlpark40@earthlink.net

Acknowledgement

The XIth International IUPAC Symposium on Mycotoxins and Phycotoxins was organized by committees headed by Dr. Douglas L. Park, Director of the Division of Natural Products, U.S. Food and Drug Administration (FDA), College Park, Maryland in collaboration with the FDA-led international outreach program on mycotoxins. The compositions of the organizing and scientific committees are listed below.

On behalf of the organizing and scientific committees, the editors wish to express their gratitude to invited speakers, oral and poster presenters and all symposium participants. Particular appreciation is extended to the keynote speaker, Mr. Ezzeddine Boutrif, Senior Food Standards Officer, Food Quality and Standards Service, Food and Agriculture Organization of the United Nations. They are also indebted to the many reviewers who assisted in the selection of quality manuscripts.

Significant support for the international mycotoxins outreach program and international participants, hence the symposium as a whole, was provided by the U.S. Food and Drug Administration, the U.S. Department of Agriculture Foreign Agricultural Service and the Food and Agriculture Organization of the United Nations. The symposium was also made possible by the financial and material support of the AOAC INTERNATIONAL, Gaithersburg, Maryland, USA; VICAM® LP, Watertown, Massachusetts, USA; Romer Labs Inc., Union, Missouri, USA; Agilent Technologies, Palo Alto, California, USA; Neogen Corporation, Lansing, Michigan, USA; R-Biopharm, Darmstadt, Germany and the Almond Board of California/ International Treenut Council, Modesto, California, USA.

The organizers wish to thank the AOAC INTERNATIONAL's Director, Dr. E. James Bradford and his staff for the effective management of the symposium and the symposium hosts, the Natcher Center and the National Institutes of Health, Bethesda, Maryland, USA, for their contributions to the success of the symposium. Gratitude is also extended to the secretariat of the IUPAC and the IUPAC Subcommittee on Food Chemistry for affording the organizers the opportunity to put together an insightful symposium.

Organizing Committee

Douglas L. Park	Chair, Food and Drug Administration, College Park, MD, USA
Elke Anklam	European Commission Directorate-General Joint Research Centre, Geel, Belgium
Magda Carvajal	National Autonomous University of Mexico (UNAM), Mexico D.F., Mexico
John Gilbert	Central Science Laboratory, Sand Hutton, York, United Kingdom
Marina Miraglia	Istituto Superiore di Sanità, Rome, Italy
Samuel W. Page	IPCS/World Health Organization, Geneva, Switzerland
Albert E. Pohland	AOAC INTERNATIONAL, Gaithersburg, MD, USA
Myrna Sabino	Instituto Adolfo Lutz, São Paulo, SP, Brazil

Scientific Committee

Douglas L. Park	Chair, Food and Drug Administration, College Park, MD, USA
Jean-Marc Fremy	Agence Française de Sécurité et Sanitaire des Aliments, Maisons-Alfort, France
Ana Gago-Martinez	Universidad de Vigo, Vigo, Spain
John D. Groopman	The Johns Hopkins University, Baltimore, MD, USA
Leonardo Guzmán	Instituto de Fomento Pesquero, Puerto Mont, Chile
Sherwood Hall	Food and Drug Administration, Laurel, MD, USA
Patrick Holland	Cawthron Institute, Nelson, New Zealand
Paul C. Howard	Food and Drug Administration, Jefferson, AR, USA
Russell J. Molyneux	United States Department of Agriculture, Albany, CA, USA
Henry Njapau	Food and Drug Administration, College Park, MD, USA
Mark A. Poli	U.S. Army Medical Research Institute of Infectious Diseases, Fort Detrick, Maryland, USA
Michael A. Quilliam	National Research Council, Halifax, Nova Scotia, Canada
Vildes Maria Scussel	Federal University of Santa Catarina, Florianopolis, SC, Brazil
Gordon Shephard	Medical Research Council, Tygerberg, South Africa
Gerrit J.A. Speijers	National Institute for Public Health and the Environment (RIVM), Bilthoven, The Netherlands
Joerg Stroka	European Commission Directorate-General Joint Research Centre, Geel, Belgium
Socrates Trujillo	Food and Drug Administration, College Park, MD, USA
Hans P. van Egmond	National Institute for Public Health and the Environment (RIVM), Bilthoven, The Netherlands
Thomas B. Whitaker	United States Department of Agriculture, North Carolina State University, Raleigh, NC, USA

Table of contents

Section IV: Toxicology of mycotoxins

Section V: Occurrence of mycotoxins

Section VI: Management of mycotoxins

Section I:

Overview of mycotoxins

Mycotoxins and mycotoxicoses: A 2004 update

J.L. Richard

Romer Labs, Inc., 1301 Stylemaster Dr., Union, MO 63084, USA; johnr@romerlabs.com

Abstract

A compilation of some current activities in the area of mycotoxins and mycotoxicoses is presented. Important aspects of current research in pre-harvest strategies for the control of mycotoxins in crops include development of resistant crops using proteomics and genomic approaches for the fungal pathogen. Transgenic plants providing reduced levels or elimination of mycotoxins are some possibilities being investigated. The use of bio-control agents is providing success in maize, cottonseed and peanuts. Further understanding of the relationships of certain mycotoxins to human and animal disease is being elucidated and continues to be a need within the medical/veterinary communities. The conceptual development and implementation of a HACCP program for the control of mycotoxins from "Farm to Fork" is an important facet of some international agencies. Standardised international regulations based on sound risks for mycotoxins are a continuing need to equilibrate trade among nations. Development of rapid methods for analysis using new, currently available technologies will increase the usefulness of such tests worldwide. Litigation costs involving mycotoxins within certain countries have increased significantly because the lack of exposure data in animals allows for cases to continue into the courtroom. The true economics of the mycotoxin problem are not understood and likely is far more significant than we currently estimate. The knowledge base accrued in developed countries must be used in advancing the education of those in developing countries to reduce mycotoxin occurrence in commodities used in traditional foods, beverages and medicines. Finally, with most mycotoxin problems, the consideration for exposure involves ingestion or the oral/alimentary route of exposure. However, there is now sufficient information to consider respiratory exposure to mycotoxins to be an area of concern for human and animal health and possibly of significance to a safe food supply of animal origin. Information as an indication of the importance of this kind of exposure involves aflatoxins, *Stachybotrys* toxins and ochratoxin.

Keywords: mycotoxins, mycotoxicoses, bio-control, control, HACCP, litigation, testing, economics, respiratory exposure

Introduction

The investigational area of mycotoxins and mycotoxicoses is receiving a considerable amount of activity with the overall thrust of preventing mycotoxin occurrence and, because this is currently not achieved, to prevent them from contaminating our food/feed products and causing mycotoxicoses. An overview of some major current activities regarding mycotoxins and mycotoxicoses is presented. If mention is made of an investigational area in which the reader is currently working, then it may mean agreement with the selection of the area as a current "hot issue". However, bias may appear in the selection of topics discussed here. The selection was based primarily on those issues that are receiving considerable funding and those where a considerable need for additional information was evident. This overview attempts to only highlight the specific areas of activity and not present an overall summary of each activity.

Controlling the occurrence of mycotoxins in crops

While the order of presentation of topics is not meaningful insofar as importance, however, this topic would be at or near the top of the list of important areas of investigation for mycotoxins.

The development of resistant crops is receiving considerable funding from the United States Department of Agriculture and selected commodity groups in the United States, and this area is being investigated in a few other countries as well. In this endeavour we have moved from the green revolution to the gene revolution whereby the genetics controlling resistance in crops is being investigated through both proteomic and transgenic approaches. However, this activity is also being approached from the pathogen or fungal side of the issue whereby the genomics of the fungus is being investigated with development of the understanding of the biosynthetic pathway for mycotoxin biosynthesis. To be able to control the biosynthesis of a given mycotoxin could be crucial in the elimination of their occurrence.

The use of bio-control measures whereby selected non-toxigenic strains of fungi are applied to crops allowing for the non-toxigenic strain to compete in the field with the "wild type" toxigenic strains is another option being actively pursued. Currently, this approach has been applied to the control of aflatoxin in cottonseed, peanuts and maize. Cooperative efforts of the USDA and the Arizona Cotton Research and Protection Council have resulted in use of this bio-control of aflatoxin in cottonseed. Other companies are involved in utilisation of non-toxigenic strains of non-toxigenic fungi for bio-control of aflatoxin as well.

Relationship of mycotoxins to disease

Considerable effort is currently being applied to the study of selected mycotoxins in human disease. Epidemiologic investigations on the relationship of aflatoxin to human liver carcinoma are currently ongoing in several countries. The potential involvement of the fumonisins in human esophageal cancer and neural tube defects is of interest to several investigational groups. Similarly, the involvement of zearalenone in thelarche and human breast cancer has received considerable attention among several investigative teams throughout the world. The etiology of ochratoxin in Balkan Endemic Nephropathy has been a question of concern for numerous epidemiologic investigations and continues to be suspect but not proven.

The involvement of mycotoxins in animal disease has been known for a long time but there remain numerous problems both in the diagnostic arena as well as actual cause effect relationships. While mouldy silage has been incriminated in animal disease the actual mycotoxin(s) remain elusive. This may be because multiple toxins are involved or the amounts found in silage are of low concentrations and the observed disease cannot be explained by these small amounts of mycotoxins present in the examined samples. Also, many of the observed effects such as loss of production or unthriftiness are non-specific and could be the result of one or more of numerous toxic compounds. Current efforts are aimed at determining the cause of these signs of animal disease by assessing the occurrence of biomarkers as an aid to more specific causes of the malady being investigated.

The HACCP approach to controlling mycotoxins

Several international organisations such as the World Health Organisation (WHO) and the Food and Agriculture Organisation (FAO) of the United Nations are involved in identifying mycotoxin control points through a "Farm to Table" approach. This approach includes the gathering of food from the sea, as well. Because of the extremely complex variety of processes in the manufacture of food and/or feed that occur after the commodities leave the farm and before the food reaches the table, this is a large and difficult task. Identification of the critical control points is always an ongoing process in the manufacture of food/feed. Also, in this time of modern agricultural practices, all or almost all of the produced grains leave the farm and often come back to the farm as bulk or processed feed. Thus, farmers recognise that they cannot send contaminated grain to market as it may come back to them as contaminated feed and affect the production or effect losses of their livestock.

Standardised international regulations

Currently, worldwide regulations exist for mycotoxins with approximately 100 or more countries having such regulations. However, these regulations must be examined by international scientific bodies and must be based on sound exposure data and risk assessments that are agreeable to the examining bodies. For some developing countries currently trying to become involved in international trade, the regulations pose a trade barrier as well as a health risk when, in an effort to try to meet strict regulations of importing countries they ship their highest quality grain and retain the contaminated lots for consumption by their own populace. The harmonisation of regulations from country to country is needed to equilibrate trade and effect a better quality of trade negotiations, unbiased by non-scientific approaches that are often exacerbated by the popular press and other media.

Development of rapid analytical methods

New technologies are rapidly becoming available such as the use of biosensors, array technology, molecular imprinted proteins, novel immunology, and rapid, one pass cleanup columns. Because of the demurrage being assessed by shipping companies to grain processors and manufacturers of food and/or feed, the requirement for development of rapid tests is highly economically motivated. It is also important to maintain a rapid flow of commodities in the marketing channels. Currently there are a number of companies and research organisations involved in the development of test kits and technologies for measuring mycotoxins in multiple matrices and the simultaneous analysis of multiple mycotoxins in a single matrix. It is important in this endeavour that the tests maintain adequate accuracy, precision and ruggedness, as well as other qualities for testing for mycotoxins in complex matrices.

Assembling of animal exposure data

Because funding is lacking for much needed research on exposure of animals to mycotoxins, there is a lack of sufficient data required to keep litigation out of the courtroom; thus litigation costs have skyrocketed, especially in the United States, and cases are being won on rather frivolous claims. Currently, gathering of animal exposure data is being accomplished by many law firms involved in such litigation. More exposure data from both natural and experimental intoxications of animals by mycotoxins is needed especially for long term, low level studies in livestock. Additionally, this data can be used in improving risk assessments due to mycotoxins.

True economics of the mycotoxin problem

A team of investigators from the United States Food and Drug Administration (FDA), attempted an economic assessment model to estimate losses in the United States due to three selected mycotoxins (CAST, 2003). The mean economic annual losses in crops from aflatoxins, fumonisins, and deoxynivalenol were estimated at $932 million. The true economic costs of mycotoxins are impossible to determine accurately because sufficient information for other mycotoxins is not available to determine crop, livestock and human losses. The assessment of such losses from other countries as well as the United States must be undertaken to be able to understand the economic impact of mycotoxins on a worldwide basis.

Transfer of knowledge base to developing countries

To reduce or eliminate mycotoxin occurrence in commodities used in the preparation of traditional food, beverages, and medicines in developing countries, the knowledge base accrued for mycotoxins must be transferred. This would include knowledge of adequate sampling and testing for mycotoxins in the various matrices and developing an adequate cadre of personnel in these countries to serve as resource persons and trainers within the specific countries. This was the basis of the International Workshop on Mycotoxins held at the FDA in Maryland in July of 2002. A follow up of this workshop with representatives from numerous developing countries was an addendum to this IUPAC meeting in Bethesda, Maryland in 2004. Furthermore, the Donald Danforth Plant Science Center in St. Louis, MO along with the United States Department of Agriculture, the Rockefeller Foundation and the USAID program, held an International Workshop on Mycotoxins in November of 2003 with similar objectives. Cooperative efforts among agencies involved in the transfer of this knowledge base are needed for maximum efficacy.

Respiratory exposure to mycotoxins

We often consider that exposure to mycotoxins is through the gastrointestinal route but considerations of exposure by both contact and respiratory route are necessary as well. Interest in the latter area has surged because of numerous reports in the popular press concerning the dreaded "black mould" that is found in indoor habitats that have been dampened because of a number of reasons. However, this is a problem area in which more solid information is needed to confirm many people's suspicion that a mould toxin is causing their illness. Nevertheless, this aspect of certain mycotoxicoses may be more important than we currently regard. In the brief comments herein, the known or potential aspects of respiratory exposure of humans and other animals to only aflatoxins, *Stachybotrys* toxins, and ochratoxins is addressed.

Aflatoxins

When looking back at the proceedings from past IUPAC symposia, it was noticed that Dvorackova and Kralove (1979) from Czechoslovakia (CSSR) had reported at the IVth International IUPAC Symposium, held in Lausanne (1979), on the relationship between aspergilloma and pulmonary carcinoma. They reported the finding of aflatoxin B_1 in lung tissue from an agriculture worker, who had developed an aspergilloma, and who was likely exposed through the respiratory route. Dvorackova (1976) reported on another incidence whereby aflatoxin inhalation resulted in alveolar cell carcinoma in an engineer who was sterilising Brazilian peanut meal. His lung contained aflatoxin B_1. Dvorackova and Pichova (1986) described the occurrence of aflatoxin B_1 in lung tissue of two agricultural workers and one textile worker suffering from pulmonary interstitial fibrosis, again likely of respiratory exposure.

Epidemiologic studies have shown that workers in peanut and linseed oil processing plants exposed to the dust from press residues had a 3 times higher incidence of liver cancer than that of a matched control group (Van Nieuwenhuize *et al.*, 1973; Hayes *et al.*, 1984). Concern for agricultural workers for respiratory exposure is warranted based on the fact that aflatoxin B_1 can occur at significantly high levels (up to 200 ng/g) in grain dust or 25 ng/cubic meter of air containing maize dust (Burg *et al.*, 1982). Selim *et al.*, (1998) investigated the airborne aflatoxin B_1 generated by various on-farm activities on farms in Iowa (USA). High levels of aflatoxin were found in certain operations such as bin cleaning (Table 1).

Table 1. Concentration of airborne aflatoxin B_1 (ng/cubic meter) found during three different farm activities.

Harvesting (24)	Animal feeding (22)	Bin cleaning (14)
0.04-92	5-421	124-4849

The number between brackets is the number of sites investigated.

Certainly, these would be levels of concern for individuals exposed to them on a routine basis.

It is well known that aflatoxins affect the immune system (Richard, 1991). Moreover, Richard and Thurston (1975) demonstrated that cellular defence mechanisms in the lung were diminished even by feeding rabbits various levels of aflatoxin B_1. The ingestion of aflatoxin B_1 caused a reduction (as high as 43%) of the phagocytic activity of the alveolar macrophages when harvested from the lungs of the exposed rabbits. It was not determined if direct exposure of aflatoxin B_1 to the alveolar macrophages via the respiratory route was more efficacious than the oral or gastrointestinal route.

Stachybotrys *toxins*

The exposure of animals, including man, to *Stachybotrys* has been considered to be a problem from both gastrointestinal exposure and respiratory exposure. The current "Black Mould," made famous quite recently in the popular literature is *Stachybotrys* and has raised a cause for concern of respiratory exposure to one or more of the toxic metabolites produced by this organism, known more correctly as *S. chartarum*. Historical evidence may place this organism as a causative agent of disease as far back as the time of the Dead Sea Scrolls where "Houses of Mildew" were discussed as causing problems of inhabitants. In more recent times, this organism caused problems in animals in such countries as Russia, affecting primarily horses, but this was not necessarily an aerosol phenomenon. However, human disease occurring in the same localities may have resulted from respiratory exposure as descriptions of these diseases included chest pains and inflamed respiratory passages.

Some evidence supporting respiratory exposure to this organism was provided by the report of Croft and co-workers (1986) where pulmonary irritation, headaches, fatigue and diarrhoea were found in residents and workers in *Stachybotrys*-contaminated houses. Samples taken from these houses yielded various trichothecenes known to be produced by isolates of *Stachybotrys* and the diseases subsided when contaminated materials were removed from the premises.

Fairly recent outbreaks of idiopathic pulmonary hemorrhage/hemosiderosis were described in 10 infants under 6 months of age in Cleveland, Ohio (Dearborn *et al.*, 1999; Robbins *et al.*, 2000). This syndrome was considered to be due to the presence of *Stachybotrys atra* (syn. *chartarum*) in the homes of the patients where water damage had occurred; an event usually preceding the growth and establishment of this fungus in a building. A similar outbreak had occurred in Chicago, IL at about the same time but the mould was not corroborated.

While we know that *Stachybotrys* strains are capable of producing a variety of toxic metabolites including the macrocyclic trichothecenes, allergens and volatiles, the current illnesses in humans occupying *Stachybotrys*-contaminated buildings are not yet proven to be causally related to the toxic products of this fungus but are extremely suspect.

Ochratoxins

Although the occurrence data of ochratoxin A would indicate that it is found in numerous foods and commodities, the concentrations are not considered to be very high by most investigators. However, this may not be of importance insofar as ochratoxin A is concerned because it apparently has a rather lengthy half life in the human body and likely is extended in many other animal species as well. Therefore, the accumulation of small amounts from a variety of foods would perhaps eventually reach a toxic threshold level in the body. Of concern also is that it may be important as an airborne toxin resulting in toxicity by the respiratory route. Breitholz-Emanuelsson and co-workers (1995) demonstrated that ochratoxin A was absorbed very efficiently from the lungs of rats given doses intratracheally. Indication that ochratoxins cause human disease was presented by DiPialo and colleagues (1993) whereby a husband and wife suffered acute renal failure from apparent inhalation of ochratoxin after handling *Aspergillus ochraceus* contaminated wheat. When rabbits and guinea pigs were exposed by the respiratory route with air passed through the contaminated wheat, renal disease typical of ochratoxicosis was established in both animal species.

Ochratoxin A has been found in exceedingly high levels (>1500 ng/g) in house dust collected from a problem household whereby both the human residents and pet animals (dogs and guinea pigs) had some symptoms typical of ochratoxicosis (Richard *et al.*, 1999). Although no cause and effect relationship was established, the residents did improve physically once they vacated the premises. Subsequent to this finding we have found ochratoxin A in carpet sweepings from 3 homes in Texas, however the levels were considerably lower (6-17 ng/g) than that found in the above residence. Similarly, Skaug and colleagues (2000) found ochratoxin (0.2-70 ng/g) in airborne dust in 6 of 14 dust samples collected from cowsheds in Norway.

The presence of ochratoxin A in serum collected from workers in three different industries (coffee, cocoa beans and spices) at the end of a shift was reported by Italian investigators (Iavicoli *et al.*, 2002). The levels of ochratoxin A in these workers ranged from 0.94 to 3.28 ng/ml compared to controls whose serum contained only 0.03 to 0.95 ng/ml. Presumably their exposure occurred by the respiratory route.

Thus it would appear from just the few studies that have been presented here, that respiratory exposure to mycotoxins should be a concern for public health reasons and possibly for animals as well. It is likely that, especially agricultural workers and workers in processing facilities utilising commodities susceptible to ochratoxin A contamination, would be at risk because of airborne dust from the commodities that is often generated in the handling.

Conclusion

As mentioned earlier in this discussion, a bias may be evident in the author's view of the current investigations ongoing in the area of mycotoxins and mycotoxicoses. However, it is hoped that it would be evident that there certainly are some exciting investigations being conducted as is evident by a perusal of the information presented at this current XIth IUPAC Symposium. The results of these investigations will assist in leading to an increased knowledge base to understand the true risks associated with mycotoxins as they occur in nature and as they gain entry through the possible portals into the bodies of humans and other animals.

References

Breitholz-Emanuelson, A., Fuchs, R., and Hult, K., 1995. Toxicokinetics of ochratoxin A in rat following intratracheal administration. *Natural Toxins* 3: 101-103.

Burg, W.R., Shotwell, O.L., and Saltzman, B.E., 1982. Measurements of airborne aflatoxins during the handling of 1979 contaminated corn. *American Industrial Hygiene Association Journal* 43: 580-586.

CAST (Council for Agricultural Science and Technology) Report No. 139., 2003. *Mycotoxins: Risks in Plant, Animal, and Human Systems.* Task Force Report. Council for Agricultural Science and Technology, Ames, Iowa, USA.

Croft, W.A., Jarvis, B., and Yatawara, C.S., 1986. Airborne outbreak of trichothecene toxicosis. *Atmospheric Environment* 20: 549-552.

Dearborn, D.G., Yike, I., Sorenson, W.G., Miller, M.J., and Etzel, R.A., 1999. Overview of investigations into pulmonary hemorrhage among infants in Cleveland, Ohio. *Environmental Health Perspectives* 107: 495-499.

Di Paolo, N., Guarnieri, A., Loi, F., Sacchi, G., Mangiarotti, A.M., and Di Paolo, M., 1993. Acute renal failure from inhalation of mycotoxins. *Nephron* 64: 621-625.

Dvorackova, I., 1976. Aflatoxin inhalation and alveolar cell carcinoma. *British Medical Journal* 1: 691-693.

Dvorackova, I., and Kralove, H., 1979. Relation between aspergilloma and pulmonary carcinoma. Abstract 712—4th International IUPAC Symposium on Mycotoxins and Phycotoxins, Lausanne.

Dravorackova, I., and Pichova, V., 1986. Pulmonary interstitial fibrosis with evidence of aflatoxin B1 in lung tissue. *Journal of Toxicology and Environmental Health* 18: 153-157.

Hayes, R.B., Van Nieuwenhuize, J.P., Raatgever, J.W., and Ten Kate, F.J.W., 1984. Aflatoxin exposures in the industrial setting: An epidemiological study of mortality. *Food Chemistry and Toxicology* 22: 39-43.

Iavicoli, I, Brera, C., Carelli, G., Caputi, R., Marinaccio, A., and Maraglia, M., 2002. External and internal dose in subjects occupationally exposed to ochratoxin A. *International Archives of Occupational and Environmental Health* 75: 381-386.

Richard, J.L., 1991. Mycotoxins as Immunomodulators in Animal Systems. In: *Mycotoxins, Cancer, and Health*, Bray, G.A. and Ryan, D.H., (Eds), Pennington Center Nutrition Series, Vol. 1, Louisiana State University Press, Baton Rouge, pp 197-220.

Richard, J.L., and Thurston, J.R.,1975. Effect of aflatoxin on phagocytosis of *Aspergillus fumigatus* spores by rabbit alveolar macrophages. *Applied Microbiology* 30: 44-47.

Richard, J.L., Plattner, R.D., May, J., and Liska, S.L., 1999. The occurrence of ochratoxin A in dust collected from a problem household. *Mycopathologia* 146: 99-103.

Robbins, C.A., Swenson, L.J., Nealley, M.L., Gots, R.E., and Kelman, B.J., 2000. Health effects of mycotoxins in indoor air: a critical review. *Applied Occupational and Environmental Hygiene* 15: 773-784.

Selim, M.I., Juchems, A.M., and Popendorf, W., 1998. Assessing airborne aflatoxin B1 during on-farm grain handling activities. *American Industrial Hygiene Association Journal* 59: 252-256.

Skaug, M.A., Eduard, W., and Stormer, F.C., 2001. Ochratoxin A in airborne dust and fungal conidia. *Mycopathologia* 151: 93-98.

Van Nieuwenhuize, J.P., Herber, R.F.M., Debruim, A., Meyer, I.P.B., and Duba W.C., 1973. Aflatoxins-epidemiological study on the carcinogenicity of prolonged exposure to low levels among workers of a plant. *Tijdschrift voor Sociale Geneeskunde* 51: 754-759.

Section II:

Fungi-substrate interaction (ecology)

Fungal ecology and the occurrence of mycotoxins

J.I. Pitt

Food Science Australia, P.O. Box 52, North Ryde, NSW 1670, Australia; John.Pitt@csiro.au

Abstract

It is commonly assumed that if a mycotoxin is found in a particular commodity in one location, it will occur everywhere. The true picture is much more complex. Some cereal crops are grown over very wide areas of the globe, with marked differences in both temperature and rainfall patterns. Wheat and barley grown in Northern Europe and parts of Canada, with short cool, growing seasons, are often contaminated with ochratoxin A due to growth of *Penicillium verrucosum*, a species which grows only below 30 °C. However, in warmer climates such as the United States and Australia, *P. verrucosum* is at most very uncommon, so small grains from those areas do not contain ochratoxin A. The ecology of the *Fusarium* species which grow and produce trichothecenes in small grains is also complex. *F. graminearum* and relatives, the producers of deoxynivalenol, nivalenol and zearalenone, grow over a wide temperature range, and can be expected to produce toxins wherever and whenever cool damp harvest seasons predispose the crops to infection. Only in Australia, where prime quality wheat normally matures at temperatures around 40 °C, are trichothecenes not a commercial problem. Some kinds of mycotoxins are more universally produced. *Fusarium verticillioides* is associated with maize, so fumonisins can occur wherever maize is grown, and levels are mainly influenced by local climatic conditions near harvest. Crops also vary in their susceptibility to the growth of particular fungi. Aflatoxins are produced by two common fungal species, *Aspergillus flavus* and *A. parasiticus*, especially in nuts and oilseeds in the tropics. However, while *A. flavus* is associated with maize, peanuts and cottonseed, and produces aflatoxins in all three crops, *A. parasiticus* produces significant aflatoxin levels only in peanuts. Moreover *A. parasiticus* does not occur in one large peanut growing area of the world, Southeast Asia. These and other examples of the influence of ecology on fungal growth and mycotoxin production are described in this paper.

Keywords: aflatoxin, ochratoxin, deoxynivalenol, nivalenol, fumonisin, zearalenone, *Aspergillus, Penicillium, Fusarium*

Introduction

Mycotoxins are insidious poisons, produced when, and only when, common fungi grow in foods and feeds. Only a proportion of common fungi are toxigenic, and specific toxins are produced only by specific fungi. The most important mycotoxins are aflatoxins, ochratoxin A, fumonisins, the trichothecenes nivalenol and deoxynivalenol, and zearalenone (Miller, 1995), and in practice each is produced by only a small number of species, as detailed below.

However, the detection in a food or feed of a particular mycotoxigenic fungus, known to produce an important toxin, does not automatically mean that that particular toxin will be formed in appreciable quantities. Growth of each fungus is governed not only by environmental factors, in a manner specific to that fungal species, but also by the relationship that fungus has with a particular crop plant or plants, or whether such a relationship exists. This paper provides an outline of some of these factors and relationships as they relate to the production of the important mycotoxins.

Pre-harvest versus post-harvest mycotoxin formation

The most important basic consideration controlling mycotoxin production is whether the particular fungus has an association with a crop plant species. Association of a fungus with a crop occurs while the crop is growing, *i.e.* pre-harvest, and the association may be as a commensal, or perhaps as a symbiote, or as a pathogen. However, some major mycotoxin producing fungi have no association with a plant, in which case invasion of a crop, and mycotoxin production, occur post-harvest, *i.e.* after the crop is picked or pulled, during drying, transport and storage. This distinction is an extremely important one, all too frequently overlooked. The control of pre-harvest toxin formation is - must be - based on entirely different strategies from the control of toxins which occur post-harvest.

Ochratoxin A is a good example of a post-harvest toxin. Control of ochratoxin A formation in foods is basically a food technology problem. By that is meant that it can be controlled by good agricultural practice and good manufacturing practice. This will be discussed in detail later.

The control of pre-harvest mycotoxin formation, however, is a much more difficult problem. Good examples of pre-harvest toxins are fumonisins, trichothecenes and zearalenone. Uncontrollable factors - crop, climate and environmental - dictate whether the fungi which produce these toxins grow, and hence whether mycotoxins are likely to be formed. Two very important generalisations can be made. First, growth of these plant associated fungi occurs only in specific crops, where there is a definite plant–fungus association. Second, the fungi which produce these toxins grow only at high water activity, so these toxins which are formed pre-harvest are rarely formed during storage. Increases in concentrations of fumonisins, trichothecenes or zearalenone rarely if ever occur in storage. This important point is often overlooked in discussions of mycotoxin occurrence.

The mycotoxin which defies categorisation into pre- or post-harvest formation is aflatoxin. This is a very important exception and will be discussed first, as a number of important principles can be introduced.

Aflatoxins

More than 40 years after their discovery, aflatoxins remain the most important mycotoxins in the world. It is well known that they are principally produced by two related fungi, *Aspergillus flavus* and the closely related species *A. parasiticus*. It is also well known that aflatoxins are of major importance in three crops, maize, peanuts and cottonseed. What is less well known is that aflatoxin formation in two of these crops, maize and cottonseed, is due almost entirely to *A. flavus*, while aflatoxin in peanuts often comes from both *A. flavus* and *A. parasiticus*. This is due primarily to the fact that *A. flavus* is a commensal fungus in cottonseed (Klich *et al.*, 1984; Klich and Chmielewski, 1985) and probably also in maize (Lillehoj *et al.*, 1980). This association enables the fungus to grow in the plant and in developing seeds and kernels before harvest, providing a big ecological advantage over fungal competitors. It should be emphasised that *A. flavus* is not a parasite on these crops, but a commensal, so that the plant does not react to the presence of the fungus, and the fungus produces no visible damage to the plant or to the seed. In peanuts, *A. parasiticus* is a commensal as well as *A. flavus*: both fungi are able to grow in peanut plants and developing peanuts (Pitt *et al.*, 1991; Pitt, 2004).

It is often important to be able to distinguish between these two fungi. While *A. flavus* produces only B aflatoxins, *A. parasiticus* produces both B and G toxins; only about 50% of naturally occurring *A. flavus* isolates produce aflatoxins, while virtually all known isolates pf *A. parasiticus* are toxigenic. *A. flavus* is of universal occurrence in food crops in the tropical and warm temperate zones of the world. *A. parasiticus* has a much more limited distribution. It occurs wherever peanuts are grown, and is also a natural pathogen on the sugar cane mealy bug *Saccharicoccus sacchari* (Drummond *et al.*, 1991), so is of common occurrence in sugar cane fields. *A. parasiticus* is uncommon in other natural environments. It is also rare – at most – in some large geographic regions, including Southeast Asia (Pitt and Hocking, 1996).

In summary, aflatoxins in peanuts are believed to be derived from growth of *A. flavus* and *A. parasiticus* in areas of the world outside Southeast Asia, but principally or exclusively only from *A. flavus* in Thailand, Indonesia, the Philippines and Vietnam, and also perhaps other Southeast countries where this has not been studied. A small contribution from *A. nomius* in Southeast Asia also seems likely. Aflatoxins in maize and cottonseed are mostly or almost entirely produced by *A. flavus*. So while peanuts usually contain both B and G aflatoxins, the other major crops where only *A. flavus* is a commensal contain mostly or exclusively B aflatoxins.

The forgoing discussion relates to crops with an association with *A. flavus* and *A. parasiticus*, where pre-harvest infection takes place. In all other crops where aflatoxin occurs, and this includes tree nuts such as Brazil nuts, walnuts, pecans and pistachios, plus oil seeds including dried coconut, and small grains such as wheat and barley, aflatoxins are formed essentially only post-harvest, due to poor drying or inadequate storage. These two fungi grow down to about 0.80 a_w and at temperatures up to 37 °C or more, so are capable of growth, and aflatoxin formation, in crops stored under unsatisfactory conditions. This occurs particularly in the humid tropics.

Control of aflatoxin formation in crops

For crops where infection by *A. flavus* and *A. parasiticus* occurs only post-harvest, control of aflatoxin formation after harvest involves good agricultural and manufacturing practice. Rapid and thorough drying and effective maintenance of dry storage are the principal mechanisms (Highley *et al.*, 1994).

Control of aflatoxin formation before harvest, in susceptible crops, is difficult. The use of conventional plant breeding techniques, of great value for controlling the growth of pathogenic fungi, has been largely unsuccessful against the fungi which produce aflatoxins. This is undoubtedly because they are commensals, and do not exhibit classic host pathogen relationships. In maize, reduction in insect damage by the use of genetically modified cultivars shows promise. In cotton, some success has been achieved in the development of cultivars which lack the nectaries which attract insects and so assist *A. flavus* infection. One technique which has shown promise, however, is the use of bio-control techniques.

Bio-control

The type of bio-control currently coming into use against aflatoxins is more correctly called competitive inhibition. High numbers of non-toxigenic strains of *A. flavus* or *A. parasiticus* are introduced into fields where susceptible crops are being grown, where they compete against the naturally occurring toxigenic strains for invasion sites on seeds, nuts or kernels. Considerable success has been achieved by these means in recent years. Approval for commercial processes using non-toxigenic *A. flavus* strains for control of aflatoxin production in cottonseed (Cotty personal communication, 2005) and peanuts (Anon, 2004) has recently been granted in the United States, while an application has been submitted for regulatory approval to use non-toxigenic *A. parasiticus* on peanut crops in Australia (Pitt, 2004).

Ochratoxin A

In one way, the ecology of ochratoxin A formation in foods and feeds is more complex than that of aflatoxin, because ochratoxin A is formed in crops by three different fungi (and some of their close relatives) which have differing ecological niches. However, in another way the picture is simpler: no good evidence exists that the fungi concerned are associated with particular crops before harvest. This greatly simplifies control, at least from the theoretical or ecological viewpoint. The three fungi and the crops affected will be discussed separately.

Penicillium verrucosum and the formation of ochratoxin A in cereals

P. verrucosum occurs commonly in cereals grown in cool temperate climates, ranging across Northern and Central Europe, Canada and Northern Asia. The occurrence of this species in cool temperate cereal crops has two important consequences: ochratoxin A is present in European, Canadian and Japanese cereal products, especially bread and flour based foods, and in the meat of animals which eat cereals as a major dietary component. Ochratoxin A was detected in Danish pig meats 25 years ago (Krogh *et al.*, 1974) and its implications for human and animal health were recognised at the same time. As bread, other cereal products and pig meats are major components of European type diets, the further consequence is that many people from the northern cool temperate zone have shown appreciable concentrations of ochratoxin A in their blood (Breitholtz-Emanuelsson *et al.*, 1993; Scott *et al.*, 1998).

It is important to note that *Penicillium verrucosum* occurs only in cool climates, as it has a maximum temperature for growth near 30 °C (Pitt and Hocking, 1997). This species is responsible for ochratoxin formation only in cereals in the northern cool temperate zone.

Aspergillus ochraceus and the formation on ochratoxin in stored products

Ochratoxin A was originally described as a metabolite of *Aspergillus ochraceus* from laboratory experiments (van der Merwe *et al.*, 1965). It was subsequently reported from several related *Aspergillus* species (Ciegler, 1972; Hesseltine *et al.*, 1972). Until recently this species was not known to have any affinity with a particular crop or foodstuff, being known mainly as a source of ochratoxin A in long stored cereals (Pitt and Hocking, 1997). However it has recently been shown that *A. ochraceus* is a major source of ochratoxin A in green (dried) coffee (Taniwaki *et al.*, 1999, 2003). Infection has been shown to occur during drying (Taniwaki *et al.*, 1999), and no association between the fungus and the coffee tree has been demonstrated.

Aspergillus ochraceus is a mesophilic xerophile. Growth occurs between 8 and 37 °C, and optimally between 24-31 °C, with a minimum a_w for growth near 0.80 (Pitt and Hocking, 1997). It is therefore not likely to infect crops either in the cool temperate zones or drying under very hot conditions.

Aspergillus carbonarius and the formation of ochratoxin A in grapes and grape products

The possibility that ochratoxin A can be produced by *A. carbonarius* and occasionally by its much more common relative, *A. niger*, was discovered only in the past few years (Horie, 1995; Téren *et al.*, 1996; Heenan *et al.*, 1998). These species have dark hyphae and spores, so are common in vineyards and fruit orchards, where high temperatures and strong sunlight provide a competitive advantage. However, they are not primary invaders of grapes, growing only as the result of damage by plant pathogenic fungi or of skin splitting due to unseasonal rain. Growth of *A. carbonarius* in grapes before harvest causes production of ochratoxin A in grape juice and wines. Ochratoxin A occurs in wines from the warmer growing areas throughout the world, but levels are usually low, as the fermentation process positively stops growth of the fungus. In dried vine fruits, however, the possibility of mechanical damage during harvesting and the prolonged time available for fungal growth during drying increase the probability of ochratoxin A being formed.

Control of ochratoxin production

None of the fungi producing ochratoxin A are known to occur in crops before harvest, *i.e.* as systemic invaders or pathogens. Although strictly speaking *A. carbonarius* does occur before harvest, it only infects damaged grapes. In consequence, control of the formation of ochratoxin A relies on good practice. If coffee beans, grapes or cereals are dried rapidly at the time of harvest, and stored dry, little ochratoxin A will be formed. If grapes for wine production are substantially sound, *i.e.* free from rotting fruit, and are crushed quickly after harvest, leading rapidly to anaerobic conditions, ochratoxin A can be held to low limits. If grains are stored under safe conditions, ochratoxin A formation by *A. ochraceus* will also be minimal.

Fusarium *mycotoxins*

The most important general observation to be made about *Fusarium* mycotoxins is that all *Fusarium* species grow only at high water activities, above about 0.9 (Pitt and Hocking, 1997) so that toxin production in crops occurs only before harvest or during the early stages of drying. Synthesis of toxins ceases long before the crops are fully dry, and only occurs during storage under catastrophic conditions, such as flooding. Production of *Fusarium* mycotoxins occurs as the result of growth of the fungus in the living plant and seed.

Fumonisins

One major group of *Fusarium* mycotoxins, the fumonisins, are produced by *F. verticillioides* (previously called *F. moniliforme*) and the closely related species *F. proliferatum*. These species are systemic in maize worldwide, being always present in the plants, and even in healthy kernels (Miller, 1994). *F. verticillioides* has been reported to suppress the growth of other ear fungi (Reid *et al.*, 1999) and heat treated kernels lacking the fungus germinate but do not thrive (Foley, 1962), indicating the fungus has some role in plant growth. Fumonisins appear to be produced only when drought stress or other unfavourable conditions disturb the balance between fungus and plant. Fumonisin production in maize is favoured by relatively high temperatures (Munkvold, 2003). Fumonisins are found only in maize and sorghum, as *F. verticillioides* and *F. proliferatum* rarely infect other crops.

Trichothecenes

The important trichothecenes are deoxynivalenol and nivalenol, produced by *Fusarium graminearum* (often listed as *Gibberella zeae*, its sexual stage), and *F. culmorum*. *F. graminearum* occurs in maize, and both species in small grains especially wheat and barley. Unlike *F. verticillioides*, these species are rank pathogens, invading plants and grains by causing diseases, known as Gibberella ear rot and Fusarium head blight, in developing grain. Epidemics of Gibberella ear rot require the congruence of three factors: airborne or insect-borne spores, inoculum at the correct time, and, appropriate moisture and temperature (Miller, 1994; Munkvold, 2003). Hence these diseases are prevalent in north temperate climates especially in wet years, and much less common in the southern USA, South Africa or Australia.

Zearalenone

Zearalenone is produced by the same fungi that produce deoxynivalenol, and usually at the same time. The ecology of zearalenone production mirrors that of deoxynivalenol, at least in general terms.

Control of *Fusarium* mycotoxins

As with other pre-harvest mycotoxins, control of the occurrence of *Fusarium* mycotoxins is not easy. Breeding for resistance to Fusarium kernel rot has not been effective so far, although some improvements in disease expression have been seen (Munkvold, 2003). However, good agricultural practice, the control of insects, development of resistance to other ear diseases, and development of cultivars adapted to drought and temperature tolerance are all important in reducing the risk of fumonisin accumulation in maize. Some progress has been made in breeding cultivars resistant to ear rot (Miller, 1994; Munkvold, 2003).

As noted earlier, the fact that *Fusarium* species grow only at high a_w, means that once grains are dried, increases in levels of *Fusarium* mycotoxins are unlikely to occur.

References

Anon., 2004. A partner in biological control of aflatoxin contamination. http://www.circleoneglobal.com (December, 2004).

Breitholtz-Emanuelsson, A., Olsen, M., Oskarsson, A., Palminger, I., and Hult, K., 1993. Ochratoxin A in cow's milk and in human milk with corresponding human blood samples. *Journal of AOAC International* 76: 842-846.

Ciegler, A., 1972. Bioproduction of ochratoxin A and penicillic acid by members of the *Aspergillus ochraceus* group. *Canadian Journal of Microbiology* 18: 631- 636.

Drummond, J., De Barro, P.J., and Pinnock, D.E., 1991. Field and laboratory studies on the fungus *Aspergillus parasiticus*, a pathogen of the pink sugar cane mealybug *Saccharicoccus sacchari*. *Biological Control* 1: 288-292.

Foley, D.C., 1962. Systemic infection of corn by *Fusarium moniliforme*. *Phytopathology* 68: 1331-1335.

Heenan, C.N., Shaw, K.J., and Pitt, J.I., 1998. Ochratoxin A production by *Aspergillus carbonarius* and *A. niger* isolates and detection using coconut cream agar. *Journal of Food Mycology* 1: 63-72.

Hesseltine, C.W., Vandergraft, E.E., Fennell, D.I., Smith, M.L., and Shotwell, O.L., 1972. Aspergilli as ochratoxin producers. *Mycologia* 64: 539-550.

Highley, E., Wright, E.J., Banks, H.J., and Champ, B.R., eds., 1994 *Stored Product Protection*. CAB International, Wallingford, UK.

Horie, Y., 1995. Productivity of ochratoxin A of *Aspergillus carbonarius* in *Aspergillus* section *Nigri*. *Nippon Kingakkai Kaiho*: 36, 73-76.

Klich, M.A., and Chmielewski, M.A., 1985. Nectaries as entry sites for *Aspergillus flavus* in developing cotton bolls. *Applied and Environmental Microbiology* 50:602-604.

Klich, M.A., Thomas, S.H., and Mellon, J.E., 1984. Field studies on the mode of entry of *Aspergillus flavus* into cotton seeds. *Mycologia* 76: 665- 669.

Krogh, P., Hald, B., Englund, P., Rutqvist, L., and Swahn, O., 1974. Contamination of Swedish cereals with ochratoxin A. *Acta Pathologia Microbiologia Scandinavia*, Sect. B, 82: 301-302.

Lillehoj, E.B., Kwolek, W.F., Horner, E.S., Widstrom, N.W., Josephson, L.M., Franz, A.O., and Catalano, E.A., 1980. Aflatoxin contamination of pre-harvest corn: role of *Aspergillus flavus* inoculum and insect damage. *Cereal Chemistry* 57: 255-257.

Miller, J.D., 1994. Epidemiology of Fusarium ear diseases. In: *Mycotoxins in grain: compounds other than aflatoxin*, Miller, J.D., and Trenholm, H.L. (Eds.), Eagan Press, St. Paul, MN. pp. 19-36.

Miller, J.D., 1995. Mycotoxins in grains: issues for stored product research. *Journal of Stored Product Research* 31: 1-16.

Munkvold, G.P., 2003. Epidemiology of Fusarium disease and their mycotoxins in maize. *European Journal of Plant Pathology* 109: 705-713.

Pitt, J.I., and Hocking, A.D., 1996. Current knowledge of fungi and mycotoxins associated with food commodities in Southeast Asia In: *Mycotoxin contamination in grains,* Highley, E. & Johnson, G.I. (Eds.). Australian Centre for International Agricultural Research, Canberra. pp. 5-10.

Pitt, J.I., 2004. Biocontrol of aflatoxins in peanuts. In: *Meeting the Mycotoxin Menace,* Barug, D., van Egmond, H., Lopez-Garcia, R., van Osenbruggen, T., and Visconti, A. (Eds.). Wageningen Academic Publishers, Wageningen, Netherlands. pp. 141-152.

Pitt, J.I., and Hocking, A.D., 1997. *Fungi and Food Spoilage.* Blackie Academic and Professional, London.

Pitt, J.I., Dyer, S.K., and McCammon, S., 1991. Systemic invasion of developing peanut plants by *Aspergillus flavus. Letters in Applied Microbiology* 13: 16-20.

Reid,L.M., Nicol, R.W., Ouellet, T., Savard, M., Miller J.D., Young, J.C., Stewart, D.W., and Schaafsma, A.W., 1999. Interaction of *Fusarium graminearum* and *F. moniliforme* in maize ears: disease progress, fungal biomass, and mycotoxin accumulation. *Phytopathology* 89:1028-1037.

Scott, P.M., Kanhere, S.R., Lau, B.P-Y., Lewis, D.A., Hayward, S., Ryan, J.J., and Kuiper-Goodman, T., 1998. Survey of Canadian human blood plasma of ochratoxin A. *Food Additives and Contaminants* 15: 555-562.

Taniwaki, M.H., Pitt, J.I., Urbano, G.R., Teixeira, A.A., and Leitao, M.F.F., 1999. Fungi producing ochratoxin A in coffee. Proceedings of the 18th International Scientific Colloqium on Coffee, Helsinki, Finland, 2-6 August, 1999. pp. 239-247.

Taniwaki, M.H., Pitt, J.I., Teixeira, A.A., and Iamanaka, B.T., 2003. The source of ochratoxin A in Brazilian coffee and its formation in relation to processing methods. *International Journal of Food Microbiology* 82: 173-179.

Téren, J., Varga, J., Hamari, Z., Rinyu, E., and Kevei, É. 1996. Immunochemical detection of ochratoxin A in black *Aspergillus* strains. *Mycopathologia* 134: 171-176.

Van der Merwe, K.J., Steyn, P.S., Fourie, L., Scott. D.B., and Theron, J.J., 1965. Ochratoxin A, a toxic metabolite produced by *Aspergillus ochraceus* Wilh. *Nature* 205: 1112-1113.

Effect of fungal infection and insect infestation interaction, moisture content, type of earhead and grain on mycotoxin production in field sorghum

B.N. Reddy and C.R. Raghavender

Mycology and Plant Pathology Laboratory, Department of Botany, Osmania University, Hyderabad - 500 007, A.P., India; reddybn1@yahoo.com

Abstract

Three different types of earheads which are generally found in sorghum such as compact panicles with open grains, open panicles with enclosed grains and normal panicles with normal grains were examined for earhead insect pests and also for the relationship between insect pest damage and fungal infection on the production of mycotoxins. A total of 15 species of earhead insect pests belonging to 14 genera were isolated from one or the other varieties of sorghum from the stage of flowering to harvest. Caterpillars accounted for 53% of the insect pests followed by bugs (40%) and flies (7%). Among all the pests collected from the earheads, the sorghum midge (*Contarinia sorghicola*), the earhead bug (*Calocoris angustatus*), and the yellow peach moth (*Dichocrosis punctiferalis*) were of major importance in view of the great damage and heavy yield losses. Altogether 35 fungal species belonging to 13 genera were isolated in this study. Species belonging to *Aspergillus, Curvularia, Fusarium* and *Penicillium* were the most frequently isolated. In general compact headed varieties of sorghum were more highly infested with insect pests and fungi than the normal and open earheads. Most mycotoxin contamination was observed in the open grains of compact earheads containing high moisture. The least contaminated were the enclosed type of grains in the open panicles, which were found to contain less moisture. The multiple factors such as nature of the panicle, glume covering of the grain, insect pest infestation, fungal invasion and moisture content of the grain greatly influenced the production of mycotoxins in the field sorghum.

Keywords: sorghum, earhead, moistures, glume, interrelationship, mycotoxin

Introduction

The association between insects and plant pathogens including toxigenic fungi is of great importance. The insects act as disseminators and frequently as inoculators of numerous fungi, thus helping the latter spread from plant to plant. Insect-fungal interactions focusing on the effects of insect infestation on seed production (Leach, 1940) and diseases transmission (Agrios, 1980) have been studied in the past. Although these previous studies identified insects as a factor in the transmission of various fungi, none provided specific information on mycotoxin contamination in pest damaged and fungal infected earheads of sorghum (*Sorhum bicolor* L. Moench).

Sorghum is one of the four major cereal crops of the world. In the USA and South America, sorghum is principally used for animal feed while in Africa and Asia, particularly in the tropical developing countries like India, small scale farmers use sorghum mainly as human food either directly or in the form of an alcoholic beverage. More than 55% of the world's sorghum production comes from semi-arid tropical (SAT) zones. More than 90% of India's sorghum production occurs in a SAT zone, the Deccan Plateau, covering parts of Maharastra, Karnataka, Andhra Pradesh, Madhya Pradesh, Tamil Nadu Uttar Pradesh and Rajasthan. Approximately $^{2}/_{3}$ of India's sorghum production occurs during the rainy season (*Kharif*) and the remaining $^{1}/_{3}$ in winter (*Rabi*).

The purpose of this study was to determine sorghum earhead pests in general and insect-fungal associations in relation to type of earhead and variety of sorghum. Furthermore, the study sought to establish any relationship between insect infestation, fungal infection and the production of mycotoxins in field sorghum. In this paper, the term *panicle* is used as a synonym for earhead.

Materials and methods

Sorghum earhead samples, at various stages of maturity, from flowering to harvest, were collected at random from the fields of Andhra Pradesh Agricultural University, Rajendranagar, Andhra Pradesh, India. In order to collect all the diurnal pests, sampling was carried out at 5 hour intervals between 4 AM and 7 PM. Samples were collected by earhead type (compact, open and normal panicles) and grain arrangement on the earhead (fully or partially covered by glumes). Consequently the varieties IS-14338 (open headed variety with fully enclosed grain; N= 35), Y-75 (compact headed variety with open grain; N= 30), and CSH-1, CSH-5 and CSH-9 (normal earheads and grains; N=35) (Table 1, Figure 1) were included in the sample pool.

In general, most toxigenic fungi require >13% of moisture content for the growth and subsequent production of mycotoxins in different agricultural commodities. In view of this, the samples were categorised into two groups *i.e.* samples having up to 13% of moisture and above 13%. The seed samples were analysed for the presence of different mycotoxins using a multi-mycotoxin method (Bamburg *et al.*, 1969; Ullstrup, 1953).

Table 1. Different types of earheads and grain covering found in sorghum.

Sl.No.	Type of earhead	Entry	Glume covering of the grain
1	Compact (close)	(Y-75)	Partially covered
2.	Open (loose)	(IS14338)	Fully covered
3	Normal	(CSH – 1, 5 & 9)	Normal

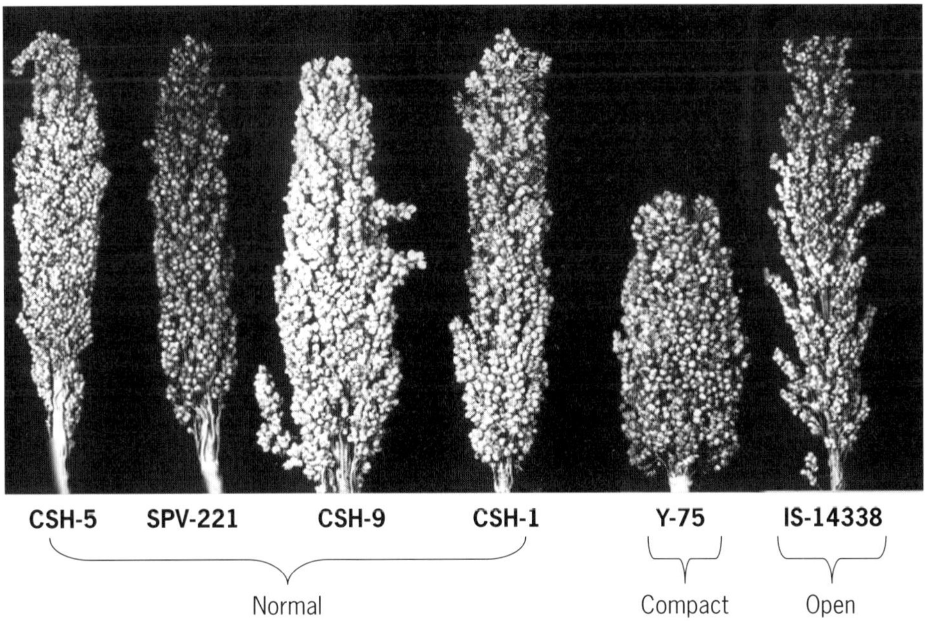

Figure 1. Different types of earheads found in sorghum.

Chemical confirmation and quantitative estimation of these mycotoxins was carried out by UV-aided visual thin layer chromatography (Stoloff *et al.*, 1971).

Results and discussion

Fifteen earhead insect pest species belonging to fourteen genera were identified from the sorghum samples (Table 2). The majority (53%) of the insect pests were caterpillars (*Calocoris angustatus, Dysdercus singulatus, Leptoglossus phyllopus,*

Table 2. Important insect pests of sorghum earheads and their nature of damage.

Common name	Scientific name	Status	Nature of damage
FLY			
Sorghum midge	*Contarinia sorghicola* coq.	K.P.	Damage developing ovary
BUGS			
Jowar earheads bug	*Calocoris angustatus* Leth.	K.P	Feed on developing seed causing shriveled seed resulting in susceptiblility to head-moulds
Pyrrhocorid bug	*Dysdercus cingulatus* Fb.	O.P.	Feed on developing seed causing shriveled seed resulting in susceptiblility to head-moulds
Leaf-footed bug	*Leptoglossus phyllopus*	O.P.	Feed on developing seed causing shriveled seed resulting in susceptiblility to head-moulds
Lygaeid bug	*Nysius raphanus* Howard	O.P.	Feed on developing seed causing shriveled seed resulting in susceptiblility to head-moulds
Green vegetable bug	*Nezara viridual* Linn.	O.P.	Feed on developing seed causing shriveled seed resulting in susceptiblility to head-moulds
Pentatomid bug	*Dolicoris indicus* Stal.	O.P.	Feed on developing seed causing shriveled seed resulting in susceptiblility to head-moulds
EARHEAD CATERPILLARS			
Armywarm	*Spodoptera litura* Hubner	O.P.	Feed on developing seed causing shriveled seed resulting in susceptiblility to head-moulds
American cotton	*Heliothis armigera* Hubner	K.P.	Destruction of seeds in head bollworm
Corn earworm	*Heliothis zeae* Boddie	K.P.	Destruction of seeds in head
Castor capsule borer	*Dichocrosis punctiferalis* Guen	O.P.	Destruction of seeds in head
Sorghum webworm	*Calama sorghiella* Riley	O.P.	Destruction of seeds in head
Head Caterpillar	*Cryptoblades quidiella* Hampsn	O.P.	Destruction of seeds in head
Webworm	*Eublemma silicula* Swinh	O.P.	Destruction of seeds in head
Webworm	*Stenochronia elongella* Hampsn	O.P.	Destruction of seeds in head

K.P. = Key Pest; O.P.= Occasional Pest.

Key Pests are serious, perennially occurring, persistent species that dominate control practices in the absence of deliberate human intervention. The pest population commonly exceed the economic injury level each year, often wide areas.

Occasional Pests cause economic damage only in localised areas, or very infrequently. Such pests are usually under natural control and exceed the economic injury level only sporadically. Most pests of sorghum are occasional pests.

Nysius raphanus, Nezara viridula, Dolicoris indicus) followed by bugs (*Heliothis armigera, H. zeae, Spodoptera lithura, Dichocrosis punctiferalis, Calama sorghiella, Cryptoblades quidiella, Eublemma silicula, Stenochronia elongella*) (40%) and the only fly (*Contarinia sorghicola*) (7%). Among the insect pests collected from the earheads, the sorghum midge (*Contarinia sorghicola*), the earhead bug (*Calocoris angustatus*), and the castor capsule borer (*Dichocrosis punctiferalis*) are of major importance because they are known to cause enormous damage to the crop leading to heavy yield losses (Sharma *et al.*, 2000).

It is evident from the results that the compact headed varieties of sorghum were more prone not only to earhead insect pest infestation but also to fungal infection (Table 3) compared to normal earheads. On the other hand, open headed varieties of sorghum were less vulnerable to insect infestation and mould infection. In the majority of the cases the colour of the earheads turned to black or pinkish white due to heavy infection by *Curvularia* or *Fusarium* (Figure 2). Altogether 13 fungal genera comprising 35 species were isolated from the sorghum seed samples (Table 3).

Figure 2. Highly infected sorghum earhead with headmoulds (black colour in the infected earhead due to Curvularia / Drechslera *and the pinkish white due to* Fusarium*).*

Table 3. Percentage dominance of field fungi in relation to type of earhead in sorghum.

Mycoflora	Compact panicle	Loose panicle	Normal panicle
Alternaria alternata	3.1	-	0.4
Aspergillus candidus	2.3	4.4	0.4
A. flavus	6.3	12.1	12.2
A. fumigatus	-	5.4	1
A. glaucus	-	-	5.7
A. nidulans	3.7	3.6	4
A. niger	4.9	6.4	5.2
A. ochraceous	1	-	1.8
A. terreus	-	-	3.7
Cephalosporium acremonium	4.8	4.3	3.1
Chaetomium indicum	3.9	3.7	2.3
Cladosporium herbarum	2.9	-	2
Curvularia clavata	3.2	-	1.7
C. geniculata	2.8	4.7	1
C. lunata	11.9	13.2	14.1
C. maculans	-	-	2.7
C. pallecens	4.5	4.3	3.6
C. tetramera	2.1	-	-
Drechslera halodes	4.1	5.7	5.1
D. hawiiense	1.8	-	0.5
Fusarium equiseti	3.1	-	2.3
F. incarnatum	2.9	2.8	1.8
F. moniliforme	5.7	6.8	5.8
F. oxysporum	5.1	4.1	5.6
F. roseum	4.1	3.7	3.8
F. solani	-	-	0.9
Nigrospora oryzae	1.3	2.7	-
Penicillium candidum	3.7	4.9	2.8
P. citrinum	1	3.1	1
P. funiculosum	-	-	2.7
P. tardum	4.3	4.1	3
P. viridicatum	1	-	-
Phoma sorghiana	1.7	-	2.5
Rhizopus nigricans	1.1	-	2.4
Trichothecium roseum	1.7	-	0.3

Aspergillus, Curvularia, Fusarium and *Penicillium* were the predominant fungal pathogens. Heavy insect infestation by piercing sucking insects such as mealybugs, leafhoppers, scales and aphids frequently leads to the production of sticky sugary honeydew, which is often colonised by sooty mould (dark coloured fungus) that affects the plant adversely. Honeydew contains sugars, amino acids, proteins, minerals, and vitamins, which serves as the growth medium for sooty moulds.

Results in Table 4 show that the highest percentage of aflatoxin contamination was reported in compact earheads with open grains whereas the least contamination was in open panicles with enclosed grains. The moisture content of the seed sample was directly proportional to the percentage of mycotoxin contamination and the amount of toxin accumulated.

The higher incidence of fungal infection in compact panicles may be due to the presence of high moisture and optimum temperature prevailing inside the panicle. At the same time, the compact earheads protect the earhead pests from the high pressure caused by the strong winds and heavy rains. However, Doering and Randolp (1960) reported that there is no direct relation between the pest population and type of earheads. According to their report, the compact panicles were relatively less prone to pest infestation, which may be due to the presence of relatively less internal

Table 4. Relationship between the incidence of mycotoxins and type of earhead found in sorghum.

Type of earhead and grain	No of samples			% of toxin contamination	No. of samples +ve for (Concentration in µg/kg)				
	assayed	+ve for toxins	with above 13% moisture		B1	B2	G1	Zearalenone	T-2
Compact panicle with open grain	35	9	21	26	6 (35-425)	4 (20-285)	1 {25}	3 (18-115)	2 (20&55)
Open panicle with enclosed grain	30	5	17	17	3 (10-85)	2 (10 & 45)	-	1 {15}	- -
Normal panicle with normal grain	35	7	19	20	5 (20-185)	3 (15-65)	-	2 (10&35)	-

area exposed in compact heads when compared to open headed ones. But it is a known fact that even the compact panicles have enough space for the pests to enter and move around inside the panicles, through a wide gap at the base of the panicle in between two compound spikelets.

Insect damage may simply provide a fortuitous access for subsequent fungal infection (Ullstrup, 1953). Obligate relationships between insect-vectors and fungal pathogens have also been described earlier by Leach (1940). Interaction between insect and fungi has traditionally been studied from the perspective of the disciplines of entomology and mycology with much emphasis on insect transmission of plant pathogens in field and storage grain.

Fennell *et al.* (1975) reported that the incidence of *Aspergillus flavus* was higher on kernels from insect damaged ears. The highest percentage of mycotoxin contamination was observed in compact panicles with open grains due to the greater moisture content held up in the closely fitted panicles, that enhance the chances for colonisation by a large number of fungi in general and toxigenic strains in particular which further increase the production of different mycotoxins.

Conclusion

From the results of the present study, insect and other pest damage of standing sorghum is greater in close or compact panicles than normal ones. The data also shows that the probability of fungal infection is increased by insect damage. Pest damage in the field facilitates further invasion of fungi (Aucam, 1969; Reddy *et al.*, 1983, Wilson *et al.*, 1981) leading to the production of mycotoxins. The greater fungal infection in compact panicles may also be due to the closely arranged earhead parts. This type of earhead arrangement leads to incomplete drainage or evaporation of water and that further leads to an increase in the moisture content of the grain sorghum, which creates suitable conditions for fungal colonisation of the earhead (Austwick *et al.*, 1963). Furthermore, fungal spores, which reach the compact panicle, cannot easily be blown out of the earhead due to its compactness. In the case of open earheaded varieties, the percentage of fungal contamination was less due to the exposure of the earhead parts to sunlight, maximising water evaporation thereby reducing the moisture content of the grain. In addition to the type of earhead, the grain in the open panicle is protected by complete enclosure by glumes forming a physical barrier to fungal infection. Other workers have attributed resistance to the fungal invasion to higher phenolic content in the grains and glumes of open panicles of sorghum (Gangadharan *et al.*, 1978)).

This study has demonstrated the relation between earhead insect pests and fungal colonisation, and the interrelationships between various parameters such as type of earhead, arrangement of grain, moisture content of the grain on the production of mycotoxins under field conditions (Figure 3)

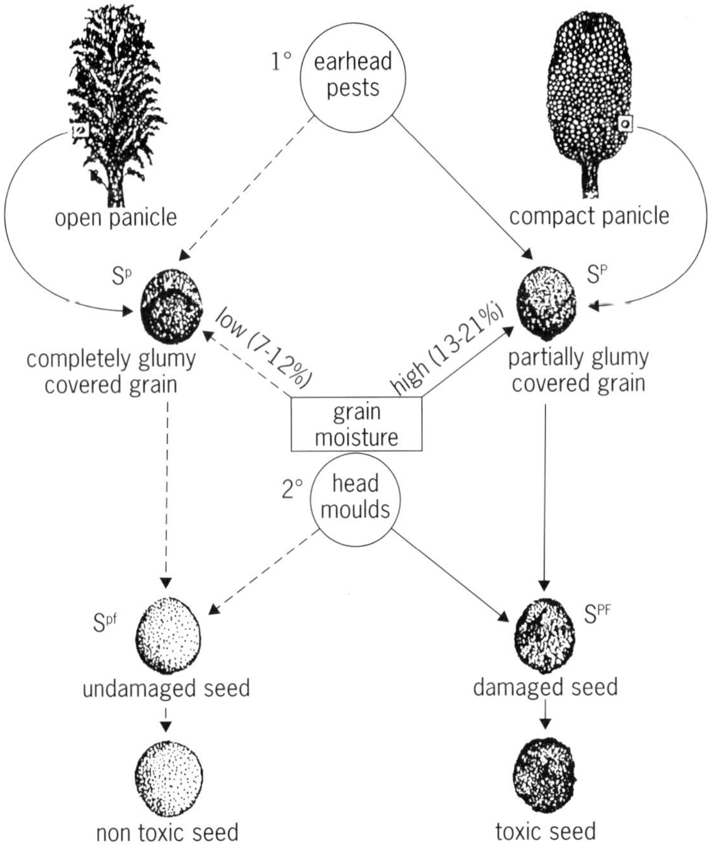

Figure 3. Schematic representation of multiple factors i.e. type of earhead and grain, grain moisture, and pest damage influencing fungal invasion and mycotoxin production in standing sorghum.
Arrows = greater infestation; dashes = lesser or no infestation. The symbols 1° and 2° denote primary and secondary invaders; S^P = seed with pest damage, S^{PF} = seed without (little) pest and fungal damage.

References

Agrios, G.N., 1980. Insect involvement in the transmission of fungal pathogens. In: *Vectors of Plant Pathogens,* K.F. Harris and K. Maramorosch, eds. Academic Press, New York. 467 pages. pp. 293-324.

Aucam, J.L., 1969. Role of mite vectors in the development of aflatoxin in groundnut. *J. Stored Prod. Res.* 5: 245–249.

Austwick, P.K.C., and Ayerst, G., 1963. Toxic products of groundnut. Groundnut microflora and toxicity. *Chem. Ind.* (London) 21: 51–61.

Bamburg, J.R., Strong, F.M., and Smalley, E.B., 1969. Toxins from moldy cereals. *J. Agris. Food Chem* 17: 443–450.

Doering, G.W., and Randolph, N.M., 1960. Field methods to determine the infestation of the sorghum webworm and the damage by the sorghum midge in grain sorghum. *J. Econ. Entomol.* 53: 749–750.

Fennell, D.I., Lillehoj, E.B., and Kwolek, W.F., 1975. *Aspergillus flavus* and other fungi associated with insect damaged field corn. *Cereal Chem.* 52: 314–321.

Gangadharan, K., Ramaraj, B., Prasad, M.N., and Palaniswamy, S., 1978. Studies on head moulds of sorghum in Tamil Nadu. *Sorghum News Letter* 20: 70–71.

Leach, J.G., 1940., *Insect transmission of plant diseases.* Mc Grawhill, New York.

Reddy, B.N., and Nusrath, M., 1983. Aflatoxin contamination in normal and insect damaged kesari (*Lathyrus sativus* L.) in Andhra Pradesh. *Curr. Sci.* 52: 370–371.

Sharma, H.C., Satyanarayana, M.V., Singh, S.D., and Stenhouse, J.W., 2000. Inheritance of resistance to headbugs and its interaction with grain molds in *Sorghum bicolor. Euphytica* 112: 167–173.

Stack, M.E., and Pohland, A.E., 1975. Collaborative study of a method for chemical confirmation of the identity of aflatoxin. *J. Assoc. Off. Anal. Chem.* 58: 110–113.

Stoloff, L., Nesheim, S., Yin, L., Rodricks, J.V., Stack, M., and Campbell, A.D., 1971. A multimycotoxin detection method for aflatoxins, ochrataxins, zearalenone, sterigmatocystin and patutin. *J. Assoc. Off. Anal. Chem.* 54: 91–97.

Ullstrup, A.J., 1953. Several ear rots of corn. In: *Plant Diseases,* the year book of Agriculture, 390 pages. U.S. Dept. Agric.

Wilson, D.M., Widstrom, N.W., Marti, L.R., and Evans, B.D., 1981. *Aspergillus flavus* group, aflatoxin and bright greenish-yellow fluorescence in insect damaged corn in Georgia. *Ceral Chem.* 58: 40–42.

Relationship between the condition of home-stored peanuts and aflatoxin B_1 contamination

A. Kane[1], N. Ba-Diop[1], T.S. Diack[1], M. Sembene[2], A. Delobel[3] and A.T. Guiro[1,2]

[1]Institut de Technologie Alimentaire, P.O.Box 2765, Dakar, Sénégal; ita@sentoo.sn
[2]Faculté des Sciences et Techniques, Université Cheikh Anta DIOP, Avenue Cheikh Anta DIOP, Dakar, Senegal
[3]Institut de la Recherche pour le Développement, Hann, Dakar Sénégal

Abstract

A study was undertaken to evaluate the relationship between different types of pest damage to peanuts at harvest and during storage and the occurrence of aflatoxin B_1. Samples of the peanut variety 73-33, were collect from peasant farmers at harvest and during storage over a 7 months period. Hand-shelled samples were assigned to various categories depending on the type and determined source of damage throughout the study and sorted out into different categories (without visible damage (healthy kernels), wrinkled or immature, discoloured, attacked by millipedes, attacked by termite and/or weevils (*Caryedon serratus*), black, contaminated by *Aspergillus flavus*, contaminated with other fungi, and infected by *Rhizopus* and/or *Mucor)*. Each category was analysed for aflatoxin using AOAC Method 970.45 coupled with HPLC. Health kernels had the least aflatoxin (mean 0.22 µg/kg), followed by wrinkled kernels (49.32 µg/kg). The highest contamination was recorded in kernels infected with *Aspergillus flavus* (208,000 µg/kg) followed by kernels attacked by millipedes (5,400 µg/kg) and termites (4,000 ug/kg). The results show that meadow harvest (preharvest) contamination is the major source of aflatoxin contamination in the study area.

Keywords: peanut, aflatoxin B_1, segregation, *Caryedon serratus*, *Aspergillus flavus*, termites, weevil, *Mucor*, *Rhizopus*, black kernels

Introduction

Contamination of peanuts by aflatoxin is a real public health problem especially in developing countries. In Senegal, peanut production is concentrated in the "Bassin arachidier". Until recently, the focus of research was on peanut plant resistance to *Aspergillus flavus* infection, determination of the different types of infections (Waliyar and Roquebert, 1979) and on detoxification procedures (Kane *et al.*, 1993).These previous studies showed that pre-harvest infection was predominant in the driest zones of the Basin, whereas post-harvest infection was of concern in the most humid parts of the basin where the rainfall season is longer (Waliyar and Zambettakis, 1979). It has also been shown that attacks by arthropods in the field encouraged infection of peanuts by *A. flavus*. The goal of the present study was to reassess the relationship between different forms of damage to the peanut, notably that caused by insects, and the level of contamination by the aflatoxins.

Materials and methods

Study area and agricultural practices in the study area

The study was carried in an area covering Keur Baka village and up to 2 km before Ndoffane in the south of the "Peanut Basin". The study area is part of the Sudan-Sahel zone with an annual rainfall of 800 mm that falls between July and October. Peanuts are harvested by hand during the month of November. The harvest process involves digging out the entire peanut plants and cleaning off loosely attached soil. The peanut plants are placed in rows (swaths) and left to dry for about 10 days. The swaths are then composted into bigger heaps (millstones) and left to continue drying for an additional 14 days. Once the farmer determines that the peanuts are sufficiently dry, the millstones are beaten and the peanut pods are put in sacks for storage at the homestead.

Sampling and sample preparation

Approximately 5 kg in-shell peanuts (variety 73-33) were collected from each of 10 peasant farmers at the beginning of the storage period and monthly thereafter for a period of seven months. The samples were transported to the laboratory, hand-shelled and sorted into different categories as shown in Table 1: For each damage category, samples from farmers 1 to 5 were combined into one composite sample and those from farmers 6 to 10 were combined into a second composite sample. As a result there were two composite samples per category that were processed further for analysis. Each composite sample was finely ground to pass a 20 mesh sieve and immediately analysed or stored frozen at -5 °C until analysis.

Chemical analysis

Analysis for aflatoxin was carried out according to the BF method (AOAC 970.45) and High Pressure Liquid Chromatography (HPLC). In some categories, the quantity of finely ground sample obtained was less than the BF method stipulated

Table 1. Classification of shelled peanut samples by type of damage.

Category	Classification	Description
A	Healthy	Kernels without visible damage
B	Wrinkled	Seeds of abnormally small size or with wrinkled cuticles
C	Discoloured	Seeds with cuticle presenting slightly coloured areas
D	Millipede	Partially eaten seeds, from a pod with an opening for entry or exit of millipedes
E	Termite attacked	Partially eaten seeds, from a pod with an opening for entry or exit of termites
F	Weevils	Presenting visible damage by *Caryedon serratus*
G	Black seeds	Seeds with dark coating of unknown origin
H	Moldy seeds	Seeds with a white coloration
I	*Rhizopus/Mucor*	Seeds presenting a characteristic dark mycelium
J	*Aspergillus flavus*	Seeds presenting a characteristic yellow mycelium

50 g quantity for aflatoxin analysis. Consequently different amounts (50, 25, 10 or 5 g) were used for the analysis. The solvent and other reagent ratios were adjusted accordingly. Briefly, 4 g NaCl was added to a 50 g, peanut meal sample and blended with 250 ml methanol+water (60:40, v/v) and 100 ml hexane in a laboratory benchtop Waring blender (Waring, Torrington, CT) for 3 minutes. The slurry was centrifuged (Eppendorf 5810R, Eppendorf AG, Hamburg, Germany) at 2,000 rpm for 3 minutes. The chloroform was evaporated to near dryness using a rotary evaporator.The residue was reconstituted in 2 ml of chloroform and evaporated under a stream of nitrogen. The final extract was derivatised with trifluoracetic acid before injection into a Waters 474 HPLC unit equipped with a fluorescence detector.

Results and discussion

Mean aflatoxin B1 contamination

Table 2 shows the mean aflatoxin contamination of the two composite samples per condition category per month. Healthy peanuts were the least contaminated (mean =0.22 μg/kg) over the 7 months period. The second least contaminated category was immature and/or wrinkled peanuts, except for the month of December when the mean value was much higher than the rest of the levels within the category. Where enough sample material was available for the category for peanuts contaminated with *A. flavus*, the contamination level was the highest recorded among all the categories.

Table 2. Aflatoxin B_1 level (μg/kg) by month and average contamination of 73-33 peanut variety.

Sample	December	January	February	March	April	May	June	Number of month of storage	Mean aflatoxin content (μg/kg)
Healthy	0	1	0.23	0	0	0.17	0.13	7	0.22
Imm + wrink	292	29	1	1	1	21	0.23	7	49.32
Discoloured	111	297	1965	520	1070	588	563	7	730.57
Mouldy	802	4,761	3,124	625	1,658	1,215	1,560	7	1963.57
Mucor	NS	5,390	7,560	25	2,534	8	0.45	6	2586.24
Black	284	1,021	625	1,022	239	24	590	7	543.57
A. flavus	NS	69,455	308,041	246,240	NS	NS	NS	3	207,912.00
Caryedon	NS	NS	52	NS	763	515	314	4	411.00
Termites	NS	NS	2,392	NS	3,671	1,017	8,779	4	3964.75
Cary + Ter	5,918	4,012	NS	644	NS	NS	NS	3	3524.67
Millipede	4,275	4,051	4,274	4,045	10,336	5,400	5,424	7	5400.71

NS: No sample.

In addition, consistently high (>100 μg/kg) contamination was recorded in samples infested by millipedes or termites, and in samples that were mouldy or discoloured. These results, while confirming that *A. flavus* is the first source of contamination, also show that millipedes and termites contribute significantly to high levels of aflatoxin in peanuts. On the other hand, attack by *Caryedon serratus* doesn't seem to be a major factor. Because we used visual classification, it is possible that those samples classified as *Mucor/Rhizopus*, discoloured and mouldy may have been infected by *Aspergillus* spp. as well. It would seem therefore, that the major source of contamination in the study area is pre-harvest contamination, since millipedes and termites are known, in Senegal, to attack peanuts whilst if the field.

Where samples were available on every sampling occasion, there is some notable variation in the contamination level from month to month in almost all categories. We attribute this variation to heterogeneity of contamination normally observed in lots contaminated with aflatoxin. Overall, there is neither a decreasing nor increasing trend in the content of aflatoxin across the sampling period. This pattern indicates contamination before storage, perhaps during meadow harvest.

Influence of the sorting

We encountered problems with sorting that could have influenced the observed results. Initially, it was easy to sort the kernels into the various categories. With time of storage however, multiple forms of damage were often simultaneously present on one kernel resulting in classification difficulties. Only kernels that were distinctly healthy and those infected by *A. flavus* could be classified with confidence.

Conclusion

Results from this study have shown that healthy seeds and those wrinkled and/or immature are the least contaminated by aflatoxin B_1. Kernels visibly infected by *A. flavus*, weevil infestation and occurrence of other moulds such as *Mucor* and *Rhizopus* can be used as indicators of potential presence of aflatoxin. However, the true contribution of each of the analysed categories to the total aflatoxin load of a lot of peanuts may be influenced by their weight ratio within the lot.

A previous survey in the same zone (Delobel *et al.*, 1996) had shown that the infection by *A. flavus* was predominant in meadow-harvest peanuts. Our results concur with this observation. Although the near ubiquitous presence of *Caryedon serratus*, and the increase in damage it causes during storage contributes to the aflatoxin presence in peanuts globally, arthropods (weevils and termites), in presence of favourable conditions for contamination are the more significant vectors of infection and aflatoxin contamination if conditions are favourable in meadow-harvest. This is an important observation and could explain the consistent aflatoxin levels after the month of January.

References

Anonymous, 1995.Aflatoxins in peanut and peanut products. *Official methods of Analysis of AOAC International*, 16th Ed., vol 2, chap. 49, pp. 10.

Kane, A., Ba–Diop, N., and Diack, T.S., 1993. Unrefined peanut oil, a source of human exposure to aflatoxins. Detoxification procedures. *Africa Newslett. On Occup. Health and Safety* Suppl. 2/93: 43-47.

Waliyar, F., and Roquebert, M.F., 1979. Mycoflore des gousses et des graines d'arachide au Sénégal. *Revue de Mycologie* 43: 169–186.

Waliyar, F., and Zambettakis, C., 1979. Etude de la mycoflore des gousses et des graines d'arachide au Sénégal. *Oléagineux* 34: 191–198.

Section III:

Determination of mycotoxins

Indirect competitive ELISA for ochratoxin A detection in coffee and molecular identification of ochratoxin A producing *Aspergillus* strains

S. Fujii[1], E.Y.S. Ono[2], M.H.P Fungaro[3], E.N. Itano[4], T.C.R.M. Oliveira[1], C.E.C. Prete[1], M.H. Taniwaki[5], O. Kawamura[6], Y. Ueno[7] and E.Y. Hirooka[1]

[1]Department of Food and Drug Technology, State University of Londrina, P.O. Box 6001, 86051-990, Londrina, Parana, Brazil; hirooka@uel.br
[2]Department of Biochemistry, State University of Londrina, P.O. Box 6001, 86051-990, Londrina, Parana, Brazil
[3]Department of Biology, State University of Londrina, P.O. Box 6001, 86051-990, Londrina, Parana, Brazil
[4]Department of Applied Pathology, State University of Londrina, P.O. Box 6001, 86051-990, Londrina, Parana, Brazil
[5]Institute of Food Technology, P.O. Box 139, 13073-001, Campinas, Sao Paulo, Brazil
[6]Department of Biochemistry and Food Science, Faculty of Agriculture, Kagawa University, Ikenobe, Miki-cho, Kita-gun, Kagawa 761-0765, Japan
[7]Bioresearch Institute, 5-10-4, Takashimadaira, Itabashi-ku, Tokyo 175-0082, Japan

Abstract

Ochratoxin A (OTA) is of major concern in international coffee trade due the risk to human health. An indirect competitive enzyme-linked immunosorbent assay (IC-ELISA) was developed for OTA detection in non-roasted, roasted and instant coffee. The immunoassay showed a detection limit of 0.20 μg/kg and a linear range of 0.20-10 μg/kg. Results of spiked non-roasted, roasted and instant coffee samples showed good correlation (correlation coefficients, 0.86-0.98) with high performance liquid chromatography (HPLC). OTA recoveries by IC-ELISA from samples spiked at 5-70 μg/kg were 81.5% (non-roasted), 46.8% (roasted) and 64.4% (instant coffee) with coefficients of variation of 8.7%, 5.1% and 4.7%, respectively. Therefore, in its current form, the IC-ELISA method may not be an appropriate assay for OTA in roasted and instant coffee. In addition, *Aspergillus* strains isolated from some coffee samples were identified by sequencing of ITS regions amplification products. This molecular tool characterised *Aspergillus ochraceus, A. niger, A. tubingensis, A. carbonarius, A. flavus* and *A. tamari* isolated from freshly harvested coffee samples and *A. ochraceus,*

A. niger, *A. tubingensis*, *A. tamarii*, *A. caelatus*, *A. flavus* and *A. oryzae* isolated from stored coffee.

Keywords: ochratoxin, ELISA, HPLC, coffee, *Aspergillus*

Introduction

The natural occurrence of ochratoxins in field crops has been a topic of concern due to potential human and animal health hazards (Petzinger and Weidenbach, 2002; Grosso *et al.*, 2003). Ochratoxin A (OTA) may cause several ailments including nephrotoxicity, teratogenicity, cytotoxicity, immunotoxicity, and genotoxicity (Hussein and Brasel, 2001). Ochratoxin A in coffee was first reported by Levi *et al.* (1974) but has since been extensively studied as a mycotoxin of public health and international trade interest (Bucheli *et al.*, 1998). Previously known to be predominantly produced by *A. ochraceus* and *P. verrucosum* (Petzinger and Weidenbach, 2002), other fungal species mainly belonging to the *Aspergillus* group have recently been associated with the occurrence of OTA (Mouyna *et al.*, 2002).

Detection of OTA may be achieved by thin layer chromatography (TLC) (Pittet and Royer, 2002), high performance liquid chromatography (HPLC) (Sibanda *et al.*, 2002) and enzyme-linked immunosorbent assays (ELISA) (Sibanda *et al.*, 2001). Immunoassays are of relatively low cost, are rapid and are highly sensitive. However, immunoassays may yield false positives due to non-specific food component-antibody interactions (Hefle, 1995). This matrix interference can be minimised by prior dilution of the sample extract, or incorporation of an adequate clean-up step (Ono *et al.*, 2000). In addition, where molecular biology techniques for advanced taxonomy are available, identification of relevant fungal species in the food commodity provide further safety nets for prevention of OTA intake.

In this study, the performance of a monoclonal antibody (anti-OTA.7 mAb) based IC-ELISA (indirect competitive enzyme-linked immunosorbent assay) for detection of OTA in green coffee produced in Paraná State, Brazil, and roasted and instant coffee, was evaluated by comparison with HPLC. Furthermore, *Aspergillus* sp. strains were identified by sequencing ITS regions of rDNA.

Material and methods

Sample preparation and IC-ELISA and HPLC analysis

Sixteen freshly harvested green coffee (cultivar IAPAR-59) samples were obtained from a university farm. Fourteen stored (non-roasted) green coffee samples were obtained from regional farms in northern Paraná State - Brazil, while roasted and instant coffee samples were acquired from local commercial sources. Five grams of ground coffee was extracted with acetonitrile+water (3:1), centrifuged, and the supernatant mixed with chloroform and 0.5% NaCl solution. The chloroform was dried and the residue dissolved in methanol-PBST (phosphate buffer saline with 0.5% Tween 20) (1:9) for IC-ELISA analysis. Ochratoxin A was analysed by IC-ELISA according to Kawamura *et al.* (1989). The procedure included standardising the OTA-BSA conjugate and horseradish peroxidase labelled anti-mouse IgG (anti-IgG-HRP) and, determining the cross-reactivity of anti-OTA.7 mAb with OTA analogues. An ochratoxin A standard curve by IC-ELISA was obtained using a range of concentrations (0.2 to 10.0 ng/ml) of OTA in methanol-PBST. The OTA standard was obtained from Sigma (Sigma, St. Louis, MO).

For HPLC analysis the chloroform residue was dissolved in 2 ml methanol+1% sodium bicarbonate (70:30) for green coffee or 2 ml 1% sodium bicarbonate (other coffee types and products). The dissolved residue was added to 8 ml PBS-0.01% Tween 20 (PBST) and filtered through a glass microfiber filter paper (GF/C, Whatman International Ltd., Maidstone, England). The filtrate was subjected to immunoaffinity column (Ochratest, Vicam Inc., Watertown, MA) cleanup as described by (Pittet *et al.*, 1996). Twenty microliters (20 μl) of the column eluate was injected into a reversed phase HPLC set up and eluted isocratically. The HPLC unit was a LC-10 AD (Shimadzu, Japan) fitted with Nucleosil 100 C18 column (0.4 x 25 cm, pore size 5 μm) (Merck, Germany) and a fluorescence detector (RF 535) (Shimadzu, Japan). The mobile phase was acetonitrile+water+acetic acid (99:99:2, v/v/v) pH 3.0 at a flow rate of 0.8 ml/min.

The standard curve for HPLC quantification was obtained using higher OTA concentrations (10.0 to 100 ng/ml).

Recoveries and matrix interference

Ochratoxin A recoveries were determined by spiking the various types of coffees with a range of OTA concentrations in methanol (5.0 to 70.0 ng/g) and analysis by both HPLC and IC-ELISA. Matrix interference on IC-ELISA was determined using non-spiked samples (OTA negative non-roasted, roasted and instant coffee) extracted as above, and diluted from 1:10 to 1:2000 in methanol-PBST (1:9).

IC-ELISA x HPLC correlation analysis

Immunoaffinity column-ELISA x HPLC correlations for OTA analysis in the various types of coffees tested were performed by Pearson's correlation test (Statistica software version 5.0, StatSoft Inc., USA, 1984-1995) to obtain linear regression equations and correlation coefficients.

Molecular identification of Aspergillus *strains*

Coffee beans were externally disinfected with 1% sodium hypochlorite. Fungal species were isolated according to Blotter test (Neergard, 1979). The genera identification of isolated fungi was carried out as described by Samson *et al.* (1995). Genomic DNA was determined by cultivating *Aspergillus* sp. in liquid medium (Pontecorvo *et al.*, 1953) and extracting the DNA according to Azevedo *et al.* (2000). The DNA analysis was performed by a fluorimetric method (DynaQuant 200, Hoefer Pharmacia Biotech, Inc., San Francisco, California). Fragments of ITS1, gene 5.8S and ITS2 regions were amplified by ITS1 and ITS4 primers (White *et al.*, 1990), and analysed by gel electrophoresis in 1 x TEB buffer (Sambrook *et al.*, 1989). Fragments of rDNA amplified by PCR were sequenced using ITS4 primer (Invitrogen- 3.0 pmol/ml), premix of DYEnamic ET terminator (Amersham Biosciences, Piscataway, New Jersey) and PCR-rDNA. The sequencing reaction was analysed by MegaBACE™ 1000 sequencer (Amersham Biosciences, Piscataway, New Jersey), and the data were analysed using the Phred/Phrap/Consed program (Ewing *et al.*, 1998) by comparing with the National Center for Biotechnology and Information (NCBI) - BLAST data bank.

Results and discussion

Appropriate dilutions of immunologic reagents for IC-ELISA were determined as shown in Table 1. Dilutions of OTA-BSA conjugate (1:1000 and 1:2000), anti-OTA mAb (1:1000 and 1:2000) and anti-IgG-HRP (1:1000 to 1:3000) were tested to select a combination yielding 50% maximal binding (absorbance of 0.7 to 0.9) (Kawamura *et al.*, 1989). Inspection of results shown in Table 1 shows that experiment 2a achieved the desired maximal binding [*i.e.* OTA-BSA (1:1000), anti-OTA mAb (1:2000) and anti-IgG-HRP (1:1000), which produced an absorbance of 0.903]. The conditions determined in experiment 2a were therefore selected and used in subsequent tests. Total protein in OTA-BSA conjugate was 4.76 mg/ml.

Table 1. Dilutions of OTA-BSA conjugate, anti-OTA monoclonal antibody and HRP labelled anti-IgG tested for OTA IC-ELISA standardisation.

Experiment		Dilutions			IC-ELISA
		OTA-BSA	anti-OTA MAb	anti IgG-HRP	(A450nm) [a]
1	1a	1:1000	1:1000	1:1000	0.913
	1b	1:1000	1:1000	1:1500	0.727
	1c	1:1000	1:1000	1:2000	0.676
	1d	1:1000	1:1000	1:3000	0.581
2	2a	1:1000	1:2000	1:1000	0.903
	2b	1:1000	1:2000	1:1500	0.692
	2c	1:1000	1:2000	1:2000	0.555
	2d	1:1000	1:2000	1:3000	0.500
3	3a	1:2000	1:1000	1:1000	0.453
	3b	1:2000	1:1000	1:1500	0.318
	3c	1:2000	1:1000	1:2000	0.322
	3d	1:2000	1:1000	1:3000	0.338
4	4a	1:2000	1:2000	1:1000	0.431
	4b	1:2000	1:2000	1:1500	0.433
	4c	1:2000	1:2000	1:2000	0.374
	4d	1:2000	1:2000	1:3000	0.314

[a]average of values in triplicate.

Considering that the specificity of an antibody is essential for reproducibility and sensitivity of IC-ELISA, cross-reactivity of anti-OTA MAb was analysed using OTA analogues. The amounts of analogues that inhibited 50% antibody binding were calculated from an inhibition curve of mAb using IC-ELISA. Ochratoxin A 7 antibody cross-reacted with ochratoxin C (79.4%) but only minimally reacted with (4 R)-4-hydroxy-ochratoxin A (1.24%), ochratoxin B (0.07%) and ochratoxin α (0.06%). Since the natural occurrence of ochratoxin C has been uncommon, its contribution to data generated using an assay performed with anti-OTA.7 mAb would be minor.

An IC-ELISA standard curve for OTA was thus developed using the standardised dilutions of anti-OTA.7 mAb, OTA-BSA and anti-IgG-HRP shown in Table 1. Figure 1 shows that the curve was linear between 0.20 and 10 ng/ml, and had a detection limit

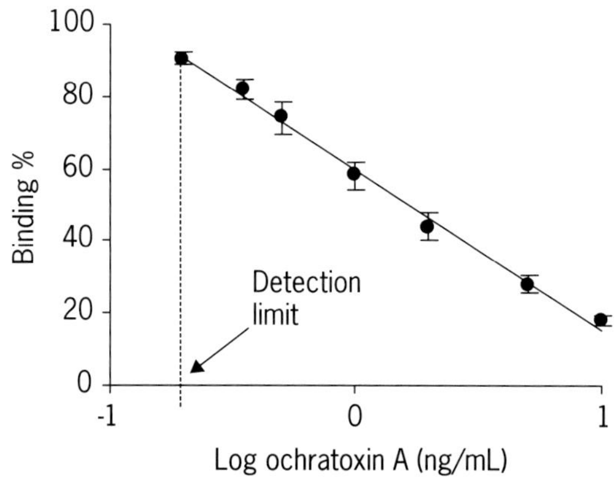

Figure 1. Standard curve of OTA detection by IC-ELISA. Each data point is a mean of seven binding data carried out on different days during 1 month. The detection limit was 0.20 ng/ml, which was the minimum concentration over 9.3% inhibition (3-fold SD at 0 ng/ml OTA) detected by IC-ELISA, corresponding to 3.73 ng/g OTA in coffee beans.

of 0.20ng/ml (3.73ng/g). The low detection limit confirms the sensitivity of this IC-ELISA method for OTA analysis.

The applicability of the developed IC-ELISA procedure was evaluated using natural and artificially OTA contaminated coffee. Figure 2 shows matrix interference of the various types of coffee samples tested in the analysis of OTA by IC-ELISA. Dilutions of the extracts of two logs (1/100) or more minimised the undesired interference for each coffee type. A single log (1/10) dilution resulted in interference of slightly more than 20%.

Table 2 shows a comparison of IC-ELISA and HPLC results for artificially OTA contaminated coffee samples. At less than 4 ng/g OTA, IC-ELISA overestimates the concentration by as much as 50% in non-roasted coffee. The overestimation may be due to probable loss of OTA at the immunoaffinity clean-up step for HPLC or matrix interference in the ELISA. Overall mean IC-ELISA/HPLC agreement among coffee types ranged from 1.02 to 1.24 with correlation coefficients of 0.98, 0.98 and 0.86 in non-roasted, roasted and instant coffee, respectively.

Table 3 shows OTA recovery by IC-ELISA and HPLC using different matrices. Artificially contaminated non-roasted coffee with 5 to 70ng/g OTA showed recovery values from 55 to 106.1% by IC-ELISA (average, 81.53%) as compared to HPLC

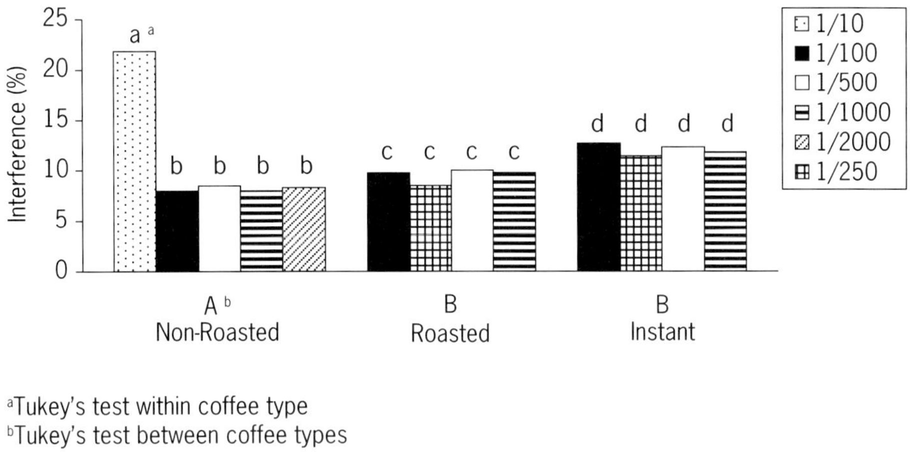

Figure 2. Effect of dilution on matrix interference during OTA analysis in coffee (non-roasted, roasted and instant) by IC-ELISA. Similar letters on bars indicate lack of significance between dilutions as determined by Tukey's test at the 5% level.

(average, 80.54%). These data demonstrate that IC-ELISA could reliably be used for analysis of OTA in non-roasted coffee. However OTA recoveries from the other coffee types were low. Mean recoveries from roasted and instant coffee spiked at 10-50 ng/g were 46.73% and 64.35%, respectively for IC-ELISA and 45.91% and 55.15%, respectively for HPLC. These results indicate that both methods may be inappropriate for these matrices.

Nevertheless, the IC-ELISA/HPLC ratios of 1.02 and 1.24 (Table 2), in addition to correlation coefficients of 0.98 and 0.86 for roasted and instant coffee, respectively, indicated that IC-ELISA and HPLC are of comparable performance for these matrices.

Among isolated fungi identified by classical taxonomy, approximately 70% belonged to the genus *Aspergillus*. The molecular analysis of sequences of amplified fragments of rDNA identified *A. flavus/oryzae* (Section *Flavi*), *A. niger* (Section *Nigri*), *A. ochraceus* (Section *Circundati*), *A. tubingensis* (Section *Nigri*), *A. tamarii* (Section *Flavi*) and *A. carbonarius* (Section *Nigri*). Section *Nigri* represented 35% of the total *Aspergillus* spp. identified. Section *Nigri* comprised *A. niger* (24%), *A. tubingensis* (10%), and *A. carbonarius* (1%). *A. tubingensis* and *A. niger* are morphologically similar and could only be distinguished using molecular tools.

Table 2. IC-ELISA/HPLC correlation of OTA analysis in coffee.

Coffee		OTA (ng/g)		IC-ELISA/HPLC		
		HPLC	IC-ELISA	Relation	X[a]	r[b]
Non-Roasted	1	3.94	5.75	1.46		
	2	3.85	4.86	1.26		
	3	8.19	8.96	1.09		
	4	6.99	7.0	1.00		
	5	18.49	20.22	1.09	1.02	0.98
	6	18.89	20.50	1.09		
	7	37.99	25.0	0.66		
	8	42.63	30.0	0.70		
	9	65.69	60.0	0.91		
	10	65.09	59.4	0.91		
Roasted	1	4.57	5.0	1.09		
	2	6.23	6.0	0.96		
	3	12.12	13.5	1.11	1.02	0.98
	4	13.19	13.1	0.99		
	5	16.55	16.0	0.97		
Instant	1	6.52	7.1	1.09		
	2	6.89	8.6	1.25		
	3	16.17	17.95	1.11	1.24	0.86
	4	9.80	17.8	1.82		
	5	23.17	21.53	0.93		

[a]X average correlation IC-ELISA/HPLC.
[b]r-coefficient of correlation IC-ELISA/HPLC.

A. ochraceus section *Circundati* represented 17% of the total isolates. Taniwaki *et al.* (2003) reported that 75% of *A. ochraceus* isolates were potentially OTA producers.

Conclusions

The results obtained in this study demonstrate the effectiveness of an anti-OTA 7 monoclonal antibody-based IC-ELISA for detection (and quantification) of OTA in non-roasted coffee grains and by-products without a clean-up step. The immunoassay showed reliability and reproducibility equivalent to HPLC. The possible use of

Table 3. Comparison of HPLC and IC-ELISA OTA recoveries in non-roasted, roasted and instant coffee.

Coffee		**OTA (ng/g)**		**Recovery (%)**	**CV (%)**	**Average recovery (%)**
		added	**recovered**			
HPLC [a]	Non-roasted	5	3.90 ± 0.06	78.00 ± 1.30	1.4	
		10	7.59 ± 0.80	75.90 ± 8.50	10.1	
		25	18.69 ± 0.30	74.76 ± 1.13	1.6	80.54 ± 7.53
		50	40.31 ± 3.28	80.62 ± 6.58	8.1	
		70	65.39 ± 0.40	93.41 ± 0.64	0.6	
	Roasted	10	5.40 ± 1.17	54.00 ± 11.70	20.9	
		25	12.66 ± 0.76	50.64 ± 3.03	5.9	45.91 ± 11.22
		50	16.55 ± 0	33.10 ± 0	0	
	Instant	10	6.71 ± 0.26	67.10 ± 2.62	3.7	
		25	13.00 ± 4.50	52.00 ± 18.10	30.0	55.15 ± 10.73
		50	23.17 ± 0	46.34 ± 0	0	
IC-ELISA [b]	Non-roasted	5	5.31 ± 0.63	106.10 ± 12.59	11.9	
		10	7.98 ± 1.39	79.80 ± 13.86	17.4	
		25	20.36 ± 0.20	81.44 ± 0.80	1.0	81.53 ± 18.20
		50	27.50 ± 3.54	55.00 ± 7.07	12.9	
		70	59.70 ± 0.42	85.29 ± 0.60	0.7	
	Roasted	10	5.50 ± 0.71	55.00 ± 7.10	12.9	
		25	13.30 ± 0.30	53.20 ± 1.13	2.3	46.73 ± 12.79
		50	16.00 ± 0	32.00 ± 0	0	
	Instant coffee	10	7.85 ± 1.06	78.50 ± 10.61	13.5	
		25	17.88 ± 0.11	71.50 ± 0.42	0.6	64.35 ± 18.77
		50	21.53 ± 0	43.06 ± 0	0	

[a] average ± SD of duplicate assays.

[b] average ± SD of triplicate assays.

molecular tools to isolate potentially ochratoxigenic *A. niger, A. ochraceus* and *A. carbonarius* from the coffee samples augments analytical options and highlights the importance of OTA monitoring in coffee and cereals.

Acknowledgments

CNPq, CAPES, JICA, FINEP and Araucaria Foundation supported this research.

References

Azevedo, A.C.S., Furlaneto, M.H., Sosa-Gomez, D.R., and Fungaro, M.H.P., 2000. Molecular characterization of Paecilomyces fumosoroseus (Deuteromycotina: Hyphomycetes) isolates. *Scientia Agricola* 57: 729-732.

Bucheli, P., Meyer, I., Pittet, A., Vuataz, G., and Viani, R., 1998. Industrial Storage of Green Robusta Coffee under tropical Conditions and Impact on Raw Material Quality and Ochratoxin A Content. *J. Agric. Food Chem.* 46: 4507-4511.

Ewing, B., Hillier, L., Wendl, M.C., and Green, P. 1998. Base-calling of automated sequencer traces using phred. I. Accuracy assessment. *Genome Research* 8: 175–185.

Grosso, F., Saïd, S., Mabrouk, I., Fremy, J.M., Castegnaro, M., Jemmali, M., and Dragacci, S., 2003. New data on the occurrence of ochratoxin A in human sera from patients affected or not by renal diseases in Tunisia. *Food and Chemical Toxicology* 41: 1133-1140.

Hefle, S.L., 1995. Immunoassay fundamentals. Food Technology 49: 102-107.

Hussein, S.H., and Brasel, J.M., 2001. Toxicity, metabolism and impact of mycotoxins on humans and animals. *Toxicology* 167: 101-134.

Kawamura, O., Sato, S., Kajii, H., Nagayama, S., Ohtani, K., Chiba, J., and Ueno, Y., 1989. A sensitive Enzyme-Linked Immunosorbent assay of Ochratoxin A based on Monoclonal Antibodies. *Toxicon* 27: 887-897.

Levi, C.P., Trenk, H.L., and Mohr, H.K., 1974. Study of the occurrence of ochratoxin A in green coffee beans. *Journal of the Association of Official Analytical Chemists* 57: 866-870.

Mouyna, I., Sarfati, J., Recco, P., Fontaine, T., Henrissat, B., and Latge, J.P., 2002. Molecular characterization of a cell wall-associated beta (1-3) endoglucanase of *Aspergillus fumigatus*. *Medical Mycology* 40: 455-464.

Neergard, P., 1979. *Seed Pathology*, vol I, McMirlan, London, pp. 739-754.

Ono, E.Y.S., Kawamura, O., Ono, M.A., Ueno, Y., and Hirooka E.Y., 2000. A comparative study of indirect competitive ELISA and HPLC for fumonisin detection in corn of the State of Paraná, Brazil. *Food and Agricultural Immunology* 12: 5-14.

Petzinger, E., and Weidenbach, A., 2002. Mycotoxins in the food chain: the role of ochratoxins. *Livestock Production Science* 76: 245-250.

Pittet, A., and Royer, D., 2002. Rapid, low cost thin layer chromatographic screening method for the detection of ochratoxin A in green coffee at control level of 10μg/kg. *Journal of Agricultural and Food Chemistry* 50: 243-247.

Pittet, A., Tornare, D., Huggett, A., and Viani, R., 1996. Liquid chromatographic determination of ochratoxin A in pure and adulterated soluble coffee using na immunoaffinity column cleanup procedure. *Journal of Agricultural and Food Chemistry* 44: 3564-3569.

Pontecorvo, G., Roper, J.A., Hemmons, L.M., Macdonald, K.D., and Bufton, AW.J. 1953. The genetics of *Aspergillus nidulans*. *Adv. Genet.* 5: 141-148.

Sambrook, J., Fristsch, E.F. and Maniatis, T., 1989. *Molecular cloning, A laboratory manual.* Cold Spring Harbor Laboratory Press, New York.

Samson, R.A., Hoekstra, E.S., Frisvad, J.C., and Filtenborg, O., 1995. Identification of the common food-borne fungi. In: *Introduction to food-borne fungi,* Samson, R. A., Hoekstra, E.S., Frisvad, J.C., and Filtenborg, O. (Eds.), Centraalbureau voor Schimmelcultures, The Netherlands, pp. 3-234.

Sibanda, L., De Saeger, S., Bauters, T.G.M., Nelis, H.J., and Van Peteghem, C., 2001. Development of a Flow through enzyme immunoassay and application in screening green coffee samples foe ochratoxin A with confirmation by High Performance Liquid Chromatography. *Journal of Food Protection* 64: 1597-1602.

Sibanda, L., De Saeger, S., and Van Peteghem, C., 2002. Optimization of solid-phase clean-up prior to liquid chromatographic analysis of ochratoxin A in roasted coffee. *Journal of Chromatography A* 959: 327-330.

Taniwaki, M.H., Pitt, J.I., Teixeira, A.A., and Iamanaka B.T., 2003. The source of ochratoxin A and Brazilian coffee and its formation and relation to processing methods. *International Journal of Food Microbiology* 82: 173-179.

Van Egmond, H.P., 1996. Analytical Methodology and regulations for ochratoxin A. *Food Additives and Contaminants* 13: 11-13.

White, T.J., Bruns, T., Lee, S., and Taylor, J., 1990. Amplification and direct sequencing of fungal ribosomal RNA genes for phylogenetics. In: *PCR Protocols: A Guide to Methods and Applications,* Innis, M.A., Gelfand, D.H., Sninsky, J.J., and White, T.J., (Eds.), Academic Press, San Diego, pp. 315-322.

Simultaneous determination of aflatoxin, ochratoxin A and zearalenone by a new immunoaffinity column

N. Zabe, E. Wadleigh and B.A. Cohen

VICAM, 313 Pleasant St., Watertown, MA 02472, USA; nancyz@vicam.com

Abstract

Existing clean up methods for HPLC determination of mycotoxins require separate analysis for aflatoxin, ochratoxin A, and zearalenone. Described is a method for simultaneous determination of aflatoxin, ochratoxin A, and zearalenone using one sample clean up and one HPLC run. A sample is extracted in aqueous methanol. The extract is diluted with buffer and passed over an immunoaffinity column containing immobilised antibodies to aflatoxin, ochratoxin A, and zearalenone (AOZ HPLC™ column). The column is then washed and eluted. Using gradient flow and wavelength switching for detection, aflatoxins, ochratoxin A and zearalenone were measured in a single HPLC run of less than 30 minutes. Recoveries for all toxins were greater than 80% from samples of spiked maize, sorghum and wheat over the range of 0-100μg/kg total aflatoxin, 0-1,000 μg/kg zearalenone and 0-100 μg/kg ochratoxin A. Analysis using multiple mycotoxin immunoaffinity columns can save time and money in performing the mycotoxin analysis required by many countries worldwide.

Keywords: aflatoxin, ochratoxin A, zearalenone, LC, sample preparation, immunoaffinity chromatography, mycotoxins

Introduction

Awareness of the threat of mycotoxins to human and animal health is increasing (Bennett and Klich, 2003) and has led to the institution of regulatory limits (FAO, 1997). The aflatoxins are the most regulated mycotoxins. The European community in 2002 added a regulation for ochratoxin A (EC, 2002) and is currently evaluating possible regulation of other mycotoxins. The global transshipment of grain may lead to the blending of grain grown in different climatic regions, some conducive to the growth of different fungi and production of different mycotoxins (Scudamore and Patel, 2000). Surveys of different regions of the world have demonstrated the existence of the aflatoxins, ochratoxin A, and zearalenone in maize and sorghum (Janardhana *et al.*, 1999; Reddy and Nusrath, 1986; Micco *et al.*, 1986). In addition, many food and feed samples contain maize and sorghum. New methods are therefore, required to meet the need for faster and easier analysis, and the desire for multitoxin analysis in a single sample.

Existing methods for determining more than one mycotoxin in a single analysis include TLC (for example, Balzer *et al.*, 1978; Soares and Rodriguez-Amaya, 1989), HPTLC (Lee *et al.*, 1980), HPLC (Chamkasem *et al.*, 1988; Joshua *et al.*, 1996), LC with gel permeation chromatography (Dunne *et al.*, 1993) and LC/MS/MS (Spanjer *et al.*, 2003; Krol and Siantar, 2004). The method presented here combines the use of immunoaffinity chromatography with HPLC to analyse for aflatoxin, ochratoxin A, and zearalenone from a single sample extract. The immunoaffinity column provides a sample clean up step that is quick (less than 20 minutes), easy, and requires only methanol as solvent. The HPLC part of the method allows for detection of the mycotoxins in less than 30 minutes using HPLC instrumentation already available in many laboratories. A previous trial was done using rice, rye and pig feed (Göbel and Lusky, 2004). The work presented in this paper was carried out in maize, wheat and sorghum. Steps in the procedure that were studied were the extract dilution, filtration, immunoaffinity column washes and the immunoaffinity column elution.

Materials and methods

Reagents and solvents

Acetonitrile and methanol (both HPLC grade) were from Burdick & Jackson (Muskegon, MI). Acetic acid was HPLC grade from EMD Chemicals (Gibbstown, NJ, USA). Ultrapure water was produced by a Millipore Milli-Q system (Millipore, Bedford, MA, USA). Aflatoxin, ochratoxin A, and zearalenone standard mixture, at a concentration of 5 ng/μl aflatoxin B_1, 1 ng/μl aflatoxin B_2, 3 ng/μl aflatoxin G_1, 1 ng/μl aflatoxin G_2, 10 ng/μl ochratoxin A, and 100 ng/μl zearalenone in methanol, was purchased from Trilogy Analytical Laboratories (Washington, MO, USA). AOZ HPLC™ immunoaffinity columns and phosphate buffered saline (PBS) with 0.01% Tween 20 concentrate were obtained from VICAM LP (Watertown, MA, USA).

Apparatus

The HPLC apparatus consisted of a Waters 2695 Separations module, a Waters 474 fluorescence detector, a Symmetry C_{18} column (4.6 × 150mm, 3.5μm) fitted with a Symmetry C_{18} (3.9 × 20mm, 5 μm) guard column. The mobile phase consisted of combinations of three solvent systems (solvent A – 100% methanol, solvent B – 100% acetonitrile, and solvent C – water with 0.1% acetic acid). The gradient was 0–12 minutes isocratic 25% A, 15% B, 60% C, 12–14 minutes linear gradient to 10% A, 50% B, 40% C, 14-24 minutes held at 10% A, 50% B, 40% C, and immediately returned to 25% A, 15% B, 60% C at minute 24 followed by a 2 min delay for equilibration. The flow rate was 1 ml/minute with a column temperature of 30 °C. The wavelength settings on the fluorescence detector were excitation 365 nm, emission 455 nm from 0-14 minutes for the aflatoxins, excitation 276 nm, emission 460 nm from 14 to 21 minutes for zearalenone and excitation 330 nm, emission 460 nm from 21 to 25 minutes for ochratoxin A. All gradient and wavelength changes programming, and data management were performed using Waters Empower software (Milford, MA, USA).

A Post Column Waters reagent manager pump with saturated iodine at 0.2ml/min flow rate was controlled by a switch to go on at 0 minute, off at 14 minutes, and then back on at 25 minutes for the next injection. The iodine was added to the HPLC eluate through a "T" fitting and then passed through a circular heated reaction coil at 70 °C (Fiatron [Eppendorf] FH-40 heater and TC 50 controller) for one minute before going through the fluorescence detector.

Sample preparation and immunoaffinity clean up

Test portions of 25 g ground sample (maize, sorghum or wheat) were blended with 100 ml methanol: water (80:20) for 2 minutes. Extracts were filtered through fluted filter paper. Ten milliliters of the filtered extract was diluted with 40 ml PBS containing 0.01% Tween 20. Twenty milliliters of the diluted extract was passed through the AOZ HPLC Immunoaffinity column at a flow rate of about 3-4 ml per minute (about 1 drop per second). The immunoaffinity column was washed with 10 ml PBS containing 0.01% Tween 20, followed by 10 ml water. After the water wash, the column was dried by passing air through the column for two to three seconds. The toxins were eluted by passing 1.5 ml methanol through the column followed by 1.5 ml water containing 0.1% acetic acid. The eluate was collected in a single vial, mixed, and 100 μl injected into the HPLC unit.

Results and discussion

HPLC determination

Figure 1 shows chromatograms of a reagent blank (a) and a maize sample containing toxins below the limits of detection (b). There is a baseline shift where the excitation and emission wavelengths are changed. Figure 1 also shows example chromatograms for maize samples spiked at low (c) and high (d) levels of aflatoxin, ochratoxin A, and zearalenone. Each of the toxins can be clearly identified.

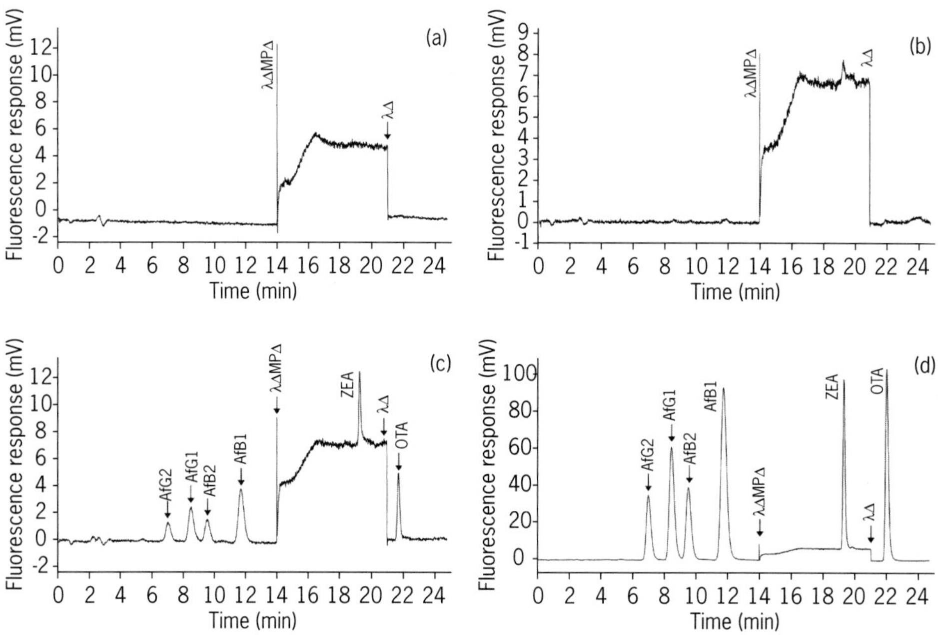

Figure 1. Aflatoxin (AF), ochratoxin A (OTA) and zearalenone (ZEA) chromatograms of (a) 100 μl injection of reagent blank (methanol: 0.1% acetic acid (50:50); (b) blank maize sample (< 0.15 μg/kg AFB_1, AFB_2, AFG_1, AFG_2, < 5 μg/kg ZEA, and < 0.25 μg/kg OTA); (c) blank maize spiked with 1 μg/kg AFB_1, 0.2 μg/kg AFB_2, 0.6 μg/kg AFG_1, 0.2 μg/kg AFG_2, 20 μg/kg ZEA, and 2 μg/kg OTA; (d) blank maize spiked with 50 μg/kg AFB_1, 10 μg/kg AFB_2, 30 μg/kg AFG_1, 10 μg/kg AFG_2, 1000 μg/kg ZEA, and 100 μg/kg OTA. Note scale change at high spike level that reduces the visibility of the baseline shift.

Method for sample preparation

A number of different steps in the analytical procedure were studied to optimise sample preparation. The steps studied were: extract dilution, column washes, extract filtration, and immunoaffinity column elution. The first step in the sample preparation method evaluated for maize was the extract dilution and the immunoaffinity column wash. Four solutions were evaluated for diluting the maize extract and for the first column wash: water, PBS, PBS with 0.01% Tween 20, and PBS with 0.1% Tween 20. Toxin recoveries for all the dilution solutions investigated averaged 75% or greater with the exception of the recovery of ochratoxin A. Ochratoxin A recovery using water dilution and water column wash averaged 49%. Further investigations (with all toxins) involved experiments using a PBS dilution followed by two 10 ml water washes, compared to a PBS dilution followed by one 10 ml PBS wash then one 10 ml water wash. The PBS immunoaffinity column wash followed by a water wash gave greater than 80% recovery of ochratoxin A.

Although recoveries for all toxins were comparable among the various combinations of PBS and PBS with Tween 20 washes, the methanol eluates from the immunoaffinity column were cloudy when using the PBS solution without Tween 20. The methanol eluates from the samples prepared using a PBS 0.01% Tween 20 or PBS 0.1% Tween 20 solution were clear, allowing for direct injection of the eluate onto the HPLC. Consequently, PBS with 0.01% Tween 20 was chosen as the solution for extract dilution and column wash.

Although filtration removed the cloudiness in the extract, it decreased toxin recovery in some samples. Filtration of the diluted extract was hence eliminated, since it was possible to load an unfiltered extract onto the column. Elimination of the filtration steps saves time and money.

The last step in the sample preparation was elution from the immunoaffinity column and drying down of the sample eluate. Aflatoxins, ochratoxin A, and zearalenone standards were dried down and reconstituted in methanol and water in the following types of vials: borosilicate culture tubes, commercially silanised tubes, tubes silanised with Pierce Surfacil, acid washed tubes, polypropylene plastic tubes, and Waters deactivated glass vials. Areas of each of the toxin peaks were compared to the area of a comparable standard that was not dried and reconstituted. Compared to standards that were not dried and reconstituted, percent recoveries for ochratoxin A and zearalenone were 90-110%. Drying and reconstitution recoveries for the aflatoxins were variable, especially for aflatoxins G_1 and G_2 where recoveries varied from 40 to 90%. Because of the inconsistent drying and reconstitution of aflatoxins G_1 and G_2, and the desire to use a single combined eluate for HPLC analysis, the eluates were not dried and reconstituted before HPLC injection.

Various elution methods were evaluated to determine the best elution to completely remove all toxins from the immunoaffinity column in the smallest volume for direct injection onto the HPLC. Eluting the immunoaffinity column with 1.5 ml methanol followed by 1.5 ml 0.1% acetic acid in water was found to give complete elution into one vial. Furthermore, the methanol+0.1% acetic acid water (50:50) is compatible with the HPLC mobile phase for direct injection.

From the observations enumerated above, the following procedure was adopted. A sample is extracted with methanol+water (80:20), filtered, and diluted with PBS with 0.01% Tween. The dilute extract (20 ml) is passed through the column and the column is washed with 10 ml PBS with 0.01% Tween followed by 10 ml water. The column is eluted sequentially with 1.5 ml methanol followed 1.5 ml 0.1% acetic acid in water. The developed method was evaluated using maize samples spiked with 0-100 μg/kg total aflatoxins, 0-1000 μg/kg zearalenone and 0-100 μg/kg ochratoxin A as shown in Table 1. The results in Table 1 show greater than 80% recovery of all toxins over the spiking ranges. Furthermore, two sets of samples, maize spiked at a low level (0.2 – 1.0 μg/kg aflatoxins, 2 μg/kg ochratoxin A, and 20 μg/kg zearalenone) and wheat spiked at a high level (2-10 μg/kg aflatoxins, 20 μg/kg ochratoxin A, and 200 μg/kg zearalenone), were each quantified ten times to determine repeatability. Repeatability data are shown in Table 2. The percent coefficient of variation (CV) for all toxins was less than 3% at the high level and less than 6% at the low level, with the exception of zearalenone (11%). At the low level, zearalenone percent CV was higher because of greater baseline variability caused by the mobile phase solvents at the lower excitation and emission wavelengths for zearalenone. The limit of detection, defined here as the smallest amount reproducibly and accurately detected with at least a 3:1 signal-to-noise ratio, was determined to be 0.1 μg/kg for aflatoxin G_2, 0.15 μg/kg for aflatoxin G_1, 0.1 μg/kg for aflatoxin B_2, 0.13 μg/kg for aflatoxin B_1, 5 μg/kg for zearalenone and 0.25 μg/kg for ochratoxin A.

Table 1. Mean percent recovery from blank maize spiked at various levels of the aflatoxins (AF), ochratoxin A (OTA) and zearalenone (ZEA) (n= 4).

Mycotoxin	2 μg/kg total aflatoxins[a] 20 μg/kg ZEA 2 μg/kg OTA	4 μg/kg total aflatoxins[a] 40 μg/kg ZEA 4 μg/kg OTA	20 μg/kg total aflatoxins[a] 200 μg/kg ZEA 20 μg/kg OTA	100 μg/kg total aflatoxins[a] 1000 μg/kg ZEA 100 μg/kg OTA
AFG_2	91	95	91	96
AFG_1	102	101	96	108
AFB_2	93	97	89	98
AFB_1	95	96	88	99
ZEA	93	101	92	94
OTA	83	89	83	88

[a]Aflatoxins added at a ratio of B_1:B_2:G_1:G_2 (5:1:3:1) by weight.

Table 2. Mean recoveries and coefficients of variation (repeatability) for aflatoxins (AF), ochratoxin A (OTA) and zearalenone (ZEA) in maize and wheat. Maize and wheat were arbitrarily spiked at two different levels.

Toxin	Spiked maize (µg/kg)	Recovery (µg/kg)	% CV	Spiked wheat (µg/kg)	Recovery (µg/kg)	% CV
AFG_2	0.2	0.18	5.5	2	1.9	1.7
AFG_1	0.6	0.53	3.3	6	5.9	0.95
AFB_2	0.2	0.18	5.2	2	1.9	1.4
AFB_1	1.0	0.84	4.1	10	9.5	0.98
ZEA	20	18	11	200	205	1.6
OTA	2	1.6	3.4	20	18	2.5

The newly developed method was also evaluated using wheat and sorghum samples. Recoveries comparable to those for maize were found when the procedure was applied to wheat and sorghum as shown in Table 3, with a small background peak in sorghum near the retention time of zearalenone (data not shown).

Table 3. Mean percent recovery of aflatoxin (AF), ochratoxin A (OTA), and zearalenone (ZEA) from spiked maize, wheat and sorghum using the AOZ HPLC immunoaffinity column.

		Matrix		
Mycotoxin	Toxin spiked (µg/kg)	Mean maize Recovery, % (n=4)	Mean wheat Recovery, % (n=10)	Mean sorghum Recovery, % (n= 2)
AFG_2	2	91	97	94
AFG_1	6	96	99	104
AFB_2	2	89	96	94
AFB_1	10	88	95	97
ZEA	200	92	102	109
OTA	20	83	90	85

Conclusion

The new AOZ HPLC immunoaffinity column clean up method improves the speed and ease of multiple mycotoxin analysis, while meeting the recovery criteria established by the CEN (European Committee for Standardization) working group "Biotoxin" for the acceptance of an analytical method (CEN, 1999). The HPLC conditions presented in this report offers a method for determining aflatoxins, ochratoxin A, and zearalenone simultaneously in one HPLC run.

Acknowledgements

We thank Dr. Stephen Powers for helpful discussions and for comments on the manuscript.

References

Balzer I., Bogdanic C., and Pepeljnjak S., 1978. Rapid thin layer chromatographic method for determining aflatoxin B_1, ochratoxin A, and zearalenone in corn. *Journal of the Association of Official Analytical Chemists* 61: 584-585.

Bennett, J.W., and Klich, M., 2003. Mycotoxins. *Clinical Microbiology Reviews* 16: 497-516

Benvenuti, M., Romano, J., Krol, J., and Sianter, D., 2004. *A Unified Method for Mixed Mycotoxin Analysis Using UV and Fluorescence Detection.* Poster presented at XI International IUPAC Symposium on Mycotoxins and Phycotoxins, May 17-21.

CEN (European Committee for Standardization), 1999. TC275-WG5 Food Analysis – Biotoxins – Criteria of analytical methods for mycotoxins. *CEN Report* CR 13505, Brussels.

Chamkasem, N., Cobb, W.Y., Latimer, G.W., Salinas, C., and Clement, B.A., 1989. Liquid chromatographic determination of aflatoxins, ochratoxin A, and zearalenone in grains, oilseeds, and animal feeds by post-column derivatization and on-line sample cleanup. *Journal of the Association of Official Analytical Chemists* 72: 336-341.

Dunne, D., Meaney, M., Smyth, M., and Tuinstra, L.G.M.Th., 1993. Multimycotoxin detection and clean-up method for aflatoxins, ochratoxin and zearalenone in animal feed ingredients using high-performance liquid chromatography and gel permeation chromatography. *Journal of Chroma*tography 629: 229-235.

EC, 2002, Commission Regulation (EC) No 472/2002 of 12 March 2002, amending Regulation (EC) No 466/2001 setting maximum levels for certain contaminants in foodstuffs. *Official Journal of the European Communities* L75: 18-20.

FAO, 1997. *Worldwide regulations for mycotoxins 1995, a compendium,* FAO Food and Nutrition Paper No. 64, Rome, pp. 1-45.

Göbel, R., and Lusky, K., 2004. Simultaneous Determination of Aflatoxins, Ochratoxin A, and Zearalenone in Grains by New Immunoaffinity Column/Liquid Chromatography. *Journal of AOAC International* 87: 411-416.

Janardhana, G.R., Raveesha, K.A., and Shetty, H.S., 1999. Mycotoxin Contamination of Maize Grains Grown in Karnataka (India). *Food and Chemical Toxicology* 37: 863-868.

Joshua, H., Joshua, M.H., Joshua, M.D., and Joshua, J., 1996. *The Simultaneous HPLC Determination of six aflatoxins, zearalenone and ochratoxin A using gradient elution, post-column photochemical derivatization and fluorescence detection.* Poster presented at the 110th AOAC International Annual Meeting and Exposition Orlando, Florida, September 8-12.

Krol, J., and Siantar, D., 2004. *Multi-analyte mycotoxin analysis using LC/MS/MS.* Poster presented at XI International IUPAC Symposium on Mycotoxins and Phycotoxins, May 17-21.

Lee, K.Y., Poole, C.F., and Zlatkis, A., 1980. Simultaneous multi-mycotoxin determination by high performance thin-layer chromatography. *Analytical Chemistry* 52: 837-842.

Micco, C., Grossi, M., Onori, R., Chirico, M., and Brera, C., 1986. Aflatoxin B1, ochratoxin A and zearalenone in Italian corn: monitoring during production 1982, 1983 and 1984 crop. *Rivista della Societa Italiana di Scienza dell'Alimentazione* 15: 113-116.

Reddy, B.N., and Nusrath, M., 1986. Mycotoxin contamination in stored sorghum. *Science and Culture* 52: 31-32.

Scudamore K.A., and Patel S., 2000. Survey for aflatoxins, ochratoxin A, zearalenone and fumonisins in maize imported into the United Kingdom. *Food Additives and Contaminants* 17: 407-416.

Soares, L.M.V., and Rodriguez-Amaya, D.B., 1989. Survey of Aflatoxins, Ochratoxin A, Zearalenone and Sterigmatocystin in Some Brazilian Foods by Using Multi-toxin Thin-Layer Chromatographic Method. *Journal of the Association of Official Analytical Chemists* 72: 22-26.

Spanjer, M., Scholten, J., and Rensen, P., 2003. *Single run LC-MS/MS Analysis of mycotoxins subject to actual and upcoming EU legislation in one sample extract.* Poster presented at Second World Mycotoxin Forum, Noordwijk, Netherlands, February 17-18.

Underestimation of fumonisin B_1 and ochratoxin A, from complex matrices using immunoaffinity columns

A. Pfohl-Leszkowicz[1], *A. Molinié*[1] *and M. Castegnaro*[1,2]

[1]*Unit of Toxicology & Food safety, Institut National Polytechnique de Toulouse, UMR CNRS N° 5503, Avenue de l'Agrobiopole, BP 107, F-31326 Auzeville-Tolosane, Toulouse, France; leszkowicz@ensat.fr*
[2]*Advisor, "Les Collanges", 07240 Saint-Jean-Chambre, France*

Abstract

This study describes our experience using immunoaffinity column approaches in the analysis of fumonisin B_1 and ochratoxin A in breakfast cereals. Reduced column binding capacity (45-25% loss) was observed when analyzing fumonisins in mixed breakfast cereals. Depending on the composition of the matrix (i.e. presence of wheat bran, lipids, and a corrector of acidity) and on the pH of the extraction medium ochratoxin A could be underestimated by (50-80%) and even not detected when using immunoaffinity columns to assay some breakfast cereals. These observations support the necessity to validate methods which use immunoaffinity columns for each new complex matrix.

Keywords: ochratoxin A, fumonisin B_1, mycotoxin analysis, immunoaffinity column, analytical problem

Introduction

From a recent survey by FAO/RIVM (FAO, 2004), there is a substantial increase in the number of countries that are setting legislation on mycotoxins. In order to reduce disputes between importing and exporting countries, the analytical data should be as comparable as possible, especially when levels of the toxin of interest are close to the regulatory limit. The present trend in analysis of mycotoxins is to use immunoaffinity columns as clean-up and enrichment devices, and the Association of Official Analytical Chemists international (AOAC INTERNATIONAL) and European Union (EU) have validated several methods which address a few food commodities. While these methods offer some advantages such as short assay times, cost, and some specificity; in complex mixtures, they may be affected by one or several compounds that may interfere with the antibodies.

Fumonisin B_1 is produced in the field by some *Fusarium* species growing on maize, oats and rice and on other cereals such as wheat, millet and sorghum (IARC, 2002; Pfohl-Leszkowicz, 1999). During storage, wheat, oats, barley, raisins, etc., contaminated by Penicillia (*Penicillium verrucosum, Penicillium aurantiogriseum, Penicillium citrinum* and *Penicillium expansum)* or Aspergilli *(Aspergillus ochraceus, Aspergillus carbonarius* and *Aspergillus niger*) can contain ochratoxin A. Both mycotoxins can be retained in the final product, such as breakfast cereals, during food processing (Molinié & Pfohl-Leszkowicz, 2002; Kim *et al.*, 2003). Currently there are no methods validated for analysis of mycotoxins in complex matrices such as breakfast cereals. There is therefore, a tendency to extrapolate the use of AOAC INTERNATIONAL or EU validated methods for the analysis of mycotoxins in such commodities.

We report here our experience with the analysis of both mycotoxins using immunoaffinity columns and more conventional clean-up/enrichment techniques (liquid-liquid partition). We analyzed fumonisin B_1 (FB_1) and ochratoxin A (OTA) in complex matrices such as breakfast cereals.

Material and methods

Ochratoxin A (free from benzene) and Ortho-phthalaldehyde (OPA) for the derivatization of FBs were obtained from Sigma Chemicals (France). Fumonisin B_1 was a generous gift from Dr D. Miller (Carleton University, Ottawa, ON, Canada). Two types of Immunoaffinity columns (IAC), Fumoniprep® and Ochraprep® were obtained from Rhône Diagnostic technologies (RDT, Lyon, France). Standard solutions of OTA, were prepared by dissolving 10 mg OTA in 1 ml of methanol. A series of working standards from 0.2 to 100 ng/ml of mycotoxin/ml were prepared by dilution. They were used to calibrate the LC detector response. The OTA stock solution was determined by measuring the absorbance at 333 nm and calculated by using the molar extinction coefficient e of 550m^2/mol.

Breakfast cereals for fumonisins and ochratoxin A analysis were collected from the French market as earlier described by Pfohl-Leszkowicz *et al.* (2004) and Molinié *et al.* (2004). Fumonisins were analysed as described by the method of Sydenham *et al.* (1992) that has since undergone an AOAC INTERNATIONAL validation (Visconti *et al.* 2001, Solfrizzo *et al.*, 2001). Ochratoxin A was analysed using a newly developed method for the simultaneous extraction and clean-up of OTA and citrinin (CIT) in breakfast cereals (Molinié *et al.*, 2004).

Extracts for each of the toxins were passed through the respective immunoaffinity columns according to the procedure provided by Rhône Diagnostics Technologies, (RDT) (1999) and Entwisle *et al.*, (2000). In order to test the performance (measured as recovery) of a method validated for maize and cornflakes (Solfrizzo *et al*, 2001; Visconti *et al.*, 2001) when applied to more complex matrices such as breakfast cereals containing fruits, sugar chocolate, oats and rice, nine samples containing the different ingredients were spiked with 300 µg/kg FB_1. The spiked samples were analysed using the Fumoniprep® column. Unpsiked samples of the same matrix were run concurrently with the spiked samples. Any toxin content naturally occurring in the sample was subtracted as background prior to the calculation of recoveries.

Results and discussion

Fumonisins

The method of Visconti and co-workers (Visconti *et al.*, 2001) stipulates the use of the Fumoniprep® column as a cleanup step in the analysis of maize and cornflakes for fumonisins. Depending on the composition of the breakfast cereal, the recoveries ranged from 54.4 to 74.9%. Only cornflakes yielded European Union (EU) study comparable recoveries of 70 to 75%. Recoveries from the other types of matrices (i.e., breakfast cereals containing, fruits, oats, rice, sugar chocolate, etc.) were low (Table 1).

By adding FB_1 to the extract of each breakfast cereal just before the immuno affinity chromatography (IAC) step, we determined that 22 to 45% of the losses could be ascribed to this step. The decreased recoveries from extracts of some types of

Table 1. Fumonisin B_1 recoveries from breakfast cereals spiked with 300 µg FB_1/kg breakfast cereal.

Sample Code	Sample composition	% FB_1 recovery
1	Maize, barley	74.9
2	Maize, barley	74.8
3	Rice, wheat flour, wheat bran	54.4
4	Maize	78.4
5	Cocoa, rice, maize, oats, wheat flour	61.7
6	Fruits, oats, wheat bran, wheat flour	79.1
7	Fruits, oats, wheat flour	59.0
8	Oats	67.0

breakfast cereals could be caused by the probable presence of some compound(s) which are recognised by the antibodies, thus blocking the fumonisin binding sites and reducing the trapping efficiency as observed with aflatoxin in maize (Castegnaro *et al.*, 2006). We suggest that the presence of high levels of unsaturated fatty acids in some of the breakfast cereal formulations may be causing the fumonisin binding interference (anti-fumonisin antibody blocking effect).

Ochratoxin A

The trend in the extraction of OTA has shifted from an acidified medium to a neutral medium (Entwisle *et al.*, 2000) or even an alkaline medium (RDT, 1999; Entwisle *et al.*, 2001). Table 2 compares the results of analysis of different samples of naturally contaminated breakfast cereals by the three methods (acidic extraction, neutral extraction and alkaline extraction). Although the number of samples analyzed was small, it can be seen that the amount of OTA detected would decrease with increasing pH of the extraction medium. We determined from previous studies that the optimum pH for good extraction of OTA is 2 (Pfohl-Leszkowicz *et al.*, 2004). In alkaline medium, some of the OTA is converted to an open-ring OTA (OP-OTA) which is not recognised by the antibodies.

Extracting from neutral rather than acidic medium may lead to about 20% loss as shown by results of samples 28, 16 and 53. Some breakfast cereal formulations contain sodium carbonate or sodium bicarbonate which correct for acidity in solution. In such instances, extraction using a neutral medium may lead to even higher losses (*e.g.* samples 16). Extracting OTA with an alkaline reagent (pH 8.4) (RDT, 1999) would therefore lead to even greater losses. To confirm this hypothesis, a solution containing 60 ng OTA in water or a solution containing 60 ng OTA in 1M sodium bicarbonate (pH 8.4) were passed through an Ochraprep® column. The OTA-1M sodium bicarbonate solution had been allowed 1 and 2 hour reaction time prior to being passed through the column. The 1 and 2-hour OTA-1M sodium bicarbonate reaction times led to 40 and 20% OTA recoveries, respectively. The loss of OTA was due to the conversion of OTA into open ring OTA (OP-OTA), which is not recognised by the antibodies on the immunoaffinity column. When pure OTA mixed with bicarbonate was analyzed by high performance liquid chromatography (HPLC), the OTA peak progressively disappeared as function of the contact time, concomitant with an increase in the peak of OP-OTA peak (data not shown).

Table 2. Comparison of OTA levels quantified using three different methods of purification

Sample code	Sample composition	Determined OTA level (µg/kg)		
		Acid extraction[a]	Neutral extraction[b]	Alkaline extraction[c]
16*	Cocoa, maize	3.4	1.6	<LOQ
25	Cocoa, maize, fruit, oats, rice, barley	<LOQ	ND	ND
28	Fruit, wheat flour, wheat bran, barley	8.8	7.2	5.6
51*	Maize	ND	ND	ND
52*	Cocoa, rice, wheat flour, oats	LOD	ND	ND
53	Oats, wheat bran, maize, raisin	1.0	0.4	ND
54	Oats, fruits	ND	ND	ND
55	Oats	<LOQ	ND	ND

[a]Acid extraction and partition (Molinié *et al.*, 2004).
[b]Neutral extraction and IAC (Entwisle *et al.*, 2000).
[c]Alkaline extraction and IAC (RDT, application note N° A9-P14.V1).
*Matrix contains a corrector for acidity (sodium carbonate or sodium bicarbonate).
LOQ, limit of quantification (0.2 µg/kg).
ND, not detected (i.e. < 0.05 µg/kg).

Conclusions

This work has demonstrated that IAC-based methods previously validated for one or two matrices may not be satisfactorily used for the analysis of the same toxins in other, sometimes very complex, matrices. The presence of some components from some formulations may interfere with the analytical process leading to the underestimation of the analyte of interest, which in our study were the mycotoxins fumonisin B_1 and ochratoxin A. In addition, in the case of OTA extraction under alkaline conditions or the presence in the matrix of a corrector for acidity (sodium carbonate or sodium bicarbonate) could induce opening of the lactone ring to produce OP-OTA. It should be realized however that the very acidic pH inside the stomach may induce the reverse conversion of OP-OTA back to OTA (Castegnaro *et al.*, 1991), thus increasing the potential hazard to consumer health arising from the underestimation of toxin content of the foodstuff. These analytical problems will have a serious impact on analytical accuracy, particularly at OTA levels close to those legislated as permissible limits by the EU.

References

Castegnaro, M., Barek, J., Jacob, J., Kirso, U., Lafontaine, M., Sansone, E.B., Telling, G.M. and Vu Duc, M., 1991. Laboratory decontamination and destruction of carcinogens in laboratory wastes : Some mycotoxins. IARC Publication N° 113, Lyon.

Castegnaro, M., Tozlovanu, M., Wild, C., Molinié, A., Sylla, A. and Pfohl-Leszkowicz, A., 2006. Advantages and drawbacks of immunoaffinity columns in analysis of mycotoxins in food. *Molecular Nutrition Food research,* 50: 480-487.

Entwisle, A.C., Williams, A.C., Man, P.J., Slack, P.T. and Gilbert, J., 2000. Liquid chromatographic method with immunoaffinity column clean up for determination of ochratoxin A in barley: collaborative study. *Journal of AOAC International,* 83: 1377-1383.

Entwisle, A.C., Williams, A.C., Man, P.J., Russell, J., Slack, P.T. and Gilbert, J., 2001. Combined phenyl silane and immunoaffinity column clean up with liquid chromatography for determination of ochratoxin A in roasted coffee: collaborative study. *Journal of AOAC International,* 84: 444-450.

FAO, 2004. FAO Food and Nutrition Paper 81, Worldwide Regulations for Mycotoxins in Food and Feed in 2003, Food and Agriculture Organization, Roma, Italy.

IARC,1993. *IARC Monographs on the Evaluation o f Carcinogenic Risks to Humans: Some Naturally Occurring Substances, Food Items and Constituents, Heterocyclic Aromatic Amines and Mycotoxins,* IARC Monograph N° 56, Lyon, pp. 445-466.

IARC, 2002. *IARC Monographs on the Evaluation o f Carcinogenic Risks to Humans, Some traditional herbal Medicines, some Mycotoxins, Naphthalene and Styrene,* IARC Monograph N° 82, Lyon, pp. 301-366.

Kim, E.K., Scott, P.M. and Lau, B.P., 2003. Hidden fumonisins in corn flakes, *Food Additives and Contaminants,* 20: 161-169. (erratum 2003 in *Food Additives and Contaminants,* 20: 417).

Lombaert, G.A., Pellaers, P., Roscoe, V., Mankotia, M., Nile, R. and Scott, P.M., 2003. Mycotoxins in infant cereal foods from the Canadian retail market. *Food Additives and Contaminants,* 20: 494-504.

Molinié, A. and Pfohl-Leszkowicz, A., 2002. Les mycotoxines dans les céréales/ les points importants de contrôle de la production au stockage, le devenir dans les produits dérivés. Les notes *ASEDIS-SO, Hors série spécial mycotoxines:* 32-42.

Molinié, A., Faucet, V., Castegnaro, M. and Pfohl-Leszkowicz, A., 2004. Analysis of some breakfast cereals on the French market for their content of ochratoxin A, citrinin and fumonisin B_1. Development of a method for simultaneous extraction of ochratoxin A and citrinin, *Food Chemistry,* 92, 391-400.

Pfohl-Leszkowicz A, editor, 1999, *Les mycotoxines dans l'alimentation, évaluation et gestion du risque,* (Paris: TEC et DOC, Lavoisier).

Pfohl-Leszkowicz, A., Molinié, A. and Castegnaro, M., 2004. Presence of ochratoxin A, citrinin and fumonisins B1 in breakfast cereals collected in French market. Comparison of OTA analysis using or not immunoaffinity clean up before HPLC. The Revista Mexicana de Micologia, 19, 7-16.

Sewram, V., Shephard, G.S., Marasas, W.F.O., Penteoda, M.F. and de Castro, M., 2003. Improving extraction of fumonisin mycotoxins from Brazilian corn-based infant foods. *Journal of Food Protection*, 66: 854-859.
Rhône Diagnostics technologies, 1999. Cereal ochratoxin A extraction method, application note for analysis of Ochratoxin A in cereal using sodium bicarbonate extraction in conjunction with Ochraprep®. Application note Ref N° A9-P14.V1.
Sydenham, E.W., G.S. Shephard and P.G. Thiel, 1992. Liquid chromatographic determination of fumonisins B_1, B_2 and B_3 in food and feeds. *J.AOAC International*, 75: 313-318.
Solfrizzo M., De Girolamo, A. and Visconti, A., 2001. Determination of Fumonisin B1 and B2 in cornflakes by high performance liquid chromatography and immunoaffinity clean-up. *Food Additives and Contaminants*, 18: 227-35.
Visconti, A., Solfrizzo, M. and De Girolamo, A., 2001. Determination of fumonisins B_1 and B_2 in corn and corn flakes by liquid chromatography with immunoaffinity column cleanup: collaborative study. *Journal of AOAC International*, 84: 1828-1837.

Cross-reactivity of six zearalenone antibodies in a hand-held fluorescence polarisation immunoassay

C.M. Maragos and E.-K. Kim

Mycotoxin Research Unit, National Center for Agricultural Utilization Research, USDA/ARS, 1815 N. University St., Peoria, IL 61604, USA; maragocm@ncaur.usda.gov

Abstract

Zearalenone is an estrogenic mycotoxin produced by several fungi, in particular *Fusarium graminearum*, and is commonly found in grain throughout the world. The native fluorescence of zearalenone (ZEN) has been used to advantage for detection of this toxin by instrumental methods such as HPLC. Antibody-based screening assays such enzyme-linked immunosorbent assays (ELISAs) have existed for a number of years for determination of ZEN and related metabolites. While convenient, the ELISAs for small molecules such as ZEN require a washing step to separate the bound and unbound enzyme label before detection. Fluorescence polarisation (FP) immunoassays do not require a separation step and therefore, with fewer manipulations, have the potential to be more rapid. Previously we have reported a FP immunoassay for detection of ZEN in maize using a commercially available antibody. In the present report we describe the cross-reactivity and kinetics of interaction of one commercially available and 5 newly developed monoclonal antibodies for the mycotoxins ZEN, zearalanone, α-zearalenol, β-zearalenol, α-zearalanol, and β-zearalanol using a miniaturised version of the previous FP device. Using the cross-reactivity and kinetic data, particular antibodies with the desired analytical characteristics can be selected for further application.

Keywords: zearalenone, zearalenol, fluorescence polarization, immunoassay, analysis

Introduction

Zearalenone (ZEN) is a mycotoxin produced by several fungi, in particular *Fusarium graminearum* (sexual state *Gibberella zeae*). The fungi and their mycotoxins are frequent contaminants of maize, barley, and other grains throughout the world (WHO, 2000). The acute toxicity of ZEN is relatively low (Hidy *et al.*, 1977). However, ZEN and related compounds (Figure 1), in particular α-zearalenol (α-ZEOL) and β-zearalenol (β-ZEOL), have potent estrogenic activity. An analog of ZEN, α-zearalanol

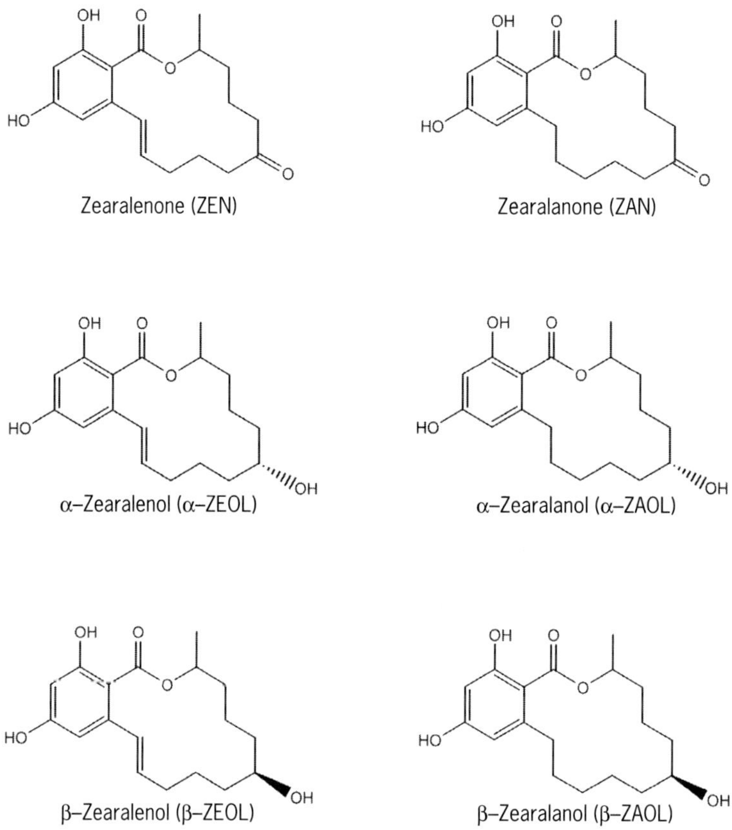

Figure 1. Structure of zearalenone and related analogs.

(zeranol or α-ZAOL), has anabolic effects and is used in some countries as a growth promoter in cattle. The estrogenic activity of ZEN and related compounds was recently reviewed (Hagler *et al.*, 2001).

Immunoassays for detection of ZEN that have been developed include radioimmunoassays and enzyme linked immunosorbent assays (ELISAs) (Thouvenot and Morfin, 1983; Liu *et al.*, 1985; Dixon *et al.*, 1987; Towers and Sprosen, 1992; Tanaka *et al.*, 1995; Park *et al.*, 2002). An ELISA for ZEN is available commercially and the performance has been validated through collaborative study (Bennett *et al.*, 1994). A dipstick immunoassay for ZEN in wheat has also been reported (Schneider *et al.*, 1995). Furthermore, polyclonal and monoclonal anti-ZEN antibodies are available from commercial sources.

While several of the existing ZEN ELISAs are very sensitive, they require a washing step to separate unbound toxin-enzyme conjugate from the solid phase, before adding a substrate for detection. Two possibilities for increasing the speed of immunoassays, relative to ELISAs, are elimination of the washing step, and elimination of the amplification (enzymatic reaction) step. Fluorescence polarisation (FP) is a technology that has been used in medical and veterinary diagnostics to detect small molecules in solution without requiring a wash step. FP immunoassay utilises the interaction of a toxin-specific antibody with a toxin-fluorophore conjugate (tracer) to effectively decrease the rate of rotation of the tracer, which increases the FP signal. In the presence of free toxin antibody binding to the tracer is reduced, decreasing the polarisation signal. In this competitive FP immunoassay format the result is a signal intensity that is inversely proportional to toxin concentration. The objective of the present work was to evaluate the performance of one commercial and 5 newly-developed antibodies for ZEN in a portable FP immunoassay format, and to characterise the cross-reactivity of the antibodies with 5 ZEN analogs.

Materials and methods

Reagents

Except where noted otherwise, deionised water (Nanopure II, Sybron/Barnstead) was used in the preparation of all reagents. All solvents were HPLC grade. ZEN, α-ZEOL, β-ZEOL, and α-ZAOL (zeranol) were obtained from International Minerals & Chemicals Corporation (Terre Haute, IN). Zearalanone (ZAN) and β-zearalanol (β-ZAOL, taleranol) were purchased from Sigma Chemical (St. Louis, MO). The commercial ZEN antibody used was a murine monoclonal, IgE isotype, and was derived from a mouse immunised with a ZEN-KLH conjugate (ZER-70 hybridoma cell line, Sigma, product Z-1751). The 6'-carboxymethyloxime of ZEN was prepared according to the method of Thouvenot and Morfin (1983) and reacted with either ovalbumin or bovine serum albumin to yield ZEN-OVA and ZEN-BSA conjugates, respectively. Ten mice were immunised with the ZEN-OVA conjugate at Harlan Bioproducts for Science (Indianapolis, IN). Procedures for immunisation of animals, testing of sera and hybridoma binding responses, and isolation of antibody producing cell lines were similar to those described previously by Maragos and McCormick (2000). Important differences were the use of ZEN-BSA as test antigen for determination of sera responses and for selection of appropriate antibody-producing hybridoma cell lines. Five monoclonal cell lines were obtained: antibodies "Ab11", "Ab21", and "Ab24" were murine IgG isotype 1, while antibodies "Ab22" and "Ab25" were murine IgG isotype 2b. The zearalenone-fluorescein tracer (ZEN-FL) was produced as described in Maragos and Kim (2004). All other chemicals and solvents were reagent grade or better and purchased from major suppliers.

Instrumentation

The fluorescence polarisation instrument was a Sentry 100 portable unit (Diachemix Corporation, Grayslake, IL). This unit is a miniaturised version of the Sentry FP unit used in our previous work. The instrument dimensions are 107 mm x 203 mm x 56 mm, and it weighs 1.1 kg. Power was supplied either with a 12V DC battery (100 mA) or with a 120V power converter.

Fluorescence polarisation immunoassay

Toxin standard solutions were prepared by diluting the appropriate toxin stock solution with acetonitrile+water (4:1, v/v) to obtain solutions over the range of 1 to 5000 ng/ml. In preliminary experiments the dilution of each antibody required to obtain maximal FP response, with minimal antibody usage, was determined. All 6 antibodies were diluted in PBS containing 0.1% ovalbumin. The antibodies and the optimal dilutions were: Sigma Mab Z-1751, 1:250, Ab11, 1:1,000, Ab21, 1:500, Ab22, 1:100, Ab24 1:1,000, and Ab25, 1:250. The working solution of the ZEN-FL was prepared fresh daily by diluting the stock solution 1:10,000 in PBS.

Glass culture tubes (10 mm x 75 mm) (VWR Scientific, West Chester, PA) were used as cuvettes for the assays. PBS (0.9 ml) was added to each tube followed by 0.1 ml of the antibody working solution and 20 μl of the sample. After thorough mixing the test solution was placed in the instrument and used as the blank. Tracer, 20 μL of the ZEN-FL working solution, was then added and mixed. The test solution was then returned to the fluorometer and the signal (mP) measured. For experiments to elucidate the kinetics of the reaction, measurements were made after holding at ambient temperature for times ranging from 15 seconds to 30 minutes. Data from standards in acetonitrile+water (4:1, v/v) were fit to a logistic dose-response curve (TableCurve software, Jandel Scientific, San Rafael, CA).

Results and discussion

Competitive FP immunoassay does not require the separation of antibody-bound label from the unbound label. In this case the label was the ZEN-FL tracer, whereas in ELISA assays the label is generally a toxin-enzyme conjugate. This format has two potential advantages. First, because separation of the label is unnecessary, the washing steps of ELISAs are not required. Second, the enzymatic amplification step is unnecessary, because a fluorescent label is used. The result is that the FP immunoassays have the potential to be more rapid than traditional competitive mycotoxin ELISAs. Whether this potential is fulfilled depends upon the kinetics of the competition between the toxin (ZEN or analogs) and the tracer for the antibody. The kinetics can be influenced by a number of factors, including the structure of the tracer and the structure of the antibody binding site. In the present work we used a tracer developed previously as part of a FP immunoassay for ZEN in maize (Maragos and Kim, 2004), and examined the effects on kinetics of using a commercially available ZEN antibody and 5 newly developed antibodies. The FP instrument used was a miniaturised, hand-held, version of the previously described instrument.

Assays were conducted by mixing antibody and sample in a tube, using the mixture to blank the instrument, then adding the tracer and holding the mixture for a brief time (15 sec to 10 min) before obtaining the FP measurement. Our previous experience with the ZEN FP immunoassay indicated that the duration of the tracer incubation was an important characteristic of assay performance (Maragos and Kim, 2004). The effect of the duration of the tracer incubation upon assay response is illustrated in Figure 2 for two of the ZEN antibodies (Sigma Mab and Ab11).

The time required for the mixture of antibody, tracer, and toxin to reach equilibrium depends upon the relative affinity of the antibody for the tracer and the toxin and their order of addition. With the format used, namely the addition of toxin before the addition of tracer, the assays were more sensitive with shorter incubation times (Figure 2). Generally this may be advantageous, however, selecting a very short incubation time may lead to the problem of having small errors in the incubation time translate into larger errors in the measured toxin content, if the mixture does not rapidly reach equilibrium. Therefore, combinations of tracer and antibody that will rapidly achieve equilibrium, while keeping the desired sensitivity for the toxin, are the most desirable (Maragos and Plattner, 2002; Maragos and Kim, 2004). For this reason 5 new ZEN antibodies were developed and evaluated for the relative kinetics of their response in the FP format.

Data for the kinetics of all 6 antibodies is provided in Table 1. Clearly the antibodies have a wide range of relative affinities for the tracer and ZEN, as manifest in the wide range of values for the concentration of ZEN required to cause a 50% reduction in FP signal (IC_{50}). For ZEN at the earliest time point (15 sec) the antibody that was most sensitive was Ab25, followed by Ab11, Sigma Mab, Ab24, Ab22, and Ab21. However, while Ab25 was the most sensitive at the earliest time point, the responsiveness of the assay was decreased dramatically over 10 min (IC_{50} of 262 ng/ml vs. 67 ng/ml). The responsiveness of the assay with the Sigma Mab also decreased dramatically in

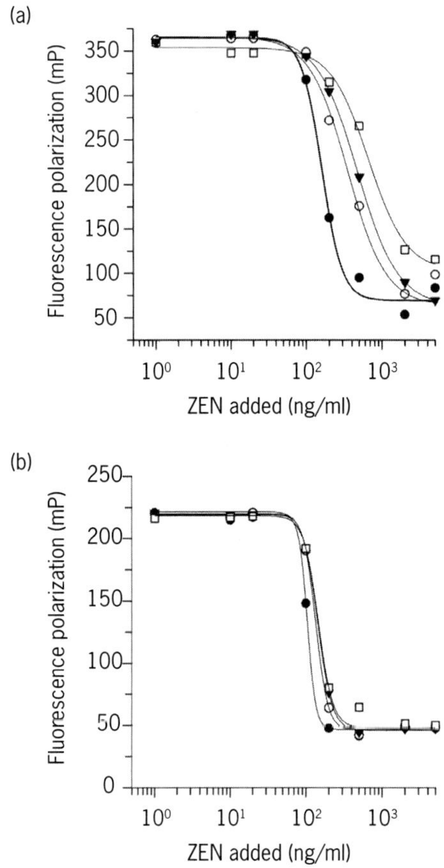

Figure 2. Kinetics of the ZEN FP Immunoassay with either Sigma Mab (A) or Ab11 (B). Calibration curves were obtained at using tracer incubations of various durations as described in the text. The duration of the tracer incubations were: 15 sec (●), 90 sec (○), 3 min (▼), 10 min (□).

this time (IC_{50} 855 vs. 172 ng/ml). Antibodies Ab21 and Ab24 also showed this effect. Two antibodies that were much less influenced by the incubation time were Ab11 and Ab22. Of these two, Ab11 was the most sensitive for ZEN.

In addition to the IC_{50}, an important parameter of the FP immunoassay is the maximum signal developed, that is, the signal generated with antibody and tracer in the absence of toxin. The maximum signal can vary dramatically, as demonstrated in Figure 3 for the 6 antibodies. For ease of comparison Figure 3 shows only data collected using a 90 sec tracer incubation. The antibodies Ab21, Ab24, Ab25, and Sigma Mab all yielded maximum signals greater than 300 mp, while the antibodies

Table 1. Summary IC_{50}'s for 6 Monoclonal Antibodies Tested with 6 ZEN analogs.

		IC_{50} (ng/ml)					
Antibody	**Time**	**ZEN**	**ZAN**	**α-ZEOL**	**β-ZEOL**	**α-ZAOL**	**β-ZAOL**
Ab11	15 s	105	143	130	156	98	108
	90 s	135	179	203	294	104	134
	3 min	144	142	193	319	119	141
	10 min	148	143	231	314	116	149
Ab21	15 s	450	148	319	598	1013	323
	90 s	635	165	469	1141	1126	494
	3 min	675	159	546	1344	1165	603
	10 min	715	173	625	1520	1275	607
Ab22	15 s	363	556	580	1634	1832	651
	90 s	357	564	629	1523	1633	652
	3 min	338	568	586	1633	1557	659
	10 min	344	597	597	1606	1651	615
Ab24	15 s	180	1314	193	316	72	402
	90 s	445	1794	348	799	116	763
	3 min	572	2112	417	1112	145	925
	10 min	627	2246	506	1330	164	1091
Ab25	15 s	67	51	279	84	86	369
	90 s	168	107	418	162	157	827
	3 min	207	127	542	226	227	1203
	10 min	262	151	598	294	257	1390
SigmaMab	15 s	172	64	689	275	52	1416
	90 s	391	113	1416	1023	94	2905
	3 min	529	170	2065	1683	130	4197
	10 min	855	258	3242	2527	233	(>5000)

Ab11 and Ab22 showed lower maximum signals (200-225 mp). Sensitivity of the assays for ZEN was not dependent upon the magnitude of the maximum signal (Figure 3). Nevertheless, a greater maximum is more convenient for assays, because it gives a greater working range of the signal.

The antibodies exhibited not only different maximum signals but also different relative affinities for ZEN and the tracer, which is manifest in the different slopes of the curves in Figure 3, and which is an expected consequence of kinetics of the

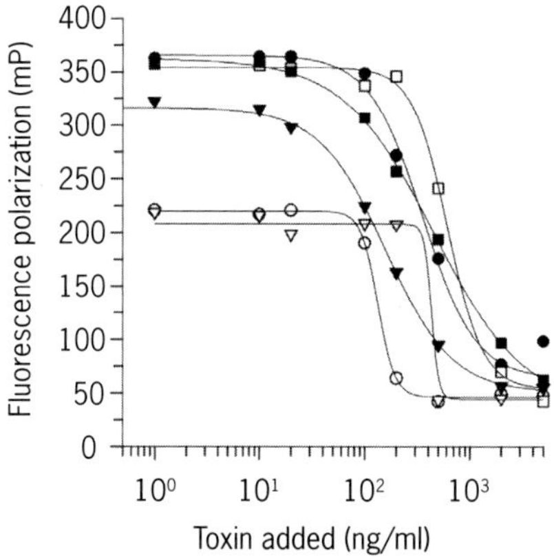

Figure 3. Response of 6 zearalenone antibodies to ZEN. All responses were measured using a 90 sec duration for the tracer incubation. Sigma Mab (●), Ab11 (○), Ab25 (▼), Ab22 (▽), Ab24 (■), Ab21 (□).

competition reaction. The usefulness of the assay depends upon the slope of the competition curve and whether a qualitative "positive/negative" response is desired or whether quantification is desired. In the former case a steep slope is preferable, giving a maximum change in signal over a minimal toxin concentration range. In the latter case a more gradual slope may be preferred, so that the changes in observed signal correspond to a broader range of toxin concentrations. Under this paradigm the three antibodies with the steepest slopes (Ab22, Ab21, Ab11) would be preferable for qualitative assays. However, because Ab21 and Ab22 are less sensitive to ZEN, the best antibody for a qualitative "positive/negative" assay would likely be Ab11. Also under this paradigm the two antibodies with high maximum signals, and broader slopes (Ab25 and Sigma Mab) would be the best antibodies for quantitative assays.

Cross-reactivity of the FP immunoassay

Cross-reactivity to structurally similar analogs is an important parameter for an immunoassay, regardless of the format. In addition to ZEN, Table 1 lists the cross-reactivity of 5 ZEN analogs (ZAN, α-ZEOL, β-ZEOL, α-ZAOL, and β-ZAOL) with each of the 6 antibodies. All 6 antibodies showed some degree of cross-reactivity to all 6 compounds. For simplicity the following comparison describes relative cross-reactivity when a 90 sec incubation time was used. The reason for selecting 90 sec is that it is empirically a convenient incubation time to use when performing the assays. Under this condition Ab11 was most sensitive for α -ZAOL followed by ZEN/

β-ZAOL, ZAN, α-ZEOL, and β-ZEOL. Ab21 was most sensitive for ZAN followed by α-ZEOL/β-ZAOL, ZEN, and α-ZAOL/β-ZEOL. Ab22 was less sensitive to all 6 compounds, but was the antibody with the greatest specificity for ZEN, followed by ZAN, α-ZEOL/β-ZAOL, and β-ZEOL/α-ZAOL. Ab24 was most sensitive to α-ZAOL, followed by α-ZEOL/ZEN, β-ZAOL/β-ZEOL, and ZAN. Ab25 was most sensitive to ZAN, followed by α-ZAOL/β-ZEOL/ZEN, α-ZEOL, and β-ZAOL. The Sigma Mab was most sensitive to α-ZAOL/ZAN, followed by ZEN, β-ZEOL, α-ZEOL, and β-ZAOL.

From these cross-reactivity data several aspects are notable. First, all of the antibodies exhibited cross-reactivity. Second, the Ab11 showed cross-reactivity to all of the analogs to a substantial degree, making this antibody a good antibody for a general screen of ZEN and related analogs. Third, the antibody with the greatest specificity for ZEN (Ab22) was also one of the least sensitive for this toxin. Fourth, several of the antibodies showed good cross-reactivity to the anabolic agent α-ZAOL (also known as zeranol), which is known to occur in the urine of pasture fed animals (Erasmuson *et al.*, 1994).

Conclusion

In summary, one commercial and 5 novel ZEN antibodies were evaluated in a portable fluorescence polarisation immunoassay using ZEN and 5 ZEN analogs. The antibodies demonstrated distinctly different abilities to rapidly achieve equilibrium, sensitivity to ZEN, and cross-reactivity to ZEN analogs. The optimum antibody would depend upon the intended use for the assay. The antibody with the best combination of rapid equilibration and sensitivity to ZEN was Ab11. The lower maximum signal developed by this antibody and the steep dose-response curve would suggest this antibody would work well in a rapid qualitative (positive/negative) assay. The Ab25 and the commercial antibody (Sigma Mab) exhibited a high maximum signal and good sensitivity for ZEN, but these assays did not attain equilibrium quickly, an important consideration for ease of use. Nevertheless, these two antibodies might be more useful in rigorously timed, quantitative, assays. All 6 antibodies showed cross reactivity to ZEN analogs, with Ab11, Ab24, Ab25, and the Sigma Mab, showing notable cross-reactivity to α-zearalanol (zeranol).

Acknowledgements

We express our appreciation to Mrs. Judith Greer, USDA-ARS-NCAUR for exceptional technical assistance.

References

Bennett, G.A., Nelsen, T.C., and Miller, B.M., 1994. Enzyme-linked immunosorbent assay for detection of zearalenone in corn, wheat, and pig feed: collaborative study. *Journal of the AOAC International* 77: 1500-1509.

Dixon, D.E., Warner, R.L., Ram, B.P., Hart, L.P., and Pestka, J.J., 1987. Hybridoma cell line production of a specific monoclonal antibody to the mycotoxins zearalenone and α-zearalenol. *Journal of Agricultural and Food Chemistry* 35: 122-126.

Erasmuson, A.F., Scahill, B.G., and West, D.M., 1994. Natural zeranol (α-zearalanol) in the urine of pasture-fed animals. *Journal of Agricultural and Food Chemistry* 42: 2721-2725.

Hagler, W.M. Jr., Towers, N.R., Mirocha, C.J., Eppley, R.M., and Bryden, W.L., 2001. Zearalenone: mycotoxin or mycoestrogen,. In: *Fusarium: Paul E. Nelson Memorial Symposium*, Leslie, J.F., Backhouse, D., Bryden, W.L., and Burgess, L.W. (Eds.), APS Press, St. Paul, Minnesota. pp. 321-331.

Hidy, P.H., Baldwin, R.S., Greasham, R.L., Keith, C.L., and Mcmullen, J.R., 1977. Zearalenone and some derivatives: production and biological activities. *Advances in Applied Microbiology* 22: 59-82.

Liu, M-T., Ram, B.P., Hart, L.P., and Pestka, J.J., 1985. Indirect enzyme-linked immunosorbent assay for the mycotoxin zearalenone. *Applied and Environmental Microbiology* 50: 332-336.

Maragos, C.M., and Kim, E.K., 2004. Detection of zearalenone and related metabolites by fluorescence polarization immunoassay. *Journal of Food Protection* 67: 1039-1043.

Maragos, C.M., and McCormick, S.P., 2000. Monoclonal antibodies for the mycotoxins deoxynivalenol and 3-acetyl-deoxynivalenol. *Food and Agricultural Immunology* 12: 181-192.

Maragos, C.M., and Plattner, R.D., 2002. Rapid fluorescence polarization immunoassay for the mycotoxin deoxynivalenol in wheat. *Journal of Agricultural and Food Chemistry* 50: 1827-1832.

Park, J.W., Kim, E.K., Shon, D.H., and Kim, Y.B., 2002. Occurrence of zearalenone in Korean barley and corn foods. *Food Additives and Contaminants* 19: 158-162.

Schneider, E., Usleber, E., Martlbaur, E., Dietrich, R., and Terplan, G., 1995. Multimycotoxin dipstick enzyme immunoassay applied to wheat. *Food Additives and Contaminants* 12: 387-393.

Tanaka, T., Teshima, R., Ikebuchi, H., Sawada, J., and Ichinoe, M., 1995. Sensitive enzyme-linked immunosorbent assay for the mycotoxin zearalenone in barley and Job's-tears. *Journal of Agricultural and Food Chemistry* 43: 946-950.

Thouvenot, D., and Morfin, R.F., 1983. Radioimmunoassay for zearalenone and zearalanol in human serum: production, properties, and use of porcine antibodies. *Applied and Environmental Microbiology* 45: 16-23.

Towers, N.R., and Sprosen, J., 1992. *Fusarium* mycotoxins in pastoral farming: zearalenone induced infertility in sheep. In: *Recent Advances in Toxinology Research*, vol. 3, Gopalakrishnakone, P., and Tan, C.K., (Eds.), pp. 272-284.

WHO, 2000. *Safety evaluation of certain food additives and contaminants*, WHO Food Additives Series: 44, Geneva. pp. 393-482.

Detection of fumonisins in maize (*Zea mays* L.) by three analytical techniques (HPLC, TLC and ELISA)

S.D. Peña Betancourt, G.H. Flores, S.J. Rosas and L.A. González

Department of Animal Production, Toxicology Laboratory, University Autonomous Metropolitan-Xoxhimilco, UAM-X. México City, Calzada del Hueso #1100, Col Villa Quietud, México, D. F. CP 04250, Mexico; spena@correo.xoc.uam.mx

Abstract

Fumonisin are mycotoxins that have been isolated from naturally contaminated maize in many countries around the world by several methods of analysis. Chromatographic techniques are widely used because they are either simple or have good precision. The aim of this study was to establish the applicability and reliability of a particular method for fumonisin screening purposes in maize. Three methods (HPLC, TLC and an ELISA) were selected for use. Fifteen samples of maize were collected from Puebla, Mexico City and Morelos during autumn 2000 and spring 2001. After analysis by all three techniques, 15/15 (100%) were found to contain fumonisins (3.69–25.03 mg/kg) by HPLC, 12/15 (80%) were shown to contain fumonisins by TLC, and 9/13 (70%) samples were shown to contain fumonisins (0.12–4.6 mg/kg) by the ELISA technique. Clearly the methods performed differently. The method of choice will depend on available equipment, the desired sensitivity and time for analysis. However, TLC is the simplest method for fumonisins screening in maize, although some improvement to the cleanup step may be necessary.

Keywords: fumonisin B_1, maize, analysis, chromatography, HPLC, TLC, ELISA

Introduction

Fumonisins are a group of mycotoxins produced principally by *Fusarium verticillioides (F. moniliforme)* and *Fusarium proliferatum*, usually occurring near harvest time of a maize crop in many regions of the world. (Jeong *et al.*, 1999; Norred, 1993; Pitt, 2000). Fumonisin B_1 (FB_1), induces many diseases in animals (Rizzi *et al.*, 2003), and has recently been associated with human esophageal cancer (Norred *et al.*, 1994). The International Agency for Research on Cancer (IARC) has classified FB_1 as a class 2B carcinogen for humans (IARC, 1987): possibly carcinogenic to humans.

The major methods used in the analysis of fumonisins include the High Performance Liquid Chromatography (HPLC), Thin Layer Chromatography (TLC), and the Enzyme-Linked Immunosorbent Assay (ELISA). These techniques have different levels of sensitivity, and also differ in cost and the time required for analysis (Betina, 1985; Sewran *et al.*, 2003). Thin Layer Chromatography (TLC) is a chromatographic technique that was initially frequently used in the analysis of mycotoxins in feeds and foods. The technique has been largely replaced by other chromatographic methods such as high performance liquid chromatography (HPLC), and gas chromatography (GC) (Rottinghaus *et al.*, 1992). Reverse phase HPLC is now more frequently used for the separation and detection of fumonisins. It has been adopted as AOAC Official Method 995.15 for the determination of FB_1, FB_2 and FB_3 in maize. (AOAC, 2000). Determination requires derivatisation with o-phthaldialdehyde (OPA) reagent before fluorescence detection, and limits of detection are in the 0.05 μg FB1/g or 50 μg/kg range (Shephard, 1998). Immunochemical methods such as ELISA are also frequently used to detect food contamination. There are a number of ELISA kits for fumonisins analysis (Scott *et al.*, 1997); one of the test kits is the RIDASCREEN®FAST Fumonisin by R-Biopharm (R-Biopharm AG, Damstadt, Germany) which has been approved for the determination of fumonisins in maize, maize meal, maize gluten meal, maize germ meal, sorghum and soy blend by the Federal Grain Inspection Service of the United States of America.

This study was initiated by the Toxicology Laboratory at the Autonomous Metropolitan University (UAM-X) to determine which of the available analytical procedures was simple and practical for fumonisin screening purposes in maize to obtain valuable information about the occurrence of fumonisin in maize from Mexico.

Materials and methods

This study was carried out using yellow, white and blue maize obtained from three states in México (Morelos, Puebla and Mexico City), during the Spring-Fall 2000-2001 harvest. Fifteen 1 kg samples were collected and ground in a disk mill. The ground material was thoroughly mixed before portions were removed for analysis by each of the three methods studied: TLC (Rottinghaus *et al.*, 1992); HPLC (Shephard *et al.*, 1996), and RIDASCREEN®FAST Fumonisin (Kit Art. R5602).

High Performance Liquid Chromatography analysis

Some slight modifications were made to the method of Shephard *et al.* (1996). Fifty grams (50 g) maize meal was extracted with 100 ml methanol+water (75:25 v/v) by shaking for 15 minutes. The slurry was filtered through Whatman No. 4 filter paper and filtrate pH adjusted to 6 with 1 M sodium hydroxide. The extract then passed through a previously conditioned Supelco SAX SPE cartridge (Sigma, St. Louis, MO) and the fumonisins were eluted with 10 ml methanol+acetic acid (99:1 v/v). The eluate was evaporated to dryness and the residue redissolved in 200 µl methanol. 25 µl of the methanolic extract was added to 225 µl of o-phthaldialdehyde reagent, mixed and 20 µl injected into the LC system.

The HPLC unit was a Varian model 9010 system with a fluorescence detector set at 334 nm excitation and 440 nm emission and an C_{18} 150 × 4.6 mm 5 µm ODS column. The mobile phase was methanol+0.1M sodium dehydrogenated phosphate (77:23 v/v), adjusted to pH 3.3 with o-phosphoric acid. Fumonisin standards were obtained from Sigma (Sigma Aldrich, St. Louis, MO). Stock solutions of FB_1, FB_2, FB_3 and FB_4 at 0.25, 0.5, 1.0, 4.0, 6.0, 8.0 and 10.0 mg/l concentrations were prepared in acetonitrile+water (75:25 v/v) and stored at 4 °C. Identification and quantification of the fumonisins was accomplished by comparing the retention times of peaks observed in samples to those for standards.

Thin layer chromatographic (TLC) analysis of the fumonisins

Some slight modification was also made to the Rottinghaus *et al.* (1992) TLC procedure. To extract fumonisins, 50 g ground analytical sample was shaken with 100 ml acetonitrile in water (1 +1) for 30 minutes using a shaker at high speed. The extract was filtered and 20 ml filtrate transferred to a 6 cm Supelco C_{18} SAX cartridge for cleanup. The fumonisins were eluted with acetonitrile+water (7:3), evaporated to near dryness and redissolved in 100 µl acetonitrile+water (1:1). 50 µl of each sample and standards were spotted on RP silica gel TLC plates and development in methanol+1%KCl (3:2). After development, spots were visualised by spraying with sodium borate buffer (pH 8.8), fluorescamine (0.4 mg/ml acetonitrile), followed by boric acid in acetonitrile (40:60), and heating at 110 °C. Fumonisins were identified by comparison of their migration (RF) and fluorescence (under ultraviolet light) to those of standards. The ultraviolet lamp used was a dual wavelength, 254/366 nm.

Fumonisins analysis using RIDASCREEN®FAST Fumonisin

The RIDASCREEN®FAST Fumonisin kit contained all materials necessary for the analysis of 43 samples plus five standards. Briefly, the extraction procedure involved weighing 5 g of ground sample into a container and adding 25 ml of 70% methanol. The slurry was manually mixed for two minutes and filtered through Whatman No. 1 filter paper. A portion of the filtrate was diluted 1:14 with distilled water and 50 µl added to microtiter well. Equal volumes of enzyme conjugate and anti-fumonisin antibody solution were added to the well and the mixture incubated for 10 minutes at room temperature. After washing, 2 drops of substrate chromogen were added to the well, mixed and incubated for 5 minutes in the dark. The reaction was stopped and absorbance measured at 450 nm.

Results and discussion

Fumonisins were found in all 15 samples analysed by HPLC (Table 1). Fumonisin B_1 was the major contaminant, with an overall incidence of 100%. Levels of FB_1 ranged from 3.7 to 25.0 mg/kg with a mean of 14.4 mg/kg. The highest incidence of fumonisins detected was in yellow maize with almost all samples containing detectable levels. Two samples of yellow maize from Puebla and Morelos had the highest levels found in this study, 23.5 and 25.0 mg/kg. Based on a 4 mg/kg guideline for acceptance for human consumption, 12 of the 15 samples contained violative levels. The TLC results showed that twelve samples contained detectable levels of fumonisin B_1. Eight of the 12 samples identified to contain fumonisins by TLC were

Table 1. Fumonisin B1 mg/kg in fifteen samples of maize using HPLC, TLC and RIDASCREEN® FAST Fumonisin methods.

Maize sample type	Origin	HPLC (mg/kg)	TLC	ELISA (mg/kg)
Yellow	Puebla	18.0	Positive	0.3
Yellow	Mexico city	28.0	Positive	7.5
White	Morelos	15.0	positive	2.3
Yellow	Morelos	3.7	negative	ND
Blue	Mexico city	16.0	positive	ND
Yellow	Mexico city	3.9	negative	0.1
Yellow	Mexico	5.6	negative	0.1
Blue	Mexico	21.0	positive	4.6
Yellow	Mexico city	23.5	ND	0.4
Yellow	Mexico city	14.0	positive	*
Yellow	Mexico city	16.0	positive	*
White	Morelos	3.8	ND	0.7
White	State of Mexico	4.3	ND	0.1
Blue	Puebla	4.8	negative	ND
Yellow	State of Mexico	25.0	positive	ND

*Not tested; ND, not detected.

Table 2. Comparison of TLC and RIDASCREEN®FAST ELISA fumonisin results to HPLC. Comparisons assume HPLC as gold standard procedure.

Test method	Correct responses at		Sensitivity/specificity	
	< 4 mg/kg	> 4 mg/kg	False positives*	False negatives
TLC	3/3	8/12	0/15	4/15
ELISA	3/3	2/10	0/13	8/13

*Method result higher than HPLC. Positive = HPLC result ≥ 4 mg/kg.

above the 4 mg/kg guideline level and classified as positive in Table 2. Nine samples were found to contain fumonisins (0.1 to 4.6 mg/kg,) using the RIDASCREEN®FAST Fumonisin test kit. The levels determined by the RIDASCREEN®FAST Fumonisin kit were much lower than those obtained by HPLC. We do not know exactly why.

Conclusion

In this study, we have demonstrated fumonisins in maize can be detected using HPLC, TLC and the RIDASCREEN®FAST Fumonisin test kit. There were differences in the concentrations of fumonisins determined by each of the three methods in most samples. Using HPLC as a gold standard, we note in Table 2 that the performance of TLC is comparable to that of HPLC, but that of the ELISA test is substantially different. The differences between the HPLC and TLC results may be due to the use of the SAX clean-up cartridge. An immunoaffinity column might have yielded better results.

Due to the variation in the results obtained, we recommended extreme care when running an analysis and that duplicate assays be carried out. The method of choice will depend on available equipment and the level of sensitivity desired and time for analysis of each purpose. Thin layer chromatography is considered a reliable and economical method for fumonisins detection in some developing countries. Our results indicate that an improvement of the clean-up step may be necessary. We also showed that Mexican maize may be contaminated by fumonisins and it may therefore be necessary to legislate permissible fumonisin B_1 in Mexico maize.

References

AOAC, 2000. Chapter 49. Natural Toxins. Subchapter 5. Fumonisins. AOAC Official Method 995.15. In: *Official Methods of Analysis of AOAC International*, 17th edition, Horwitz, W. (Ed.), Gaithersburg, Maryland, USA.

Betina, V., 1985. Thin Layer Chromatography of mycotoxins. *Journal of Chromatography* 334: 211-276.

IARC, 1987. *Monograph on the evaluation of carcinogenic risk to humans*, Suppl. 1, Lyon, France, pp. 82-87.

Jeong-A.H., and Lee, Y.W., 1999. Natural occurrence of the C series of fumonisins in moldy corn. *Appl. Environ. Microbiol.* 30: 1331-1334.

Norred, W.P., 1993. Fumonisins-mycotoxins produced by *Fusarium moniliforme*. *Journal of Toxicology and Environmental Health* 38: 309-328.

Norred, W.P., and Voss, K.A., 1994. Toxicity and role of fumonisins in animal diseases and human esophageal cancer. *J. Food Prot.* 57: 522-527.

Pitt, J.L., 2000. Toxigenic fungi and mycotoxins. *Br. Med Bull.* 56: 184-192.

Rizzi, L., Simioli, M., Roncada, P., and Zaghini, A., 2003. Aflatoxin B1 and clinoptilolite in feed for layin hens: Effects on egg quality, mycotoxin resuidues in livers and hepatic mixed function oxygenase activities. *J. of Food Protection* 66: 860-865.

Rottinghaus, G.E., Coatney, C.E., Minor, H.C., 1992. A rapid, sensitive thin layer chromatography procedure for the detection of fumonisin B1 and B2. *Veterinary Diagnostic Investigation* 4: 326-329.

Scott, P.M., Yeung, J.M., Lawrence, G.A., and Prelusky, D.B., 1997. Evaluation of enzyme-linked immunosorbent assay for analysis of beer for fumonisins. *Food Addit. Contam.* 14: 445-450.

Sewran, V., Shephard, G., Marasas, W.F.O., and Penteado, M.F., 2003. Improvement extraction of fumonisin mycotoxins from Brazilian corn-based infant foods. *J. of Food Protection* 66: 854-859.

Shephard, G.S., 1998. Chromatographic determination of the fumonisin mycotoxins. *Journal of Chromatography A.* 815: 31-39.

Shephard, G.S., Sydenham, E.W., Thiel, P.G., and Gelderbloom, W.G., 1996. Quantitative determination of fumonisin B1 and B2 by high performance liquid chromatography with fluorescence detector. *Journal Liquid Chromatography* 34: 773-775.

Shephard, G.S., Marasas, W.F.O., Leggott, N.L., 2000. Natural occurrence of fumonisins in corn from Iran. *J. Agric. Food Chem.* 48: 1860-1864.

Analytical characterisation of degradation products of fumonisin B_1 by LC-MS/MS

E. Vekiru[1], M. Täubel[2], G. Schatzmayr[2], R. Schuhmacher[1], E.M. Binder[2] and R. Krska[1]

[1]Center for Analytical Chemistry and Christian Doppler Laboratory for Mycotoxin Research, Department for Agrobiotechnology (IFA-Tulln), University of Natural Resources and Applied Life Sciences Vienna, Konrad Lorenz Strasse. 20, A-3430 Tulln, Austria; rudolf.krska@boku.ac.at
[2]Biomin GmbH, Industrie Strasse 21, A-3130 Herzogenburg, Austria

Abstract

Fumonisin B_1 (FB_1), first reported in 1988 in South Africa, is the predominant fumonisin among a group of structurally related fumonisins. It occurs in mainly maize and maize-based products. FB_1 has been shown to produce a wide range of pathological effects in animals, including leukoencephalomalacia in horses and pulmonary oedema in swine. It also disrupts sphingolipid biosynthetic pathways in animals and plant cells. It is hence understandable that there is considerable interest in the detoxification of FB_1. Decontamination using selected micro-organisms with the ability to degrade or biotransform mycotoxins has many advantages. Blackwell *et al.* (1999) showed that the fungus *Exophiala spinifera* produces enzymes that metabolise fumonisins. This fumonisin-metabolism consists of de-esterification of fumonisins tricarballylic acid esters followed by oxidative deamination of the resulting amine alcohol backbone resulting in a 2-oxo-12,16-dimethyl-3,5,10,14,15-icosanepentol (2-OP1) that undergoes internal cyclisation to form a hemiketal. Since hydrolysed fumonisin B_1 (HFB_1) can easily be adsorbed and is as toxic as or even more toxic than FB_1, a study on other micro-organisms with the capability to detoxify FB_1 and HFB_1 was conducted. As a first step, reported here, samples were analysed after degradation with *E. spinifera*. Using LC-MS/MS for characterisation of the samples, we could reconstruct the transformation of FB_1 over partially hydrolysed FB_1 to completely hydrolysed FB_1, which was then further metabolised. This metabolism led to a degradation product HFB_1, which showed a mass spectrum with similar molecular ion signal and fragmentation pattern like 2-OP1. To characterise the product we performed full scan and also product ion scan experiments in both, positive and negative ionisation mode. Detoxification of the degraded samples was also demonstrated in a bioassay.

Keywords: *Exophiala spinifera*, fumonisin B1, degradation, metabolite, detoxifying micro-organisms, liquid chromatography, tandem mass spectrometry, 2-oxo-12,16-dimethyl-3,5,10,14,15-icosanepentol

Introduction

Fumonisins are a group of mycotoxins mainly produced by *Fusarium moniliforme (F. verticillioides)* and *Fusarium proliferatum*, which are common contaminants of cereal grains, especially of maize (Thiel *et al.*, 1991). Fumonisin B_1 (FB_1), first reported in 1988 in South Africa, is the predominant fumonisin. Fumonisin B_1 not only occurs in maize and maize-based products (Shephard *et al.*, 1996) but also in an even wider range of food commodities like cereals (Rice *at al.*, 1994), beer (Hlywka *et al.*, 1999) and spices (Patel *et al.*, 1996) worldwide.

Fumonisin B_1 ingestion has been shown to cause leukoencephalomalacia in horses (Kellerman *et al.*, 1990), pulmonary oedema syndrome in pigs (Harisson *et al.*, 1990) and liver cancer in rats (Gelderblom *et al.*, 1991). In addition, esophageal cancer in humans has been associated with the toxin in distinct areas of the world, when contaminated maize was consumed (Sydenham *et al.*, 1990). Fumonisins bear a structural similarity to sphingosine and have been found to inhibit ceramide synthase, which is assumed to be the reason for inhibition of sphingolipid biosynthesis (Wang *et. al.*, 1991).

The economic implications of animal feeds contaminated with high levels of fumonisins are significant (Shepard *et al.*, 1996). Together with the high stability of these mycotoxins during storage of grains and under processing conditions, it is understandable that there is a considerable interest in the detoxification of FB_1. Appropriate FB_1 decontamination strategies are therefore necessary. Appropriate physical or chemical decontamination strategies for fumonisins are not available so far. Biological detoxification – especially the transformation of fumonisins via micro-organisms or specific enzymes – seems to be promising. Decontamination of food and feed by selected micro-organisms with the ability to degrade or biotransform mycotoxins into compounds that are no longer toxic has many advantages. The toxin can be removed under gentle conditions, without using harmful chemicals and without significant losses in nutritive value and technological properties of the detoxified food or feed. Furthermore, detoxification by micro-organisms is economically feasible.

Duvick *et al.* (1998) has already described the breakdown of FB_1 by the yeast *Exophiala spinifera* ATCC 74269 as well as by an aerobic bacterium in the course of complete metabolism of the toxin.

Under alkaline conditions, or through metabolism, the aminopentol backbone, also known as hydrolysed fumonisin B_1 (HFB_1) can be formed. Hydrolysed FB_1 can easily be absorbed and is as toxic as or even more toxic than FB_1. The aim of this investigation was to assess the metabolic pathway of fumonisin degrading micro-organism hence potential detoxification capabilities, by studying the metabolic pathway *E. spinifera*.

Materials and methods

Chemicals

The mycotoxin FB_1 was purchased from Biopure Referenzsubstanzen GmbH (Tulln, Austria). Methanol was LC grade and obtained from J.T. Baker (Deventer, the Netherlands). Formic acid (98%) was also obtained from J.T. Baker. Water for LC mobile phase was purified successively by reverse osmosis and a Milli-Q plus system from Millipore (Molsheim, France).

Biological deactivation of FB_1 by E. spinifera *DSM 1217 (Mo)*

Yeast cells were cultivated in liquid yeast medium under inductive growth conditions (25 mg/ml FB_1). After harvesting the cells by centrifugation, a degradation experiment was carried out in mineral salt solution (MSL: 1 g/l $(NH_4)_2SO_4$, 1 g/l K_2HPO_4, 1 g/l NaCl, 0.2 g/l $MgSO_4 . 7H_2O$) supplemented with 25 mg/ml pure FB_1. In parallel, a second experiment with culture control consisted of a fermentation blank in which *E. spinifera* was grown under identical conditions but without FB_1 was performed. Samples from both experiments were taken after 0, 3, 6, 20, 24, 31, 52, 72 and 96 hours of incubation at 25 °C and inactivated by heating to 95 °C. The heat inactivated samples were stored at -20 °C until analysis. After centrifugation, the samples were analysed without any other treatment.

LC-MS/MS measurements

LC-MS/MS analysis was performed with a QTrap-LC-MS/MS System (Applied Biosystems, Foster City, USA), equipped with a Turbo Ion Spray (ESI) source and a 1100 Series HPLC System (Agilent, Waldbronn, Germany) including a diode array detector (DAD). The chromatographic separation was achieved at 20 °C on a Zorbax

XDB-C8 (4.6 mm x 150mm, 5 µm; Agilent Technologies, USA) column by a gradient of methanol/mQ water including 50 µl/l of 98% formic acid.

Injection volume was 10 µl, while the flow rate was 0.4 ml/min. For metabolite identification and characterisation, we performed enhanced mass scans (+/-EMS) and also enhanced product ion scans (+/-EPI) in positive and negative modes using an electrospray ionisation source (ESI) at 400 °C source temperature. Further, MS parameters were: curtain gas (CUR) 138 kPa (20 psi), nebuliser gas (GS1) 207 kPa, (30 psi) auxiliary gas (GS2), 482 kPa (70 psi), ionisation voltage +5000 V in positive and –4200 V in negative mode, collision-activated dissociation gas (CAD) 6 (arbitrary units). Product ion spectra were gained at a collision energy (CE) of +/-30 V and a declustering potential (DP) of +/-30 V depending of the ionisation mode.

Toxicity test

To determine the toxicity of samples before and after the metabolism experiment, a OSTRACODTOXKIT F™ (MicroBioTests Inc., Nazareth, Belgium) was used. Each OSTRACODTOXKIT F™ contains all the materials to perform chronic toxicity tests with the freshwater ostracod crustacean *Heterocypris incongruens*. Standard operational procedures provided by the test kit manufacturer were followed.

Results and discussion

Characterisation of fumonisin metabolites with LC-MS/MS

LC-MS/MS analysis of FB_1 metabolites was performed with a QTrap-LC-MS/MS System. This instrument is based on a triple quadrupole mass spectrometer and is capable of performing all of the conventional tandem quadrupole scan modes as well as several high sensitivity ion trap mass spectrometer scan techniques using the final quadrupole as a linear ion trap mass spectrometer (enhanced modus).

Our first objective was to reconstruct the breakdown of FB_1 by the fungus *E. spinifera* proposed by Blackwell *et al.* (1999) (Figure 1). Using LC-MS/MS to characterise the samples enabled us to reconstruct the transformation of FB_1, through progressive de-esterification of FB's tricarballylic acid esters, via partially hydrolysed FB_1 ($PHFB_1$) to completely hydrolysed FB_1 (HFB_1) (Figure 2). Figure 3 shows changes of FB_1, $PHFB_1$ and HFB_1 during the whole experiment. Hydrolysed FB_1 was further metabolised by oxidative deamination to 2-OP1 (2-oxo-12,16-dimethyl-3,5,10,14,15-icosanepentol), which according to Blackwell *et al.* (1999) undergoes internal cyclisation to form a hemiketal. The conversion of 2-OP1 to its corresponding hemiketal is presumed

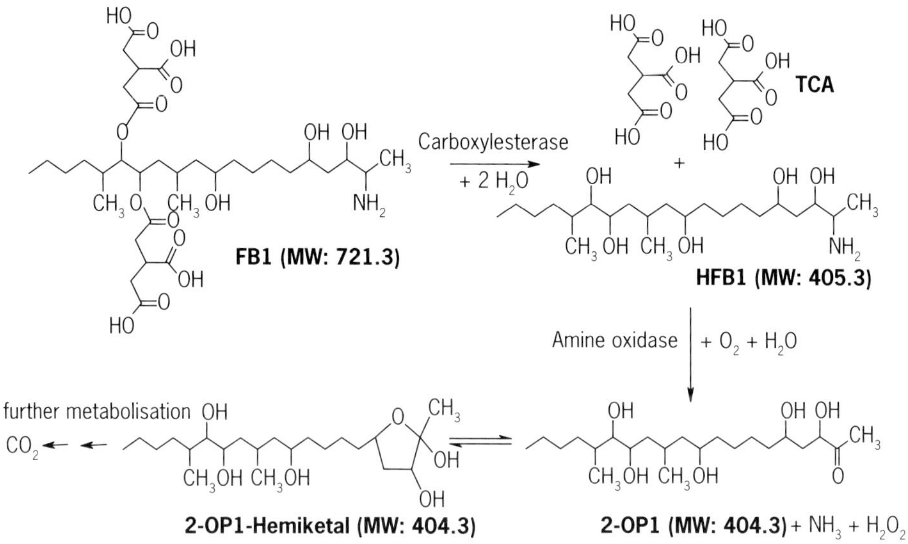

Figure 1. Pathway for FB_1-metabolism by Exophiala spinifera *according to Blackwell* et al. *(1999); TCA is tricarballylic acid.*

to occur spontaneously. In our experiments, the molecular mass of the 2-OP1 was confirmed only in the negative ion mode by detection of the deprotonated molecular ion $[M-H]^-$ (403.3 amu), which could be explained by the favored ionisation of the enol hydroxyl-group in this mode. In positive mode, the $[M+Na]^+$ (427.3 amu), $[M+Na-H_2O]^+$ (409.3 amu) and $[M+K]^+$ (443.3 amu) ions were observed. Regarding MS/MS experiments the enhanced product ion spectra (EPI) of HFB_1 and 2-OP1 were compared. The +EPI spectra of $[HFB_1+H]^+$ (406.3 amu) and $[2\text{-OP1-}H_2O+H]^+$ (387.3 amu) showed an ion series associated with the sequential loss of hydroxyl groups, as H_2O (-18 amu) (Figure 4). The same pattern was observed by the +EPI of $[2\text{-OP1-}2H_2O+H]^+$ (369.3 amu). These data are consistent with the observation of Blackwell *et al.* (1999) who reported that the base peak was corresponding to the mass of 387 amu, because the loss of water is very likely to take place in case of a hemiketal structure giving rise to the ion at 387 amu. The formation of the N-acetylated products of FB_1 and HFB_1 could also be observed.

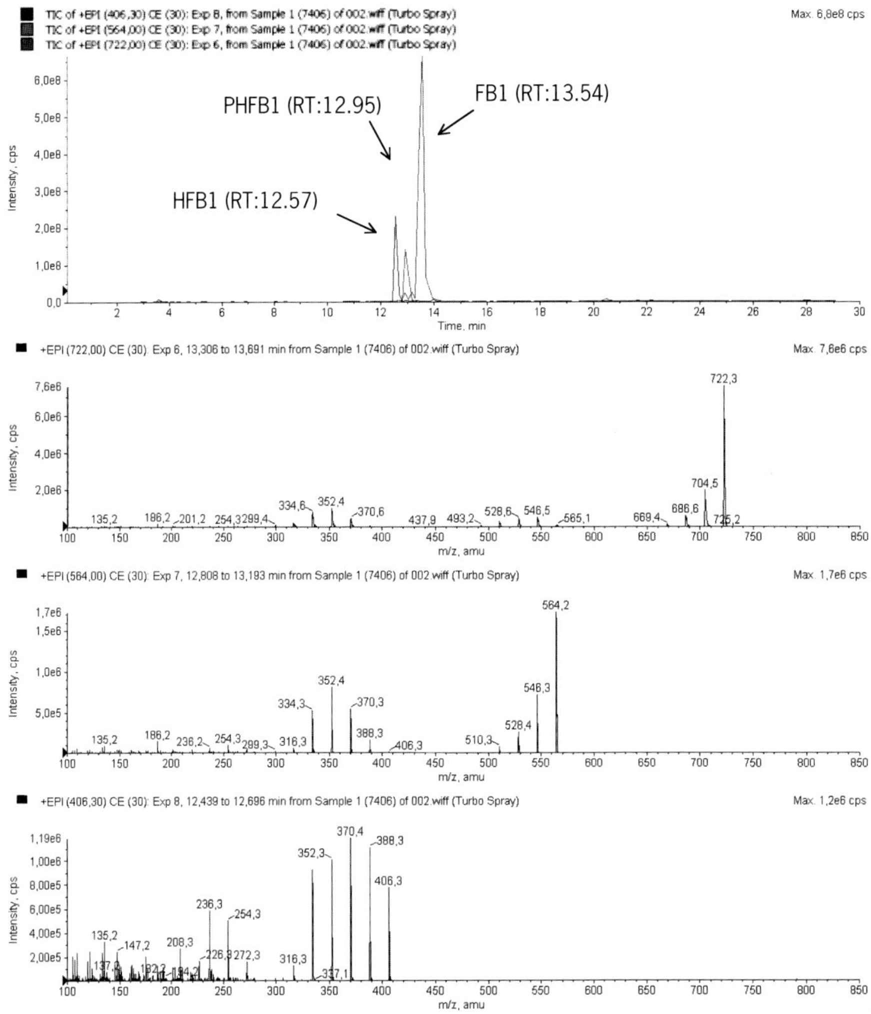

Figure 2. Occurrence of FB_1, $PHFB_1$ and HFB_1 in the sample after 3hr microbial deactivation. Total ion chromatograms and product ion MS/MS spectra of FB_1 (m/z 722), $PHFB_1$ (m/z 564) and HFB_1 (m/z 406), ESI positive ionisation mode.

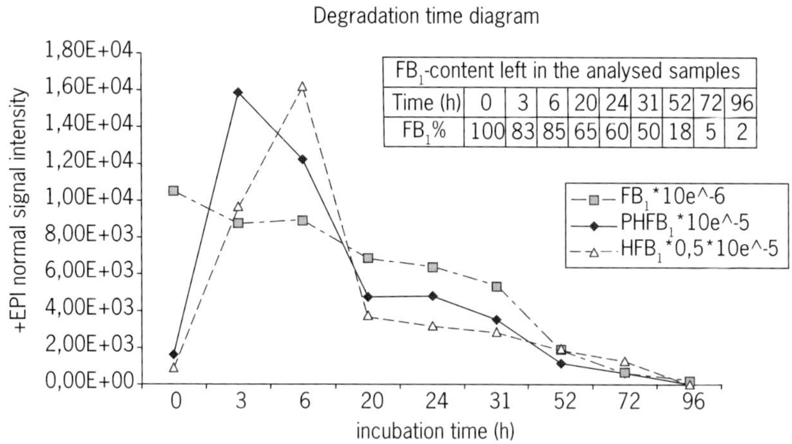

Figure 3. Changes in the content of FB_1, $PHFB_1$ and HFB_1 during the degradation experiment with Exophiala spinifera.

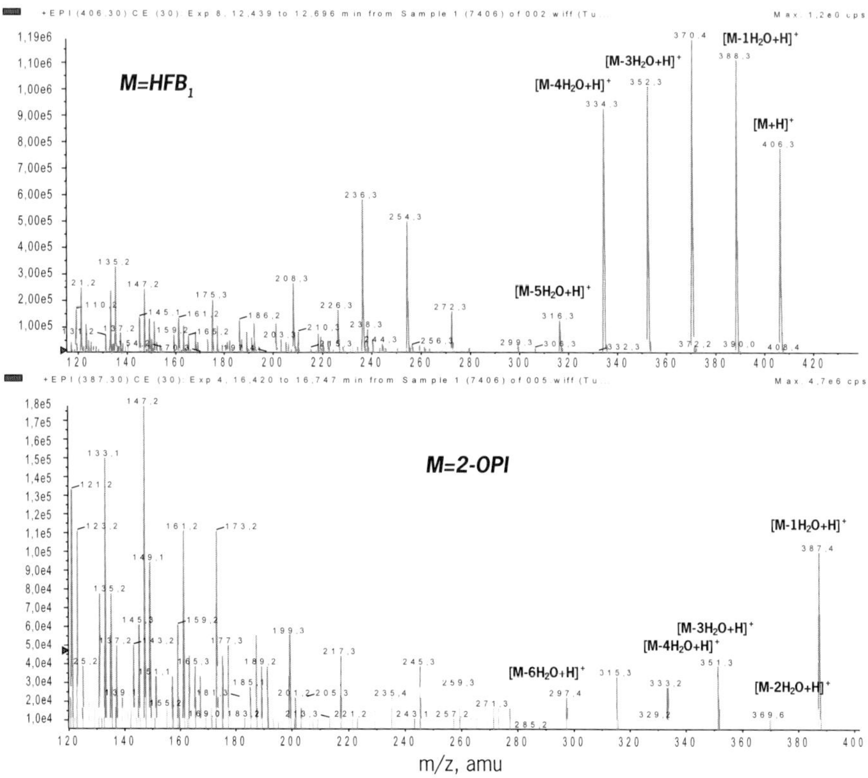

Figure 4. Comparison of the +EPI (MS/MS) spectra of HFB_1 and 2-OPI.

Toxicity test

The successful detoxification of degraded samples was demonstrated in a OSTRACODTOXKIT F™ bioassay originally designed for toxicity screening of freshwater sediments. In this study, *Heterocypris incongruens*, a benthic ostracod crustacean, was found to be sensitive towards fumonisins at concentrations as low as 5 mg/l FB_1. The test was essentially a direct contact chronic toxicity assay. Samples of mineral salt solution (MSL) with 25 mg/l FB_1 (Mo+MSL+FB_1) before (0 hr) and after (24 and 96 hr) incubation with the fumonisin-degrading microorganism, were applied to this assay. Mineral salt solution with *Exophiala spinifera* DSM 1217 (Mo) but without FB_1, was used as the reference material. The 24 and 96 hour degradation periods respectively yielded 40 and 98% degradation of the parent FB_1.

The mean mortality of the ostracods in presence of the toxin prior to microbial degradation was shown to be 100%. After 96 hr incubation with the degradation solution the mean mortality of the test organism was as low as 28% compared to 15% in MSL with the *Exophiala spinifera* (Figure 5). According to the manufacturer of the OSTRACODTOXKIT F™ bioassay, for the toxicity test to be acceptable, the mean %

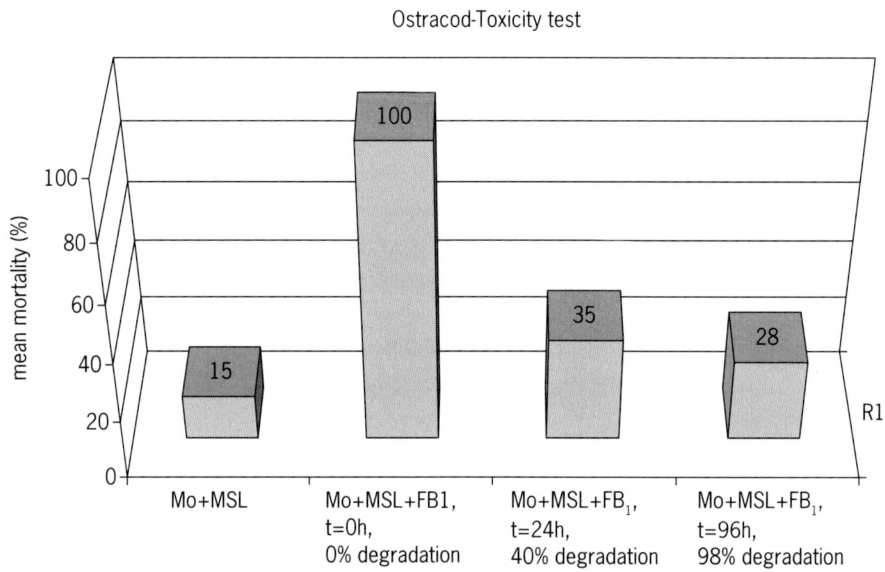

Figure 5. Mean mortality percentage of Heterocypris incongruens *measured using the OSTRACODTOXKIT F™ test kit. Mo+MSL = [reference]; Mo+MSL+FB1 (0h) = ["toxin solution" before microbial degradation]; and Mo+MSL+FB1 (24h and 96h) = ["toxin solution" after 40% and 98% microbial degradation].* E. spinifera *DSM 1217 is abbreviated as Mo.*

mortality of the ostracods in the control test (with the reference material) must not exceed 20%. Our observed 15% falls within the acceptable range.

Conclusion

The degradation-pathway by *Exophiala spinifera* according to Blackwell *et al.* (1999) could successfully be confirmed by our experiments. Furthermore, we are working on the isolation and characterisation of degradation products formed from HFB_1 by complementary analytical methods expecting that these results will help to further understand the metabolic pathway of FB_1 and to interpret the metabolites formed also by other fumonisin detoxifying micro-organisms.

Acknowledgements

The authors thank the Christian Doppler Forschungsgesellschaft and the Company Biomin GmbH for their financial support.

References

Blackwell, B.A., Gilliam, J.T., Savard, M.E., Miller, J.D., and Duvick, J.P., 1999. Oxidative Deamination of Hydrolysed Fumonisin B1 (AP1) by Cultures of *Exophiala Spinifera*. *Natural Toxins* 7: 31-38.

Duvick, J., Rood, T., Maddox, J., and Gilliam, J., 1998. Detoxification of mycotoxins in planta as a strategy for improving grain quality and disease resistance: identification of fumonisin-degrading microbes from corn. In: Kohmoto, K., and Yoder, O. (Eds.): *Molecular Genetics of Host-Specific Toxins in Plant Disease,* Kluwer Academic Publishers, pp. 369-381.

Gelderblom, W.C.A., Kriek, N.P.J., Marasas, W.F.O., and Theil, P.G., 1991. Toxicity and carcinogenecity of *Fusarium moniliforme* metabolite, fumonisin B1 in rats. *Carcinogenesis* 12: 1247-1251.

Harisson, L.R., Colvin, B.M, Green, J.T, Newman, L.E., and Cole, J.R., 1990. Pulmonary edema and hydrothorax in swine produced by fumonisin B1 a toxic metabolite of *Fusarium moniliforme*. *Journal of Veterinary Diagnostic Investigation* 2: 217-221.

Hlywka, J.J., and Bullerman, L.B., 1999. Occurrence of fumonisin B1 and B2 in beer. *Food Additives and Contaminants* 16: 319-324.

Kellerman, T.S, Marasas, W.F.O., Theil, P.G, Gelderblom, W.C.A., Cawood, M., and Coetzer, J.A.W., 1990. Leucoencephalomalacia in two horses induced by oral dosing of fumonisin B1. *Journal of Veterinary Research* 57: 269-275.

Patel, S., Hazel, C., Winterton, A., and Morty, E., 1996. Survey of ethnic foods for mycotoxins. *Food Additives and Contaminants* 13: 833-841.

Rice, L.G., and Ross,P.F., 1994. Methods for detection and quantitation of fumonisins in corn, cereal products and animal excreta. *Journal of Food Protection* 57: 536-540.

Shepard, G.S., Tiel, P.G., Stockenström, S., and Sydenham, E.W., 1996. Worldwide survey of fumonisin contamination of corn and corn-based products. *J. AOAC Int.* 79: 6671-687.

Sydenham, E.W., Thiel, P.G., Marasas, W.F.O., Shephard, G.S., Van Schalkwijk, D.J., and Koch, K.R., 1990. Natural occurrence of some *Fusarium* mycotoxins in corn from low and high esophageal cancer prevalence areas of Transkei, Southern Africa. *Journal of Agricultural and Food Chemistry* 38: 1900-1903.

Thiel, P.G., Marasas, W.F.O., Sydenham, E.W., Shephard, G.S., Gelderblom, W.C.A., and Nieuwenhuis, J.J., 1991. Survey of fumonisin production by *Fusarium* species. *Applied and Environmental Microbiology* 57: 1089-1093.

Wang, E., Norred, W.P., Bacon, C.W., Riley, R.T., and Merril, A.H., 1991. Inhibition of sphingolipid biosynthesis by fumonisins. *Journal of Biological Chemistry* 266: 14486-14490.

Multi-mycotoxin analysis by LC-MS/MS in a single sample extract

M. Spanjer, P. Rensen and J. Scholten

Dutch Food and Consumer Product Safety Authority, Primary Agricultural & Horticultural Products Department, Hoogte Kadijk 401, 1018 BK Amsterdam, The Netherlands; martien.spanjer@vwa.nl

Abstract

Mycotoxin analysis is mainly carried out using some form of an ELISA test kit or HPLC after immunoassay clean-up. This implies determination of single compounds. Aflatoxins and ochratoxin A have been subject to EU legislation for some years. Deoxynivalenol (DON) legislation has been drafted and EU member states are requested to ensure compliance with the proposed level in advance. The EC is considering legislation for *Fusarium* toxins such as fumonisins, T2, HT2, zearalenone and nivalenol. All these legal limits are set for a variety of matrices. To comply with the legal limits, it would, therefore, be preferable to routinely be able to determine all the mycotoxins in a matrix in one single extract. Application of Liquid Chromatography-Mass Spectrometry has made it possible to obtain such a method. LC-separation of the mycotoxins is done with a gradient of water and acetonitrile on a reversed-phase column within 25 minutes. The Mass spectrometer is used with an Electrospray Interface in the positive mode (ESI^+). By means of multiple reaction monitoring it is possible to analyze fragment ions, and increase specificity and sensitivity necessary for quality control. Thus a method was developed that could assay for the aflatoxins B_1, B_2, G_1 and G_2; ochratoxin A; DON; 3-acetyl-deoxynivalenol; 15-acetyl-deoxynivalenol; the fumonisins B_1 and B_2; diacetoxyscirpenol; zearalenone; T2-toxin; HT2-toxin; roquefortin and sterigmatocystin in a single extract. Applying the method to food samples showed the existence of a matrix effect from peanuts and cornflakes. This was expected due to the lack of clean-up. Method performance data (recoveries) are presented for aflatoxins in peanuts and figs, and ochratoxin A and DON in wheat. Results indicate that the multi-mycotoxin method should be suitable for use for enforcement of regulations as well as HACCP demands.

Keywords: multi-mycotoxin, matrix-effect, electrospray ionization, LC-MS/MS, multiple reaction monitoring, ion suppression, baby food, ergot

Introduction

In the recent past, mycotoxins have been determined mainly by single-compound analytical methods based on immunoaffinity clean-up. These methods were collaboratively studied for aflatoxins (Stroka *et al.*, 2000), for ochratoxin A (Entwisle *et al.*, 2000) and fumonisins B_1 and B_2 (Visconti *et al.*, 2001). The immunoaffinity clean-up was considered to be so specific that confirmation was hardly necessary. Increasing quality control demands changed these views in the nineties. Methods were developed in which mass spectrometry (MS) was applied for confirmation purposes but still based on individual mycotoxins. Hurst *et al.* (1991) applied thermospray MS for the confirmation of aflatoxins in peanuts, Xiao *et al.* (1995) used electron impact MS to elucidate ochratoxin structures and Holcomb *et al.* (1993) choose fast atom bombardment MS to confirm the presence of fumonisin B_1 in food. In our laboratory the need for confirmation emerged when use of existing deoxynivalenol (DON) methods revealed peaks in the chromatogram that could possibly interfere with the DON peak. We started to use the LCQ™ (Thermo Electron, Waltham, MA) for confirmation of DON as described by Berger *et al.* (1999). While doing so the European Union Scientific Committee of Food (2000) published opinions on zearalenone, fumonisins and trichothecenes on the EU website. Such opinions precede Commission Regulations, *i.e.* the opinion on ochratoxin A was published in 1998 and the corresponding regulation was published in 2000.

Facing the expected legislation in the future we set out to find a way to analyse as many mycotoxins as possible in one LC-MS run. This principle originates from the mycological field, where measurements preferably are made in crude fungal extracts to identify mould species (Smedsgaard and Frisvad, 1996). They applied single quadrupole equipment to *Penicillium* isolates, in which they identified 13 mycotoxins, of which ochratoxin A, citrinin, penicillic acid and roquefortine C are the most important. To perform quantification and identification at the same time a triple quadrupole is needed. In that case quantification can be carried out on the parent ion and identification is obtained from characteristic daughter ions. Kussak *et al.* (1995) applied this technique for the determination of aflatoxins in dust, Lukacs *et al.* (1996) used the procedure to identify fumonisins in maize, and Razazzi *et al.* (1999) determined nivalenol and DON in wheat.

However, in all the studies cited above, the authors still applied clean-up. Janssens and Franken (2000, Technical Laboratory Rotterdam, The Netherlands, personal communication) compiled a method for not only trichothecenes, but also aflatoxins and ochratoxin A in a single run in feed. We set out to optimise, extend and validate it for food, as lower limits exist for human consumption. Special attention was paid to the effects of the matrix. Results of the assay of the first 13 mycotoxins were presented at the Second World Mycotoxin Forum (Spanjer *et al.*, 2003). Since then the method has been expanded to cover an additional 10 mycotoxins.

Materials and methods

Chemicals

All solvents were analytical-reagent or HPLC grade. Water was obtained from a Milli-Q system (Millipore, Bedford, MA, USA). Mycotoxins standards were purchased from Sigma (St. Louis, MO, USA). Stock and standard solutions were prepared by routine procedures.

Equipment

The LC-MS consisted of a Waters Alliance 2695 HPLC system (Waters, Milford, MA, USA), coupled to a Quattro Ultima triple quadrupole mass spectrometer (Micromass, Manchester, UK), equipped with a pneumatically-assisted electrospray interface, operating in positive mode, at a capillary voltage of 2.5 kV, a desolvation temperature of 450 °C, variable cone voltage (see table), cone gas flow of 100 l/hr and a desolvation gas flow of 600 l/hr, both nitrogen gas. Collision gas was argon at 0.8 bar on pressure regulator (see Table 1). Dwell time was 0.02 sec. Separation was achieved on a 150 x 3.2 mm ID Alltima C_{18} (5 µm) column (Alltech, Breda, The Netherlands) at a column temperature of 30 °C. The gradient was composed from solvents A (0,1% formic acid in water) and B (0,1% formic acid in acetonitrile) at a flow rate of 0.3 ml/min. The solvent program started with 80% A. In 12 minutes the gradient developed to 30% A, which was held for another 5.5 minutes.

Analytical procedure

Twenty-five grams of ground sample was shaken with 100 ml (80:20, v/v) acetonitril+water for 2 hours. The slurry was filtered and the filtrate diluted four times with water. Twenty microliters (20µl) of the diluted extract was injected into the LC-MS system. For each mycotoxin the parent and two daughter ions were recorded. Calculations were performed by the Masslynx & Quanlynx software.

Results and discussion

All chromatographic and mass spectrometry data are presented in Table 1. Comparable calibration curves are presented in Figure 1. For the latter experiments aflatoxin $B_{1,}$

Table 1. Relevant mass spectrometry and recovery data. N.D. = not determined in these matrices.

Mycotoxin	parent (m/z)	daughter (m/z)		ratio prim/sec.	retention time	cone voltage	collision eV	spike range	recoveries in		fragmentation reference
		primary	sec.						solvent	matrix	
Nivalenol	313	175	295	1.9	3.4	35	19/15	n.d.	n.d.	n.d.	Razzazi *et al.* (1999)
Deoxynivalenol	297	249	231	2.3	4.5	35	10/10	50 - 800	60	91	Berger *et al.* (1999)
15-Acetyl-DON	339	231	279	2.6	9	35	15/12	n.d.	n.d.	n.d.	Berger *et al.* (1999)
3-Acetyl-DON	339	231	213	2.6	9.3	30	12/20	50 - 800	84	102	Berger *et al.* (1999)
Fumonisin B_1	722	334	352	0.8	10.5	50	40/35	10 - 160	63	58	Lukacs *et al.* (1996)
Roquefortine C	390	193	322	1.1	11.3	91	30/30	n.d.	n.d.	n.d.	Rundberget *et al.* (2002)
Aflatoxin G_2	331	245	275	3.1	11.4	100	29/29	1 - 16	125	107	Kussak *et al.* (1995)
Fumonisin B_3	706	336	318	1.7	11.5	75	36/36	n.d.	n.d.	n.d.	Lukacs *et al.* (1996)
Fumonisin B_2	706	336	318	2.1	12.0	75	38/39	10 - 160	91	88	Lukacs *et al.* (1996)
Aflatoxin B_2	315	259	287	0.9	12.2	100	30/24	1 - 16	93	106	Kussak *et al.* (1995)
Aflatoxin G_1	329	243	283	1.9	12.2	90	27/24	1 - 16	111	99	Kussak *et al.* (1995)
Diacetoxyscirpenol	367	307	247	4.2	12.3	31	10/15	50 - 800	83	104	Razzazi *et al.* (2002)
HT-2 Toxin	425	263	105	2.2	12.9	50	10/38	50 - 800	90	110	Razzazi *et al.* (2002)
Aflatoxin B_1	313	241	285	0.9	13.0	90	37/23	1 - 16	78	101	Kussak *et al.* (1995)
T-2 Toxin	467	305	245	1.3	15.8	40	10/8	50 - 800	99	111	Razzazi *et al.* (2002)
Ochratoxin A	404	239	241	2.9	16.7	50	22/22	2 - 32	98	108	Xiao *et al.* (1995)
Zearalenone	319	187	185	1.4	17.2	32	20/20	50 - 800	41	108	van Bennekom *et al.* (2002)
Sterigmatocystin	325	281	253	6	18.2	85	41/45	n.d.	n.d.	n.d.	Nielsen *et al.* (2003)

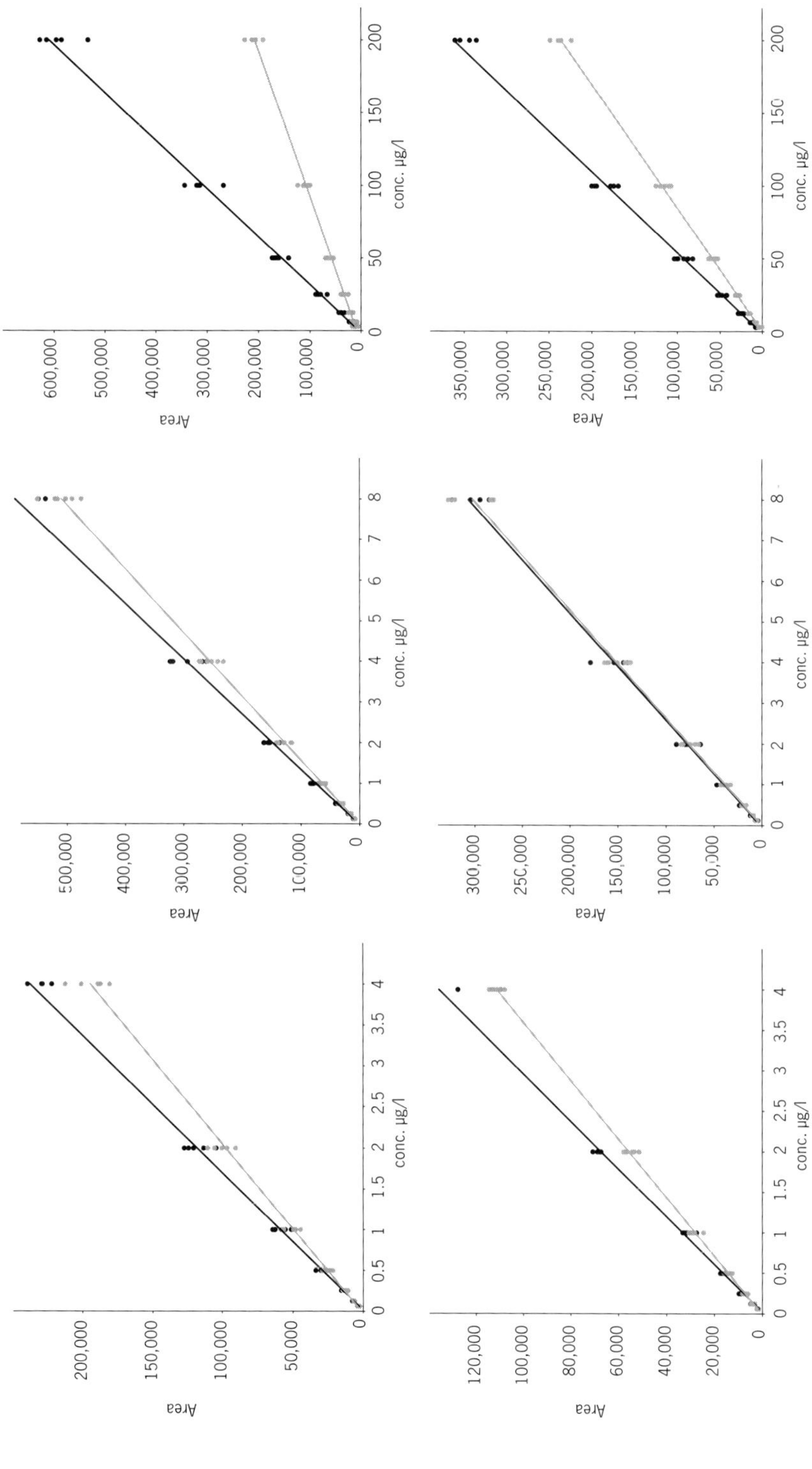

Figure 1. Ion suppression of aflatoxin B_1, ochratoxin A and DON in peanut and cornflakes (below). Upper calibration curves in solvent, lower curves in matrix. Multi reaction modes 313 → 241, 404 → 239 and 297 → 249 respectively.

ochratoxin A and DON were chosen because of the present legislation for these mycotoxins. Assessment of the performance of the method based on recoveries was carried out only in 2 matrices for 13 mycotoxins.

To separate all compounds a gradient is needed that starts with a high percentage of water, which is unfavourable for the ionization process in the ion source. In the ESI^+ mode the most prominent ion is the protonated molecular $[M+H]^+$ ion, the parent (m/z). To determine the identity of the parent, these ions were fragmented into daughter ions with argon gas in the collision cell of the triple quadrupole. Quantification was carried out on the parent ion. We did not investigate differences between positive or negative electrospray. Razzazi *et al.* (1999, 2002) considered positive ion mode to be preferred for type A and negative mode for type B trichothecenes. Tuomi *et al.* (1998) noted that the positive ion mode was clearly more effective for trichothecenes, an observation supported by Lagana *et al.* (2003) who examined data for a water/acetonitril/formic acid eluent. From the data presented, it is clear that the matrix effect, as indicated by Rundberget *et al.* (2002) for penitrem A, has to be determined for every separate matrix-mycotoxin combination. Park *et al.* (2002) reported finding several mycotoxins in a single sample run in a single assay.

We have applied this newly developed multi-mycotoxin method to several commodities: baby foods determined to have aflatoxin and ochratoxin A levels, rye samples in which ergot alkaloids were identified, maize and buckwheat. We also compared our procedure to the European Committee for Standardization (CEN). The results from these investigations will be published separately.

Conclusion

The method presented in this report has proven to be a good starting point for further development of a multi-mycotoxin method of analysis. Further work on various types of food matrices and validation needs to be carried out.

Acknowledgments

Harm Janssens and Gerard Franken, Technical Laboratory Rotterdam, are acknowledged for developing the basis for the procedure and allowing us to expand and apply it to matrices other than feeds.

References

Berger, U., Oehme, M., and Kuhn, F., 1999. Quantitative determination and structure elucidation of type A- and B-trichothecenes by HPLC/Ion trap multiple mass spectrometry. *Journal of Agricultural and Food Chemistry* 47: 4240-4245.

Commission Regulation (European Commission 472/2002) of March 12th 2002, amending Regulation EC 466/2001, setting maximum levels for certain contaminants in foodstuffs.

Entwisle, A.C., Williams, A.C., Mann, P.J., Slack, P.T., and Gilbert, J., 2000. Liquid Chromatographic Method with Immunoaffinity Column Cleanup for Determination of Ochratoxin A in Barley: Collaborative Study. *Journal of AOAC International* 83: 1377-1383.

European Union, Scientific Committee of Food, 2000. Brussels, http://europa.eu.int/comm/food/fs/sc/scf/opinions.

Holcomb, M., Sutherland, J.B., Chiarelli, M.P., Korfmacher, W.A., Thompson, Jr., H.C., Lay, Jr. J.O., Hankins, L.J. and Cerniglia, C.E., 1993. HPLC and FAB Mass spectrometry analysis of fumonisins B_1 and B_2 produced by *Fusarium moniliforme* on food substrates. *Journal of Agricultural and Food Chemistry* 41: 357-360.

Hurst, W.J., Martin, R.A., and Vestal, C.H., 1991. The use of HPLC/thermospray MS for the confirmation of aflatoxins in peanuts. *Journal of liquid chromatography* 14: 2541-2550.

Kussak, A., Nilsson, C.-A., Andersson, B. and Langridge, J., 1995. Determination of aflatoxins in dust and urine by liquid chromatography/electrospray ionization tandem mass spectrometry. *Rapid Communications in Mass Spectrometry* 9: 1234-1237.

Lagana, A., Curini, R., Dáscenzo, G., De Leva, I., Faberi, A., and Pastorini, E., 2003. Liquid chromatography/tandem mass spectrometry for the identification and determination of trichothecenes in maize. *Rapid Communications in Mass Spectrometry* 17: 1037-1043.

Lukacs, Z., Schaper, S., Herderich, M. Schreier, P., and Humpf, H.-U., 1996. Identfication and determination of fumonisin FB_1 and FB_2 in corn and corn products by high-performance liquid chromatography-electrospray-ionization tandem mass spectrometry (HPLC-ESI-MS-MS). *Chromatographia* 43: 124-128.

Nielsen, K.F., and Smedsgaard, J., 2003. Fungal metabolite screening: database of 474 mycotoxins and fungal metabolites for dereplication by standardized liquid chromatography-UV-mass spectrometry methodology. *Journal of Chromatography A* 1002: 111-136.

Park, J.W., Kim, E.K., Shon, D.H., and Kim, Y.B., 2002. Natural co-occurrence of aflatoxin B_1, fumonisin B_1 and ochratoxin A in barley and corn foods from Korea. *Food Additives and Contaminants* 19: 1073-1080.

Razzazi-Fazeli, E., Böhm, J., and Luf, W., 1999. Determination of nivalenol and deoxynivalenol in wheat using liquid chromatography-mass spectrometry with negative ion atmospheric pressure chemical ionisation. *Journal of Chromatography A* 854: 45-55.

Razzazi-Fazeli, E., Rabus, B., Cecon, B., and Böhm, J., 2002. Simultaneous quantification of A-trichothecene mycotoxins in grains using liquid chromatography-atmospheric pressure chemical ionization mass spectrometry. *Journal of Chromatography A* 968: 129-142.

Rundberget, T., and Wilkins, A.L., 2002. Determination of *Penicillium* mycotoxins in foods and feds using liquid chromatography-mass spectrometry. *Journal of Chromatography A* 964: 189-197.

Smedsgaard, J., and Frisvad, J.C., 1996. Using direct electrospray mass spectrometry in taxonomy and secondary metabolite profiling of crude fungal extracts. *Journal of Microbiological Methods* 25: 5-17.

Spanjer, M.C., Scholten, J.M., Rensen, P.M., 2003. *Single run LC-MS/MS analysis of mycotoxins subject to actual and upcoming EU legislation in one sample extract.* Poster presentation at the Second World Mycotoxin Forum, 17-18 february 2003, Noordwijk, The Netherlands.

Stroka, J., Anklam, E., Jörissen, U., and Gilbert, J., 2000. Immunoaffinity Column Cleanup with Liquid Chromatography Using Post-Column Bromination for Determination of Aflatoxins in Peanut Butter, Pistachio Paste, Fig Paste, and Paprika Powder: Collaborative Study. *Journal of AOAC International* 83: 320-340.

Tuomi, T., Saarinen, L., and Reijula, K., 1998. Detection of polar and macrocyclic trichothecene mycotoxins from indoor environments. *The Analyst* 123: 1835-1841.

van Bennekom, E.O., Brouwer, L., Laurant, E.H.M., Hooijerink, H., and Nielen, M.W.F., 2002. Confirmatory analysis method for zeranol, its metabolites and related mycotoxins in urine by liquid chromatography-negative ion electrospray tandem mass spectrometry. *Analytica Chimica Acta* 473: 151-160.

Visconti, A., Solfrizzo, M., and De Girolamo, A., 2001. Determination of Fumonisins B_1 and B_2 in Corn and Corn Flakes by Liquid Chromatography with Immunoaffinity Column Cleanup: Collaborative Study. *Journal of AOAC International* 84: 1828-1837.

Xiao, H., Marquardt, R.R., Frohlich, A.A., and Ling, Y.Z. 1995. Synthesis and structural elucidation of analogs of Ochratoxin A. *Journal of Agricultural and Food Chemistry* 43: 524-530.

Determination of masked mycotoxins using HPLC–tandem mass spectrometry

F. Berthiller[1], R. Schuhmacher[1], B. Poppenberger[2], D. Lucyshyn[2], M. Lemmens[3], G. Adam[2] and R. Krska[1]

[1]Christian Doppler Laboratory for Mycotoxin Research, Center for Analytical Chemistry, Department for Agrobiotechnology (IFA-Tulln), University of Natural Resources and Applied Life Sciences, Vienna, Konrad Lorenz Str. 20, A-3430 Tulln, Austria; rudolf.krska@boku.ac.at
[2]Institute of Applied Genetics and Cell Biology, Department of Applied Plant Sciences and Plant Biotechnology, University of Natural Resources and Applied Life Sciences, Vienna, Muthgasse 18, A-1190 Vienna, Austria
[3]Plant Production Biotechnology, Department for Agrobiotechnology (IFA-Tulln), University of Natural Resources and Applied Life Sciences, Vienna, Konrad Lorenz Str. 20, A-3430 Tulln, Austria

Abstract

The topic of conjugated mycotoxins first caught the attention of investigators in the mid eighties of the last century because in some cases of mycotoxicoses, clinical observations in animals did not correlate with the low mycotoxin content determined in the corresponding feed. High performance liquid chromatography combined with tandem mass spectrometry offers a powerful tool for identification and characterization of polar non-volatile mycotoxin conjugates. Derivatives can be identified due to their molecular mass and their fragmentation behaviour. Moreover, known conjugates can be quantified and plant extracts can be screened for unknown mycotoxin conjugates with different MS/MS techniques. We chemically synthesized deoxynivalenol-3-β-D-glucopyranoside, deoxynivalenol-15-β-D-glucopyranoside and zearalenone-4-β-D-glucopyranoside. The two deoxynivalenol-glucosides showed different fragmentation behaviours on a hybrid quadrupole – linear ion trap tandem mass spectrometer and could therefore be distinguished. It was possible to confirm the identity of deoxynivalenol-3-glucoside in deoxynivalenol treated yeast, which expressed the *Arabidopsis thaliana* glucosyltransferase *DOGT1*. Furthermore, deoxynivalenol-3-glucoside was detected in deoxynivalenol treated *Arabidopsis* and wheat. We also confirmed the formation of ZON-4-glucoside, a substance that was already known to occur in contaminated wheat. Finally, we isolated and characterized zearalenone-4-sulfate from *Fusarium* inoculated rice.

Keywords: deoxynivalenol, deoxynivalenol-3-β-D-glucopyranoside, zearalenone, zearalenone-4-β-D-glucopyranoside, zearalenone-4-sulfate, conjugated mycotoxins, masked mycotoxins, liquid chromatography, tandem mass spectrometry

Introduction

Plants can decrease the toxicity of phytotoxins, like mycotoxins, either by chemical modification and/or inclusion into the plant matrix (Engelhardt *et al.*, 1999). This detoxification process also includes the conjugation of mycotoxins to polar substances like glucose, acetate, glutathione or sulfate, which can be stored in vacuoles. Although these "masked mycotoxins" show far lower toxicities, hydrolysis in the digestive tract of mammals can transform them back to the original, unconjugated, toxins (Gareis, 1994). The trichothecene deoxynivalenol (DON) and the estrogenic toxin zearalenone (ZON) are often found in cereals and are used as key substances in the determination of *Fusarium* mycotoxins.

Ever since the work of Young *et al.*, (1984) that showed that the DON content of yeast fermented food products was higher than that of the contaminated flour used for their production, it has been speculated that a DON conjugate of some form, arising from plant metabolism, might exist. Savard (1991) was the first to chemically synthesize glucose and fatty acids conjugates of DON, while Sewald *et al.* (1992) could identify deoxynivalenol-3-β-D-glucopyranoside (DON-3-glucoside) as a DON metabolite in maize cell suspension cultures. Still, data about the toxicity, occurrence and stability of this substance in cereals is lacking.

Zearalenone-4-β-D-glucopyranoside (ZON-4-glucoside) is certainly the best studied conjugated mycotoxin. ZON-4-glucoside was originally found as a fungal metabolite (Kaminura, 1986), but Engelhardt *et al.* (1988) reported that ZON can be transformed to ZON-4-glucoside also by plant cultures. Gareis *et al.* (1990) proved that ZON-4-glucoside is cleaved during digestion in swine, releasing the estrogenic aglycone. Schneweis *et al.* (2002) developed and in-house validated a HPLC-MS method for the detection of ZON-4-glucoside, finding the substance in 42% of all ZON positive wheat samples.

Sulfate and glucuronide conjugates of zearalenone were shown to occur in the urine of ZON fed animals (Mirocha *et al.*, 1981), although the identity of the conjugates was not confirmed. Ten years later, zearalenone-4-sulfate (ZON-4-sulfate) was isolated as a fungal metabolite from *Fusarium* spp. and characterized using both mass spectrometry and NMR techniques (Plasencia and Mirocha, 1991). The same authors have also shown that the ZON-4-sulfate can be hydrolyzed easily by 1% hydrochloric acid, releasing ZON.

Materials and methods

Mycotoxins and reagents

The mycotoxins DON, ZON, α-zearalenol (α-ZOL) and β-zearalenol (β-ZOL) were purchased from Biopure Referenzsubstanzen GmbH (Tulln, Austria). Methanol and acetonitrile were LC grade and obtained from J.T. Baker (Deventer, the Netherlands).

Mycotoxin conjugates

DON-3-glucoside and DON-15-glucoside were synthesized in a two-step reaction from 1-β-bromo-1-deoxy-2,3,4,6-tetra-O-acetyl-α-D-gluco-pyranose and 15-acetyl-DON or 3-acetyl-DON (Berthiller *et al.*, 2003). Wheat ears were treated with DON (1 mg per ear) at anthesis. Whole ears were ground, weighed and extracted with acetonitrile+water (84:16, v/v) for 90 minutes using an automatic shaker set at 180 rpm. The extracts were filtered and evaporated to dryness at 50 °C. The residues were then dissolved in methanol+water (10:90, v/v), passed through 0.45 μm PVDF membrane filters and analyzed by HPLC-MS/MS.

ZON-4-glucoside was synthesized (Scharnhorst *et al.*, 2003) after a modified protocol from Grabley *et al.* (1992). For the production of ZON-4-sulfate, rice was mixed with sucrose and casein hydrolysate, sterilized, and inoculated with a *Fusarium graminearum* culture. After 3 weeks at 16-18 °C, the mouldy rice was homogenized and extracted with methanol+water (75:25, v/v).

LC-MS/MS analysis

LC-MS/MS analysis were performed with a QTrap-LC-MS/MS System (Applied Biosystems, Foster City, USA), equipped with a Turbo Ion Spray (ESI) source and a 1100 Series HPLC System (Agilent, Waldbronn, Germany) including a Diode Array Detector (DAD). Chromatographic separations were achieved at 25 °C using methanol+water (DON-3-glucoside: 25% methanol; ZON metabolites: 70% methanol) and a 100 mm x 4.6 mm i.d., 3 μm RP-18 Aquasil column (Thermo-Keystone, Waltham, USA). Injection volume was 25 μl, while the flow rate was 0.5 ml/min. UV-detection was performed at 270 nm for zearalenone derivates and at 220 nm for deoxynivalenol derivates. The ESI source was operated at 400 °C in negative ionization mode. Further MS parameters were: curtain gas (CUR) 20 psi (138 kPa), nebulizer gas (GS1) 30 psi (207 kPa), auxiliary gas (GS2) 65 psi (448 kPa), ionization voltage -4200 V, collision-activated dissociation gas (CAD) 6 (arbitrary units), multiple reaction monitoring (MRM) dwell time 25 ms, pause between mass ranges 5 ms. Product Ion spectra were gained at a collision energy (CE) of –30 V and a declustering potential (DP) of –46 V. For MRM transitions the following settings were used, ZON: m/z 317.1 -> m/z 131.0, DP –51 V, CE –38 V; α- and β-ZOL: m/z 319.1 -> m/z 160.0, DP –81 V, CE –36 V; ZON-4-glucoside: m/z 479.1 -> m/z 317.1, DP –31 V, CE –22 V; α- and β-ZOL-4-glucosides: m/z 481.1 -> m/z 319.1, DP –31 V, CE –22 V; ZON-4-sulfate: m/z 397.1 -> m/z 317.2, DP –51 V, CE –18 V.

Results and discussion

Characterization of DON-glucosides with LC-MS/MS

Despite having the same molecular mass and similar retention times, it was possible to distinguish the chemically synthesized DON-3-glucoside and DON-15-glucoside due to the loss of a "CH_2O" group (-30 amu) attached to the C-6 of the DON-3-glucoside (Figure 1). In case of DON-15-glucoside, cleavage of the bond between glucose C-6 and DON C-15 results in the loss of the whole sugar moiety (-192 amu) and no peak at m/z 427 can be observed. A more detailed explanation of the MS fragments of DON-3-glucoside is given elsewhere (Dall´Asta *et al.*, 2004). An *Arabidopsis thaliana* cDNA library screen in yeast resulted in a yeast transformant that survived on high DON concentrations in the media. The gene responsible for DON tolerance was identified as encoding a uridine diphosphate (UDP) dependant glucosyltransferase, named *DOGT1*. We previously demonstrated that the metabolite formed after DON feeding to either *A. thaliana* or to transformed yeast was identical to synthetically prepared DON-3-glucoside (Poppenberger *et al.*, 2003).

Occurrence of DON-3-glucoside in DON treated wheat

Although it was shown that DON-3-glucoside can be formed by DON treated *Arabidopsis*, it remained unclear if this substance could also occur in cereals. As a result, wheat ears were treated with DON. DON-3-glucoside was successfully detected in wheat for the first time. An enhanced product ion (EPI) chromatogram of the precursor mass 457.1 (corresponding to $[DON\text{-}3\text{-}glucoside\text{-}H]^-$) shows a single peak at the same retention time as the synthetically prepared standard (Figure 2). The MS/MS spectra were also identical and the occurrence of the m/z 427 fragment could be attributed only to DON-3-glucoside. The concentration of the DON-3-glucoside was determined in two different wheat cultivars with respect to whole ears.

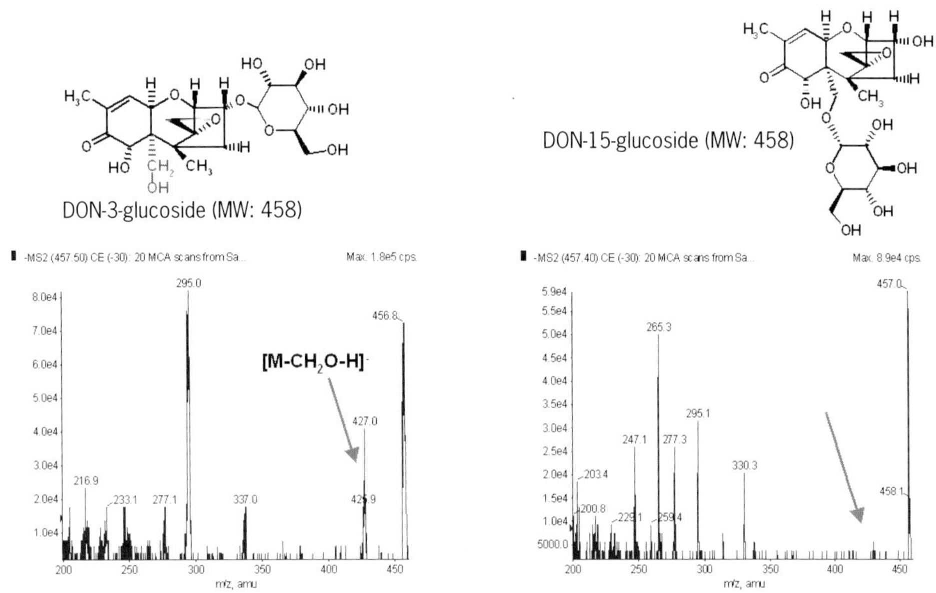

Figure 1. DON-glucosides, MS/MS spectra, ESI negative ionization mode, 10 mg/l in methanol+water (10:90, v/v), syringe pump (10 µl/min, standards).

Based on MS/MS and UV analysis, the content of DON-3-glucoside was 14-29% of the concentration of DON.

Formation of ZON-metabolites in genetically modified yeast

An LC-MS/MS method for the simultaneous detection of ZON, α-ZOL, β-ZOL and ZON-4-glucoside was established. The MRM mode was used for the method, offering high selectivity, sensitivity and linearity. With this technique, only ions of a fixed m/z ratio (*e.g.* m/z 317.1 for ZON, corresponding to the $[ZON\text{-}H]^-$ ion) could enter the collision cell of the mass spectrometer. These ions then collided with nitrogen molecules and produced smaller, characteristic fragments (*e.g.* m/z 131.0 for ZON). A chromatogram for ZON, α-ZOL, β-ZOL and ZON-4-glucoside of an 1 mg/l mixed standard solution is given in Figure 3. Similar to the *DOGT1* gene, we found a UDP dependant glucosyltransferase in the *A. thaliana* genome which is able to detoxify zearalenone. Expressing this gene in yeast resulted in the almost complete transformation of ZON into ZON-4-glucoside. In addition to the LC-MS peaks of the toxins for which standards were available, two peaks corresponding to m/z 481 were observed. The mass and fragmentation behavior indicate that these two

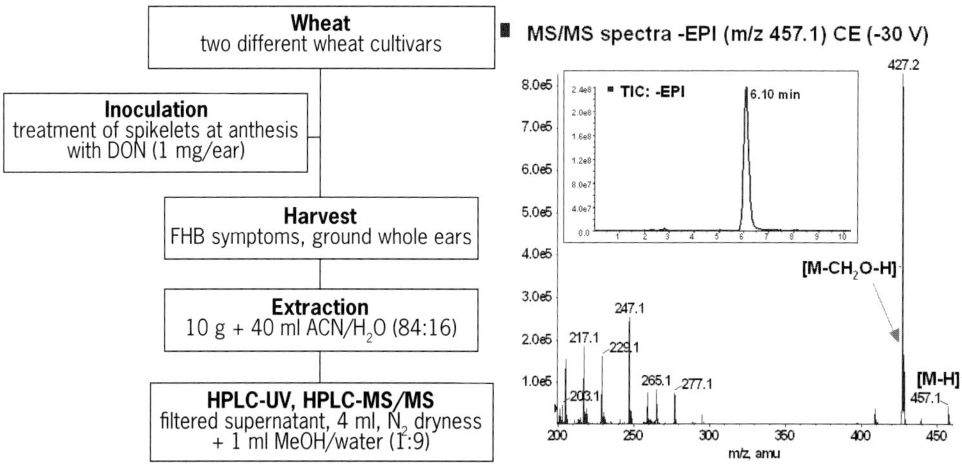

Figure 2. Occurrence of DON-3-glucoside in DON treated wheat, MS/MS spectra and total ion chromatogram, ESI negative ionization mode.

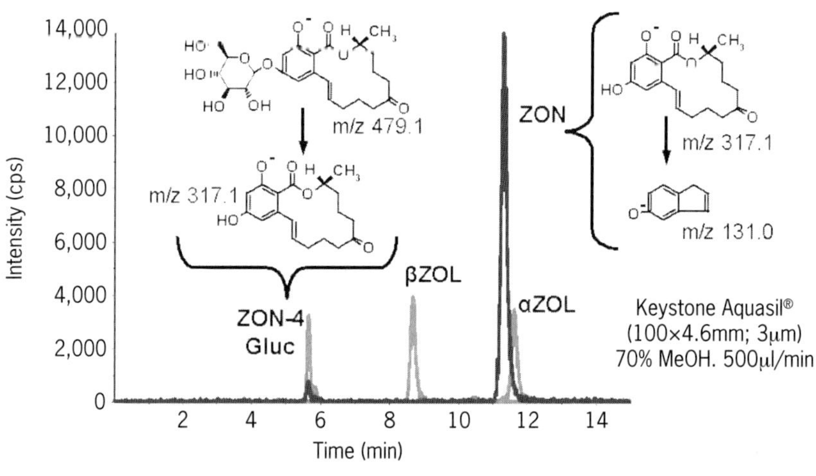

Figure 3. MS/MS (MRM) chromatogram, ESI negative ionization mode, ZON, α-ZOL, β-ZOL, ZON-4-glucoside (1 mg/l, in methanol/water (50/50, v/v)).

compounds could be α-ZOL-4-glucoside and β-ZOL-4-glucoside. We tested several commercially available ZON clean-up columns for their suitability to also purify ZON-4-glucoside. Ideally, antibodies that recognize ZON might also recognize

ZON-4-glucoside, but the recoveries of the tested immuno-affinity columns were about zero. So we changed to solid phase columns and discovered that MycoSep® #226 columns (RomerLabs Inc., Union, MO, USA) produced satisfactory results (ZON-4-glucoside recovery 92%, RSD < 10%, n=8, 40-1000 μg/kg).

Formation of ZON-4-sulfate by Fusarium graminearum *on rice*

ZON-4-sulfate and ZON can be separated by HPLC and detected by UV and MS/MS in the MRM mode. Four batches of rice were inoculated with *F. graminearum* and extracted. On average, about a quarter of the total ZON in the cultures occurred in the form of ZON-4-sulfate. This was quantified only by the UV-Signal, as we are currently working on the purification of a suitable standard for LC-MS/MS experiments. ZON-4-sulfate is far less toxic compared to ZON (Dr. Rudolf Mitterbauer, 2004, personal communication). It still remains to be investigated whether or not this compound is a relevant contaminant in cereals, as it can be hydrolyzed quickly into ZON.

Conclusion

As part of their metabolism, plants are capable of transforming phytotoxins into conjugated forms. For the first time DON-3-glucoside has been detected in wheat which was artificially contaminated with DON. Methods for the simultaneous detection of ZON and its metabolites, including α-ZOL, β-ZOL, ZON-4-glucoside, α-ZOL-4-glucoside, β-ZOL-4-glucoside and ZON-4-sulfate were developed and successfully applied for their determination in biological samples.

Acknowledgements

The authors thank the Christian Doppler Forschungsgesellschaft and the Austrian Genome Research programme (GEN-AU) for their financial support.

References

Berthiller, F., Schuhmacher, R., Buttinger, G., Freudenschuss, G., Adam, G., and Krska, R., 2003. Synthesis of Deoxynivalenol-glucosides and their Characterization using a QTrap LC-MS/MS. *Mycotoxin Research* 19: 47-50.

Dall'Asta, C., Berthiller, F., Schuhmacher, R., Adam, G., Lemmens, M., and Krska, R., 2004. DON-Glycosides: Characterisation of Synthesis Products and Screening for their Occurrence in DON-treated Wheat Samples. *Mycotoxin Research* (in press)

Engelhardt, G., Ruhland, M., and Wallnöfer, P.R., 1999. Metabolism of mycotoxins in plants. *Advanced Food Science* 21: 71-78.

Engelhardt, G., Zill, G., Wohner, B., and Wallnöfer, P.R., 1988. Transformation of the *Fusarium* mycotoxin zearalenone in maize cell suspension cultures. *Naturwissenschaften* 75: 309-310.

Gareis, M., 1994. Maskierte Mykotoxine. Übersichten zur Tierernährung 22: 104-113.

Gareis, M., Bauer, J., Thiem, J., Plank, G., Grabley, S., and Gedek, B., 1990. Cleavage of zearalenone-glycoside, a „masked" mycotoxin, during digestion in swine. *Journal of Veterinary Medicine B* 37:236-240.

Grabley, S., Gareis, M., Böckers, W., Thiem, J., 1992. Glycosylation of Mycotoxins. *Synthesis* 11: 1078-1080

Kamimura, H., 1986. Conversion of zearalenone to zearalenone glycoside by *Rhizopus sp*. *Applied and Environmental Microbiology* 52: 515-519.

Mirocha, C.J., Pathre, S.V., and Robison, T.S., 1981. Comparative metabolism of zearalenone and transmission into bovine milk. *Food and cosmetics toxicology* 19: 25-30.

Plasencia, J., and Mirocha, C.J., 1991. Isolation and characterization of zearalenone sulfate. *Applied and Environmental Microbiology* 57: 146-150.

Poppenberger, B., Berthiller, F., Lucyshyn, D., Sieberer, T., Schuhmacher, R., Krska, R., Kuchler, K., Glössl, J., Luschnig, C., and Adam, G., 2003. Detoxification of the *Fusarium* Mycotoxin Deoxynivalenol by an UDP-glucosyltransferase from *Arabidopsis thaliana*. Journal of Biological Chemistry 278: 47905-47914.

Savard, M.E., 1991. Deoxynivalenol fatty acid and glucoside conjugates. *Journal of Agricultural and Food Chemistry* 39: 570-574.

Scharnhorst, K., 2003. Nachweis von Zearalenon und dem Metaboliten Zearalenon-4-O-β-Glukosid in Mais und Maisprodukten. Diploma Thesis, Universität für Bodenkultur Wien und Technische Universität Wien, pp. 1-62.

Schneweis, I., Meyer,K., Engelhardt, G., and Bauer, J., 2002. Occurrence of zearalenone-4-β-D-glucopyranoside in wheat. *Journal of Agricultural and Food Chemistry* 50: 1736-1738.

Sewald, N., Lepschy von Gleissenthall, J., Schuster, M., Müller, G., and Aplin, R.T., 1992. Structure elucidation of a plant metabolite of 4-desoxynivalenol. *Tetrahedron* 3: 953-960.

Young, J.C., Fulcher, R.G., Hayhoe, J.H., Scott, P.M., and Dexter, J.E., 1984. Effect of milling and baking on deoxynivalenol (vomitoxin) content of Eastern Canadian wheats. *Journal of Agricultural and Food Chemistry* 32: 659-664.

Section IV:

Toxicology of mycotoxins

Interaction between fumonisin B_1 and pig liver cytochromes P_{450}

L. Marvasi[1], D. Marin[2], J.L. Viadère[3], J. Laffitte[3], I.P. Oswald[3], P. Galtier[3] and N. Loiseau[3]

[1]*Facolta' di Medicina Veterinaria, Dipartimento di Sanita' Pubblica Veterinaria e Patologia Animale, Alma Mater Studiorum Universita di Bologna, Via Tolara di Sopra n°50, 40064 Ozzano Emilia (BO), Italia*
[2]*Institute of Biology and Animal Nutrition, Laboratory of Animal Biology, Calea Bucuresti n°1, Balotesti, Ilfov, Roumanie*
[3]*INRA, UR66 Laboratoire de Pharmacologie-Toxicologie, 180 Chemin de Tournefeuille, 31931 Toulouse, France; nicolas.loiseau@toulouse.inra.fr*

Abstract

Fumonisin B_1 (FB_1) is known to cause liver damage in all animal species. The aims of the present study were to describe the molecular interactions between this toxin and hepatic cytochromes P_{450} and to measure the effects of a low exposure of piglets to FB_1 on these liver enzymes. *In vitro* spectral interaction of FB_1 with P_{450} was studied on liver microsomal suspensions from piglets pre-treated with P_{450}-inducers: β-naphtoflavone, phenobarbital, rifampicin and clofibrate, which induce respectively the P_{450} isoforms 1A, 2B, 3A and 4A. With unbuffered solution of FB_1, polar interactions (inverse type I) were observed between P_{450} apoprotein and acid functions of this mycotoxin. This type of interaction is not specific for only one isoform of P_{450}. In potassium phosphate buffered solution (pH 7.4), FB_1 interacts specifically with all isoforms according to a polar interaction. In the particular case of microsomal protein suspension induced with clofibrate, a hydrophobic interaction (type I) was observed indicating a potential capability of metabolism of this mycotoxin. This result has shown an influence of the ionisation state on the propensity of metabolism of this mycotoxin. In piglets, FB_1 treatment (0.5 mg/kg/day for 6 days of pure toxin) induced significant decreases in total cytochrome P_{450}, P_{450} 4A subfamily and ethylmorphine demethylation whereas both P_{450} 1A and ethoxyresorufin *O*-demethylation were significantly increased. There was no significant change in P_{450} 2C and 3A monooxygenases.

Keywords: mycotoxin, fumonisin B_1, cytochrome, P_{450}, metabolism, pig

Introduction

Fumonisin B_1 (FB_1) is one of the main mycotoxins produced by *Fusarium verticillioides* and *F. proliferatum*. These two fungal species are frequently found as maize pathogens (Pittet *et al.*, 1992). Fumonisin B_1 is therefore a common contaminant of maize based human food and animal feeds. Dutton (1996) reported that three quarters of the European and world maize production is variably contaminated with this mycotoxin. Fumonisin B_1 presents a specific inter species pathological panel. Except hepatotoxicity and nephrotoxicity which may occur in all species, it causes equine leukoencephalomalacia (Marasas *et al.*, 1988), porcine pulmonary edema (Harrison *et al.*, 1990); in humans it is associated with a propensity to develop esophageal cancer (Sydenham *et al.*, 2000).

Fumonisin B_1 produces cellular cycle deterioration and impacts the immune response via the cytokine network (Bhandari *et al.*, 2002). It may also interfere with cell to cell communication via a corruption of lipid metabolism (Wang *et al.*, 1991). Although the toxicological target has been identified, sufficient data to explain the pathological specificity of this mycotoxin is lacking. Only the presence of toxic metabolites produced in the liver or in a specific target tissues could explain this pathological panel, since the scientific literature describes a metabolism of FB_1 limited to the hydrolysis of ester moeties. In this study we set out to determine metabolism of FB_1 by cytochrome (CYP) P_{450} 4A (CYP4A) using clofibrate induced-liver microsomal system, and secondly the effect of FB_1 on cytochrome P_{450} regulation.

Materials and methods

Chemicals

FB_1 was purchased from The Programme on Mycotoxins and Experimental Carcinogenesis (PROMEC) (PROMEC, South African Medical Research Council, Tygerberg, South Africa); NADP, glucose 6-phosphate (G6P), glucose 6-phosphate dehydrogenase (G6PDH), clofibrate (CLO), sodium phenobarbital (PB) and β-naphthoflavone (BNF), were purchased from Sigma (Sigma, St. Louis, MO, USA) Rifampicin (RIF) was used as Rifadin pharmaceutical form.

CYP induction treatments

Male piglets were housed and treated according to French legislation, in a facility authorised by the Ministry of Agriculture. Male piglets large white (5-9 kg) were treated either with BNF (100 mg/kg i.p. in maize oil for 5 days), PB (50 mg/kg *p.o.* in sugar solution for 5 days), RIF (50 mg/kg *p.o.* rifadin formulation for 5 days) or CLO (200 mg/kg i.p. for 5 days). Piglets were sacrificed one day after the last treatment.

Preparation of liver microsomes

Microsomes were prepared from 50 g pig liver using the method of Schenkman *et al.*, (1998) for preparing rat liver microsomes. The microsomes were frozen in liquid nitrogen and stored at –80 °C until use.

Enzyme assays

The contents of cytochromes P_{450} and b_5 were determined from the difference between a microsomal sample reduced with sodium dithionite and a sample gassed with CO (Omura *et al.*, 1964). The microsomal cytochrome P_{450} (P_{450}) dependent activities of ethoxyresorufin (EROD), methoxyresorufin (MROD) and pentoxyresorufin *O*-deethylations (PROD) were determined spectrofluorometrically by the method of Lake (1987). Aminopyrine, benzphetamine and ethylmorphine N-demethylase activities were measured using 1 mg microsomal protein, 1 mM substrate and 1 mM NADPH. The formation of formaldehyde was estimated by the Nash reaction (Cochin *et al.*, 1959).

Western blot analysis

Cytochrome P_{450} isoenzymes 3A and 2B were separated by electrophoresis in 8.5% polyacrylamide gel in the presence of sodium dodecyl sulfate (Laemmli, 1970). For comparison, the amount of P_{450} was standardised for each lane at 5 pmoles. The separated microsomal proteins were transferred to nitrocellulose membranes according to the method of Towbin *et al.* (1979). Staining of the blot was obtained using diaminobenzidine or a chemiluminescence technology (ECL kit, Amersham plc, Little Chalfont, Buckinghamshire, UK) following incubations with two antibodies: a primary anti-P_{450} antibody and a secondary peroxidase-labeled antibody. Relative content of each form of cytochrome P_{450} was estimated by densitometry of the Western blots by using an Omnimedia XRS 12 CX Bioimage Scanner (Bioimmage, Roissy, France). Band intensities were compared between control and treated samples.

Substrate binding to microsomal P450s by differential visible spectroscopy

Binding of substrates to cytochrome P_{450} was studied by difference spectroscopy using 2 μM P_{450} from pig liver microsomes in 0.1 M phosphate buffer (pH 7.4). The solution was equally divided between two cuvettes (0.5 ml/cuvette). After recording the baseline, aliquots (0.5 – 1 µl) of substrate solutions were added to the sample cuvette, the same volume of solvent being added to the reference. Difference spectra were recorded over a 360–440 nm range at room temperature with a KONTRON instrument UVIKON 922 spectrophotometer (Kontron Instruments, Bletchley, UK). Ks (spectrally estimated dissociation constant) and ΔDOmax were estimated by linear regression from (1/ [FB_1]; 1/ ΔDO) values using Microsoft Excel software (Microsoft Corp., Redmond, WA, USA).

Microsomal incubations

Metabolism of FB_1 was studied at 37 °C in 0.1 M phosphate buffer (pH 7.4) with 2 µM P_{450} from pig liver microsomes using 200 µM of substrate and an NADPH-generating system (1 mM NADP+, 10 mM G6P and 4 IU G6PDH). The incubation was stopped at the indicated times by addition of the same volume of acetonitrile. The mixture was then frozen until analysis, before which it was remixed and centrifuged at 10,000 rpm for 10 min. Each sample was dried under vacuum and derivatised by 5µL of thiomethyl alcohol, 10 µl of 10^{-2} M OPA methanolic solution and 50 µl of 0.1 M sodium tetraborate solution. HPLC analysis was performed in a linear gradient mode on a LiChrospher RP18, 5 µm, 150 x 4.6 mm reverse phase column (Agilent Technologies, Massy, France). Eluants were a 0.2% acetic acid solution (A) and a 100% acetonitrile (B). The elution gradient can be described by the following sequence: at 0 min 100% A; 40 min 50% B; 45 to 60 min 90% B and 65 to 70 min 100% A; with a 1ml/min of total flow. FB_1 metabolism was estimated by UV detection at 333 nm.

Analysis of data

Means and standard deviation (SD) were calculated for all investigated parameters. Significant differences were assessed by analysis of variance and Bonferroni's test by using Instat 2.01 software (Graph PAD, San Diego, USA). In all cases, a value of $p<0.05$ was considered significant.

Results and discussion

Metabolism of FB_1

Groups of pigs were treated with several inducers for 5 days: BNF 100 mg/kg (in maize oil IP), PB 50 mg/kg (in sugar solution PO), RIF 50 mg/kg (rifadin PO) and CLOF 200 mg/kg (IP) to induce respectively, CYP 1A, 2B, 3A and 4A. The resulting induction was ascertained by Western Blot (Figure 1a) on microsomal preparation. In spite of a significant difference ($p<0.05$) in the total amount of P_{450} resulting from treatment with BNF, the proportion of each P_{450} isoform did not show a significant variation. There was just a little increase of CYP1A. Induction by PB decreased CYP2C and increased CYP4A. Since we did not measure variation in CYP2B, the increase in CYP4A may be an artefact of the omission. On the contrary, induction with RIF and CLOF produced significant increases of CYP3A and CYP4A respectively.

We tested the interaction of FB_1 with each induced liver microsomal preparation. By means of ultra-violet analysis, we have shown a hydrophobic interaction only with clofibrate induced microsome (Figure 1b). The other preparations lead to only membrane disturbances. From UV interaction between FB_1 and clofibrate induced microsomes, we could deduce, from a linear regression of the inverse of ΔDO as a function of the inverse of FB_1 concentration (Figure 1c), an apparent affinity of this mycotoxin for CYP4A around 5 μM. The interaction with clofibrate induced microsomes has been confirmed by the production of a new metabolite after prolonged incubation. The new metabolite (probably a hydroxylated analog of FB_1) is different and independent of the previously described hydrolysed FB_1 (Figure 2). Plans are underway to characterise the new metabolite using mass spectrometry and NMR.

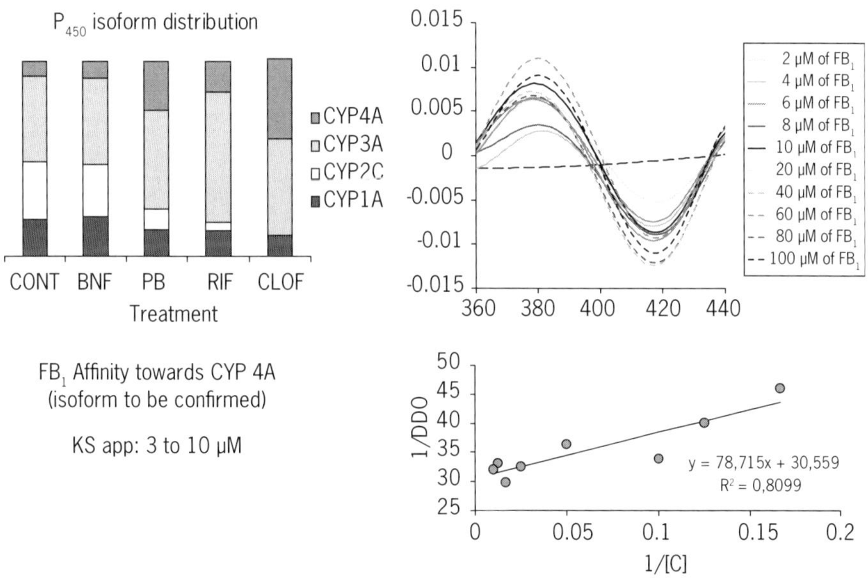

Figure 1. Upper left to bottom right: (a) P_{450} isoform distribution ascertained by western blot analysis; (b) differential UV measurement obtained by interaction between clofibrate induced microsome and several concentrations of FB_1; (c) linear regression of the reverse ΔDO in function of the reverse concentration of FB_1.

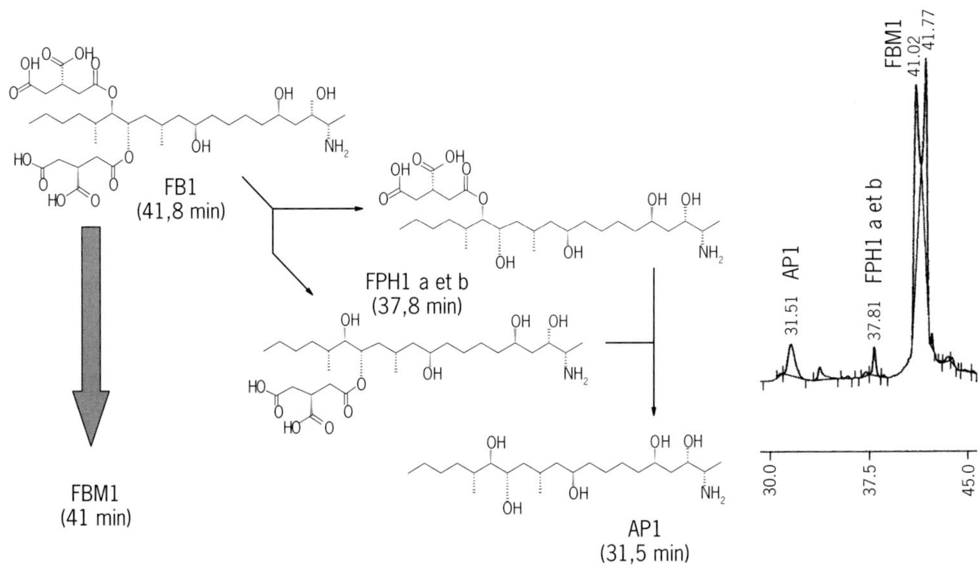

Figure 2. Metabolism scheme of fumonisin B_1 and related chromatogram.

FB_1 effect on the expression of P_{450} in liver

In vivo liver parameters (microsomal cytochrome P_{450} 1A, 2C, 3A and 4A subfamilies, EROD, PROD, MROD, benzphetamine, aminopyrine and ethylmorphine dealkylations) were measured in eight 3-week-old weaned piglets (Table 1). The animals were divided into three groups: four control piglets given a commercial starter diet, 2 piglets receiving an oral dose of 0.5 mg/kg of pure FB_1, and 2 piglets receiving a methanolic extract of FB_1. All groups were kept on their respective diets for 6 days. All animals were euthanized 24 hours after the last treatment and microsomes were prepared separately from the right and left liver lobes of each piglet. FB_1 treatment provoked a significant decrease in total cytochrome P_{450}. We observed a significant increase of CYP2C in the left liver lobe of piglets treated with the methanolic extract of FB_1 and significant decreases in both CYP3A and CYP4A in the left and right liver lobes, respectively, of piglets treated with pure FB_1 or the methanolic extract of FB_1. These results were confirmed by the demonstration of a significant increase of MROD activity in the left liver lobe treated by the extract of FB_1 and a significant decrease of ethylmorphine demethylation in the left liver lobes of piglets treated with pure or extract FB_1. There was no significant change in other liver monooxygenase activities, even though certain activities or expressions will differ between the different liver lobes.

Table 1. In vivo *liver parameters: P_{450} isoform dosage and enzymatic assays measurements in the left and right liver lobes after three different treatments.*

Liver lobes	control		Pure FB_1		Extract FB_1	
	Right	Left	Right	Left	Right	Left
CYP 1A	1.02 ± 0.49	1.38 ± 0.22	2.47 ± 1.3	1.28 ± 0.33	2.99 ± 0.59	1.67 ± 0
CYP 2C	1.21 ± 0.24	0.17 ± 0.12	1.23 ± 0.31	0.26 ± 0.16	0.84 ± 0.27	0.58 ± 0.08 **
CYP 3A	2.64 ± 0.15	1.28 ± 0.36	2.83 ± 0.28	0.48 ± 0.16 **	2.83 ± 0	0.70 ± 0.07 **
CYP 4A	3.53 ± 0.93	1.44 ± 0.21	2.13 ± 0.57 *	1.81 ± 1.07	2.04 ± 0.39 **	2.20 ± 0
EROD (pmol/min/mg)	9.49 ± 2.9	8.53 ± 1.8	9.21 ± 2.4	9.31 ± 1.8	9.55 ± 1.8	12.36 ± 1.5
PROD (pmol/min/mg)	1.72 ± 0.06	1.82 ± 0.10	1.76 ± 0.09	1.75 ± 0.14	1.87 ± 0.05	1.62 ± 0.01
MROD (pmol/min/mg)	5.52 ± 1.8	8.53 ± 1.8	5.73 ± 1.6	9.31 ± 1.8	4.7 ± 0.2	12.36 ± 1.5*
Benzphetamine demethylase (nmol/min/mg)	0.152 ± 0.01	0.153 ± 0.01	0.134 ± 0.01	0.137 ± 0.01	0.142 ± 0.01	0.150 ± 0
Aminopyrine demethylase (nmol/min/mg)	0.131 ± 0.02	0.139 ± 0.01	0.138 ± 0.01	0.133 ± 0.02	0.119 ± 0.02	0.123 ± 0
Ethylmorphine demethylase (nmol/min/mg)	0.144 ± 0.02	0.152 ± 0.03	0.134 ± 0.01	0.123 ± 0.01 *	0.136 ± 0.01	0.127 ± 0 *

* Significant difference from control group ($p \leq 0.05$).

** Significant difference from control group ($p \leq 0.01$).

Conclusion

In the first part of this study, we have shown that FB_1 is recognised by a particular isoform of P_{450} induced by clofibrate that could be CYP4A. Furthermore, this specific isoform slowly metabolised FB_1 to produce a new compound that is probably a hydroxylated form of this mycotoxin. We believe such new metabolites could be produced at different levels according to the species and the tissues. If this hypothesis were proven to be true, it would explain the specific pathology of FB_1.

In the second part of the study, the influence of FB_1 on the expression of P_{450}s has been demonstrated using *in vivo* liver parameters. Fumonisin B_1 induces a significant increase of CYP2C and a decrease of CYP3A and CYP4A in the left and right liver lobes, respectively. These results are in accordance with the hepatotoxicity induced by FB_1 in all species. Nevertheless, these results could not explain the synergistic effect of FB_1 with aflatoxin B_1 in the propensity to develop esophageal tumors.

Acknowledgments

This work was supported by the Ministère de la Recherche Française Grant N°03P475 and INRA within the framework of « Transversalité Mycotoxines » work group

References

Bhandari, N., Brown, C.C., and Sharma, R.P., 2002. Fumonisin B_1-induced localized activation of cytokine network in mouse liver. *Food and Chemical Toxicology* 40: 1483-1491.

Cochin, J., and Axelrod, J., 1959. Biochemical and pharmacological changes in the rat following chronic administration of morphine nalorphine and normorphine. *The Journal of Pharmacology and Experimental Therapeutics* 125: 105-110.

Dutton, M.F., 1996. Fumonisins, mycotoxins of increasing importance: Their nature and their effects. *Pharmacology and Therapeutics* 70: 137-161.

Harrison, L.R., Colvin, B.M., Greene, J.T., Newman, L.E., and Cole, J.R., 1990. Pulmonary oedema and hydrothorax in swine produced by fumonisin B1, a toxic metabolite of *Fusarium moniliforme*. *Journal of Veterinary Diagnostic Investigation* 2: 217-221.

Laemmli, U.K., 1970. Cleavage of structural proteins during the assembly of the head of bacteriophage T4. *Nature* 227: 680-685.

Lake, B.G., 1987. Preparation and characterization of microsomal fractions for studies on xenobiotic metabolism. In: Snell, K., and Mullock, B. (Eds.), *Biochemical Toxicology*. IRL Press, Oxford, UK, pp. 183-215.

Lowry, O.H., Rosebrough, N.J., Farr, A.L., and Randall, R.J., 1951. Protein measurement with the Folin-Phenol reagents. *Journal of Biological Chemistry* 193: 265-275.

Marasas, W.F.O., Kellerman, T.S., Gelderblom, W.C.A., Coetzer, J.A.W., Thiel, P.G., and Van der Lugt, J.J., 1988. Leukoencephalomalacia in a Horse Induced by Fumonisin B1 Isolated from *Fusarium moniliforme*. *Journal of Veterinary Research* 55: 197-203.

Omura, T., and Sato, R., 1964. The carbon monoxide-binding pigment of liver microsomes. I. Evidence for its hemoprotein nature. *Journal of Biological Chemistry* 239: 2370-2378.

Pittet, A., Parisod, V., and Schellenberg, M., 1992. Occurrence of fumonisins B1 and B2 in corn-based products from the Swiss market. *Journal of Agricultural and Food Chemistry* 40: 1352-1354.

Schenkman, J.B., and Jansson, I., 1998. Isolation and Purification of Constitutive Forms of Microsomal Cytochrome P450. In: *Methods in. Molecular Biology: Cytochrome P450 Protocols*, vol.107, Phillips, I.R. and Shephard, E.A., (Eds.), Humana Press, Totowa, New Jersey, pp.58-59.

Sydenham, E.W., Thiel, P.G., Marasas, W.F.O., Shephard, G.S., Van Schalkwyk, D.J. and Koch, K.R., 2000. Natural occurrence of some *Fusarium* mycotoxins in corn from low and high esophageal cancer prevalence areas of the Transkei, Southern Africa. *Journal of Agricultural and Food Chemistry* 38: 1900-1903.

Towbin, H., Staehelin, T., and Gordon, J., 1979. Electrophoretic transfer of proteins from polyacrylamide gels to nitrocellulose sheets: procedure and some applications. *Proceedings of The National Academy of Sciences of the United States of America* 76: 4350-4354.

Wang, E., Norred, W.P., Bacon, C.W., Riley, R.T., and Merill, A.H., 1991. Inhibition of sphingolipid biosynthesis by fumonisins. Implications for diseases associated with *Fusarium moniliforme*. *Journal of Biological Chemistry* 266: 14486-14490.

In vitro and *in vivo* effects of fumonisin B_1 on the intestine: A review

S. Bouhet and I.P. Oswald

Laboratoire de Pharmacologie-Toxicologie, INRA, 180 Chemin de Tournefeuille, 31931 Toulouse Cedex 9, France; ioswald@toulouse.inra.fr

Abstract

Fumonisins are a family of toxic and carcinogenic mycotoxins produced by *Fusarium verticillioides* (formerly *F. moniliforme*), a common fungal contaminant of maize. Contamination with fumonisin B_1 (FB_1) is of concern as this mycotoxin causes various animal diseases. The gastrointestinal tract represents the first barrier against ingested food contaminants. Following ingestion of fumonisin-contaminated food or feed, intestinal epithelial cells could be exposed to a high concentration of toxin. In this review, we summarise data on the impact of FB_1 on the intestine. Although FB_1 is poorly absorbed and metabolised in the intestine, it induces intestinal disturbances (abdominal pain or diarrhoea) and causes extra-intestinal organ pathologies (pulmonary edema, leukoencephalomalacia or neural tube defects). The main toxicological effect of FB_1 reported *in vivo* and *in vitro* is the accumulation of sphingoid bases as well as the depletion of complex sphingolipids. This disturbance of the sphingolipid biosynthesis pathway could explain the other observed toxicological effects such as alteration in intestinal epithelial cell viability and proliferation, modification of cytokine production, and modulation of intestinal physical barrier function.

Keywords: fumonisin B_1, intestine, epithelial intestinal cell, toxico-kinetic, pathology, sphingoid bases, barrier integrity, cytokine, viability

Introduction

Fumonisins are ubiquitous contaminants of maize and other grain products produced by *Fusarium verticillioides* (formerly *Fusarium moniliforme*) and several other Fusarium species. Contamination with fumonisin B_1 (FB_1) is of concern as this mycotoxin causes various animal diseases: leukoencephalomalacia in horses, pulmonary edema in pigs, nephrotoxicity in rats, rabbits and lambs and hepatotoxicity in all species examined (WHO, 2000). This toxin has also been reported to be a carcinogenic agent in rodents and a contributing factor in human esophageal cancers

(Rheeder *et al.*, 1992). It has been postulated that fumonisin exerts its toxicological effects via inhibition of the enzyme ceramide synthase, leading to an increase in the concentration of sphingoid bases (sphinganine and sphingosine) as well as the depletion of ceramide and complex sphingolipids (Merrill *et al.*, 2001).

The gastrointestinal tract has two conflicting roles: it plays a major function in the absorption of nutrients while at same time maintaining a barrier between the internal and external environments. The continuous monolayer of epithelial cells that lines the small and large intestine restricts the passage of potentially harmful molecules from the intestinal tract into the surrounding tissues. Intestinal epithelial cells are also able to secrete defense molecules such as mucus, anti-microbial peptides, immunoglobulins, and cytokines. As the first barrier against ingested chemicals, food contaminants and toxins, intestinal epithelial cells could be exposed to a high concentration of FB_1 following ingestion of food or feed contaminated by this toxin (Prelusky *et al.*, 1996). This paper reviews the effect of FB_1 on the intestinal tract and on the exposed intestinal epithelial cells (Figure 1).

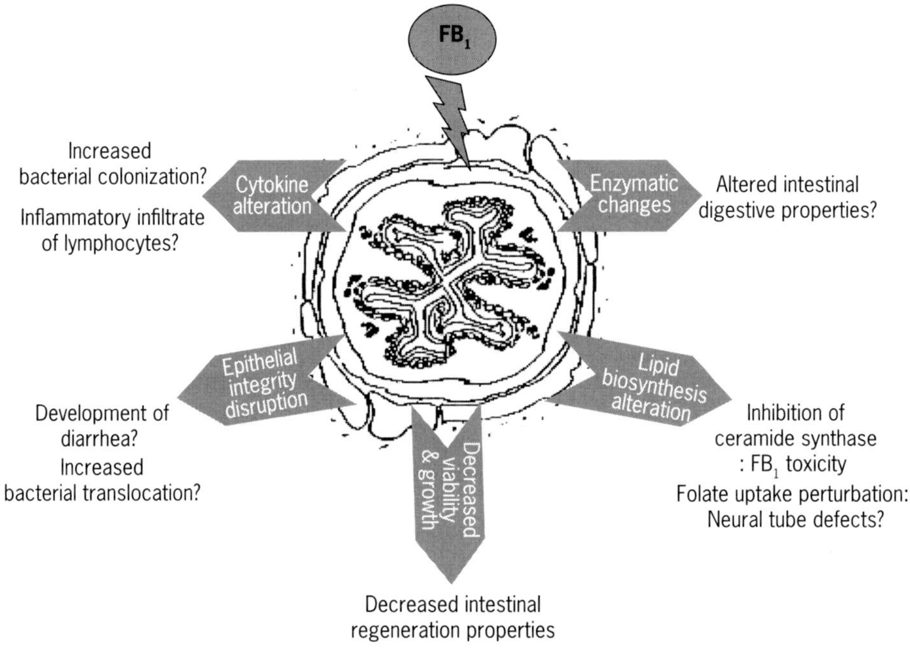

Figure 1. Summarised impacts of FB_1 on the intestine and intestinal cells. The hypothetical consequences of FB_1 in terms of pathologies and intestinal disturbances are presented with a question mark.

FB_1 toxico-kinetics

Absorption, distribution and excretion of FB_1

In rats and most other animals, the absorption kinetics of FB_1 indicates a rapid distribution and elimination that is adequately described by a two- or three-compartment model (Martinez-Larranaga *et al.*, 1999). Only a low level of FB_1 is detected in plasma and tissues after oral administration, indicating that absorption is negligible. Indeed, in cows and laying hens, systemic absorption of orally given FB_1 is less than 1% (Prelusky *et al.*, 1995; Vudathala *et al.*, 1994). In pigs, the bioavailability of FB_1 following intragastric administration is estimated to be 3-6% (Prelusky *et al.*, 1995). In rats, after intragastric administration of the toxin, up to 80% of the radiolabel is recovered in faeces and up to 3% in urine (Norred *et al.*, 1993). After prolonged animal exposure, small traces of FB_1 are found in the liver and kidneys, while no detectable quantity is reported in other organs (Shier, 2000). The poor absorption of this mycotoxin has been qualified by Shier (2000) as the "fumonisin paradox" (how can a toxin cause agriculturally significant diseases and possibly human cancers if it is not effectively absorbed after oral administration?). Whether this poor absorption and consequent low availability is due to poor transport across the epithelium of the intestine or due to the strong association of fumonisin with the intestinal content remains uncertain.

The absorption of FB_1 by enterocytes has also been studied *in vitro*. Using Caco-2 cells that, despite their colonic origin, are representative of human small intestine enterocytes, Caloni *et al.*, (2002) showed no absorption of FB_1 and partially hydrolysed metabolites by undifferentiated or differentiated cells. By contrast, absorption of totally hydrolysed FB_1 (AP1) was evident in differentiated Caco-2 cells, which expressed enzymatic and metabolic characteristics of mature enterocytes.

Intestinal cells are exposed to a substantial portion of the fumonisin that is ingested. Indeed, 24 hours after administration of radio-labelled FB_1, intestinal epithelial cells of non human primates contained 25% of the dose (Shephard *et al.*, 1995). The major route of elimination of FB_1 is via bile and the excreted toxin is still biologically active (Enongene *et al.*, 2000). Therefore, enterohepatic circulation probably increases the exposure of the intestine to the mycotoxin, at least for rats and pigs (Norred *et al.*, 1993; Prelusky *et al.*, 1996; Shephard *et al.*, 1994a). Recent data have indicated an interaction between FB_1 and cholesterol and/or bile salts, which may lead to the incorporation of dietary FB_1 into mixed micelles. This would facilitate the FB_1 intestinal absorption and its biliary excretion (Mahfoud *et al.*, 2002).

Biotransformation of FB_1

Although the liver plays a major role in drug metabolism (*e.g.* by oxidative cytochrome P450 (CYP)-dependent phase I and conjugation or phase II reactions), drug metabolising enzymes are also present at other sites. For example, the intestinal mucosa contains at least 3 CYP families (Doherty and Charman, 2002) and is able to metabolise drugs via phase I and II reactions (Krishna and Klotz, 1994). There is little or no evidence that fumonisin is metabolised *in vitro* or *in vivo*. Indeed, Cawood *et al.* (1994) failed to indicate any metabolism of fumonisins by microsomal esterases or by cytochrome P-450 monooxygenases. More recently, Spotti *et al.* (2001) reported no metabolism of FB_1 by bovine liver microsomes.

Fumonisins are excreted primarily in the faeces, either unchanged or depleted of one ester-linked tricarballylic acid (at the C14-position). Only trace amounts of the fully hydrolysed FB_1 (AP1) were found in faeces (Shephard *et al.*, 1994b). As no hydrolysed product has been found in the urine or bile, it is assumed that the hydrolysis occurs in the gut, probably performed by micro-organisms (Shephard *et al.*, 1994b, 1995).

AP1 is a substrate for ceramide synthase. The product of this reaction, N-palmytoyl-hydrolysed FB_1 also called PAP1, is at least 10 times more toxic than FB_1 or AP1 as demonstrated by *in vitro* studies on the human intestinal cell line HT-29 (Humpf *et al.*, 1998). However, it is not known if this compound is formed *in vivo*.

Pathologic manifestations in the gastrointestinal tract following FB_1 ingestion

Data obtained from human studies

In various regions of the world, an association has been established between the occurrence of *F. verticillioides* on maize and the incidence of the development of esophageal cancer in humans (Marasas, 1995). Epidemiological studies are now being conducted to establish the correlation between FB_1 concentrations in food and the development of other pathologies such as liver cancers and neural tube defects.

An epidemiological survey in India also indicated that consumption of mouldy sorghum and maize containing FB_1 (in the range of 0.14-7.8 mg/kg and 0.25-64.7 mg/kg, respectively) resulted in human gastrointestinal disease characterised by abdominal pain, borborygmi and diarrhoea (Bhat *et al.*, 1997).

Data obtained from laboratory and domestic animal studies

As in humans, FB_1 concentrations up to 8.5 mg/kg were associated with diarrhea in laying hens and 1-day-old cockerels (WHO, 2000).

Histological alterations of the intestine

Despite the fact that, after oral contamination, the small intestine is exposed to the highest concentration of FB_1, this is not the organ most affected by the toxin. In the rat, subchronic experimental mycotoxicosis (100 mg/kg FB_1, 90 days) induces lymphocytic infiltrate in the small intestine and also increases the number of mitotic figures in the intestinal crypts (Theumer *et al.*, 2002). Similarly, an inflammatory infiltrate and necrosis were observed in 50% of mice fed for 60 days with a diet contaminated with 10 mg/kg FB_1 (Casado *et al.*, 2001).

In domestic animals, hyperplasia of intestinal goblet-cells was observed in broiler chicks receiving the toxin (300 mg FB_1/kg BW for 2 weeks), (Brown *et al.*, 1992). In pigs, we recently showed that ingestion of pure FB_1 (1.5 mg/kg BW for 7 days) induced a marked proliferation of lymphatic nodules in the terminal part of the ileum and in the caecum as well as a diffuse infiltration of lymphocytes in these parts of the intestine.

Functional impairments in the intestine

Consumption of pure FB_1 or FB_1-contaminated feed can induce a reduction of body weight in piglets. Indeed, average daily gain for castrated growing pigs decreased by 8% and 11% for animals exposed for 8 weeks to feed contaminated with 1 and 10 mg/kg of pure FB_1, respectively (Rotter *et al.*, 1996). Nevertheless, a shorter exposure to the toxin (10 mg/kg for 1 or 4 weeks) does not affect the weight gain of weaned piglets (Dilkin *et al.*, 2003; Oswald *et al.*, 2003). The former data suggest that FB_1 impairs the intestinal absorption of nutrients.

In vitro treatment of CaCo-2 cells with FB_1 also inhibits folate uptake by this human intestinal cell line (Stevens and Tang, 1997). This effect is due to the increase in sphingoid bases and the subsequent inhibition of the GPI-anchored folate receptor. Because folate deficiency causes neural tube defects, fumonisins are considered as potential risk factors for embryonic development (Marasas *et al.*, 2004).

Susceptibility to intestinal infections

FB_1 can be considered as a predisposing factor to intestinal infectious diseases. Indeed, in our laboratory, we have examined the effect of dietary exposure of FB_1 on the intestinal colonisation by opportunistic pathogenic *Escherichia coli*. Weaned pigs received daily for 6 days a low concentration of FB_1 as a crude extract or as a purified toxin (0.5 mg/kg BW/day) and, on the last day of the toxin treatment, pigs were orally inoculated with *E. coli*. We observed that colonisation of the ileum, caecum and colon by the pathogenic *E. coli* strain was significantly increased (10 to 100 fold) in animals that received the toxin (Oswald *et al.*, 2003).

Intestinal toxicity of FB_1

Two major modes of action have been proposed to explain FB_1-induced animal diseases. Both involve the disruption of lipid metabolism as the initial molecular event. They lead ultimately to lipid-mediated alterations in signalling and metabolic pathways which are crucial to cell biological processes such as cell cycle regulation, differentiation, and apoptosis. One mechanism involves changes in the polyunsaturated fatty acid and phospholipid pools in the liver caused by the disruption by FB_1 of *Delta6* desaturase and cyclooxygenase metabolic pathways (Gelderblom *et al.*, 2001; WHO, 2000). The other mechanism is based on an inhibition of ceramide synthase, a key enzyme in the *de novo* biosynthesis of sphingolipids (WHO, 2000). This latter mechanism has been proposed to explain most of the effect of FB_1 on the intestine.

Disruption of the sphingolipid biosynthesis pathway in the intestine

FB_1 inhibits ceramide synthase which causes a rapid increase in the intracellular concentrations of sphinganine (Sa), and to a lesser extent in sphingosine (So). As this disruption in the sphingolipid pathway occurs before other indicators of cell injury, the Sa/So ratio has been proposed as a biomarker of fumonisin exposure (Riley *et al.*, 1994; Wang *et al.*, 1992).

Mice treated subcutaneously or orally with FB_1 (25 mg/kg BW, single dose) presented a transient increase of Sa and So in the small intestine (Enongene *et al.*, 2000). The rapid rate of cell turnover in the intestine could contribute to the transient nature of the elevation in free sphinganine in this organ. Data obtained *in vitro* indicate a direct fumonisin inhibition of ceramide synthase in intestinal epithelial cells. Indeed, studies performed on the human colonic epithelial intestinal cell line, HT-29, demonstrated an increase of free sphingoid bases by FB_1, in time- and dose-dependent manner (Schmelz *et al.*, 1998).

FB_1 metabolites also alter sphingolipid pathway in intestinal epithelial cells. The hydroxylated form of FB_1, AP1, causes an elevation in sphinganine and toxicity in HT-29, although the potency of this compound is reduced by at least fivefold

(Schmelz *et al.*, 1998). By contrast, PAP1, which also inhibits ceramide synthase in Caco-2 cells, another human colonic cell line, seems more toxic than FB_1 or AP1 (Humpf *et al.*, 1998).

Impact of FB_1 on the viability and proliferation of intestinal epithelial cells

The increase of free sphinganine observed in the HT-29 cell line induces a significant reduction in cell number via growth inhibition and induction of apoptosis (Schmelz *et al.*, 1998). By contrast, two other studies indicate that another human colonic cell line, Caco-2, is fairly resistant to FB_1 and FB_1 metabolites (Caloni *et al.*, 2002). Using the porcine intestinal epithelial cell line, IPEC-1, we have reported an inhibition of cell proliferation by FB_1, due to a blockade of cells in the G0/G1 phase of the cell cycle (Bouhet *et al.*, 2004).

At higher doses, a cytotoxic effect of FB_1 on intestinal epithelial cells has been reported (Bouhet *et al.*, 2004; Schmelz *et al.*, 1998). Both studies indicated that proliferating cells are more sensitive than confluent cells to the toxic effect of FB_1.

Impact of FB_1 on the physical barrier function of intestinal epithelial cells

Intestinal epithelial cells form a physical barrier which can be reproduced *in vitro* by the culture of intestinal epithelial cells on a porous filter. In this model, the barrier function and the monolayer integrity can be assessed by the determination of the trans-epithelial electrical resistance (TEER). Using the porcine intestinal cell line, IPEC-1, we demonstrated that a prolonged exposure of the intestinal monolayer with FB_1 prevents the establishment of TEER and alters the resistance of an already established monolayer (Bouhet *et al.*, 2004). Consequently, we showed that *Escherichia coli* bacteria added on the apical side of an FB_1-treated monolayer trans-located to the baso-lateral compartment. Such results may explain the *in vivo* translocation of experimentally inoculated pathogenic *E. coli* from the intestinal tract to the extra intestinal organs (mesenteric lymph nodes, lungs and kidneys) observed in piglets fed with FB_1.

Effects of FB_1 on intestinal epithelial cell cytokine production

Intestinal epithelial cells are major sources of cytokines and chemokines, molecules involved in the regulation of the immune system (Stadnyk, 2002). Among them, IL-8 is of particular interest because it is involved in the recruitment of polymorphonuclear neutrophils at the site of infection, mediating the non-specific acute inflammatory response.

Even if FB_1 has been shown to alter cytokine production in several organs and cell types, only few data are available concerning the impact of FB_1 on cytokine production in the intestine. We investigated the effect of FB_1 on inflammatory cytokine production in the intestine. We showed that ingestion of FB_1 (0.5 mg/kg BW, 7 days) by piglets decreased the expression of IL-8 mRNA in the ileum. These observations were confirmed by *in vitro* experiments where FB_1 decreased the expression of IL-8 from porcine epithelial intestinal cells, both at the mRNA and protein level (Figure 2). This decrease of IL-8 by FB_1 may lead to a reduced recruitment of inflammatory cells in the intestine during infection and may participate in the observed increased susceptibility of FB_1-treated piglets to intestinal infections (Oswald *et al.*, 2003).

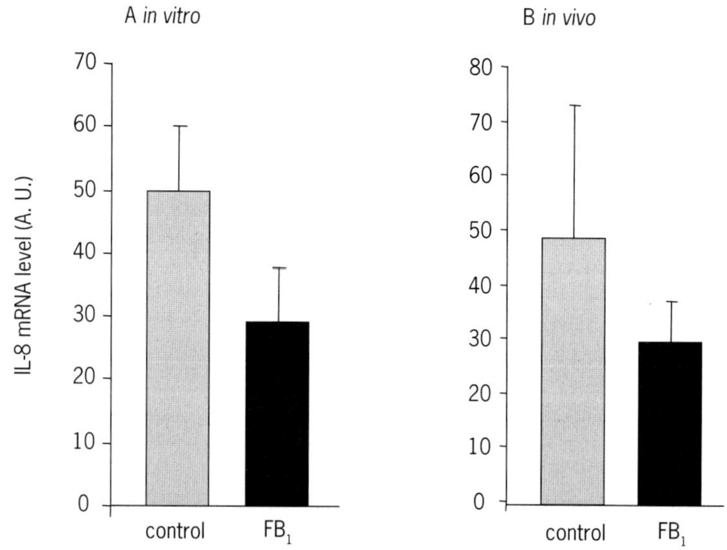

Figure 2. Effect of FB_1 on intestinal expression of IL-8. Panel A: in vitro *results obtained on porcine intestinal epithelial cells (IPEC-1) treated for 4 days with 50μM of FB_1. Panel B:* in vivo *results obtained from the ileum of piglets receiving 0.5 mg/kg b.w. of FB_1 for 7 days. IL-8 mRNA levels were analysed by semi-quantitative RT-RCR and expressed as the ratio of the band intensity of the IL-8 specific product over that of the "housekeeping" gene, cyclophilin.*

Conclusion

In conclusion, ingestion of purified FB_1 and/or contaminated cereals induces species specific pathologies in humans and animals. Even if this mycotoxin has not been associated with a specific intestinal disease, the data available indicate that it disturbs the intestinal lipid biosynthesis pathway and induces cytotoxicity on intestinal epithelial cells. FB_1 also inhibits intestinal epithelial cell proliferation and disrupts their barrier function. In addition, recent data from our laboratory indicate that FB_1 affects cytokine production at the intestinal level. All these effects may affect human and animal health and contribute to the specific pathologies observed in humans and animals as well as the increased susceptibility to intestinal infections.

Acknowledgments

Sandrine Bouhet was supported by a Fellowship from the Ministère de l'Éducation Nationale, de la Recherche et de la Technologie. This work was supported in part by the Région Midi-Pyrénées, France (DAER-Rech/99008345) and by the Transversalité INRA (Mycotoxines-P00263).

References

Bhat, R.V., Shetty, P.H., Amruth, R.P., and Sudershan, R.V., 1997. A foodborne disease outbreak due to the consumption of moldy sorghum and maize containing fumonisin mycotoxins. *J. Toxicol. Clin. Toxicol.* 35: 249-255.

Bouhet, S., Hourcade, E., Loiseau, N., Fikry, A., Martinez, S., Roselli, M., Galtier, P., Mengheri, E., and Oswald, I.P., 2004. The mycotoxin fumonisin B1 alters the proliferation and the barrier function of porcine intestinal epithelial cells. *Toxicol. Sci.* 77: 165-171.

Bouhet, S., Le Dorze, E., Pérès, S.Y., Fairbrother, J.M., and Oswald I.P., 2006. Mycotoxin fumonisin B_1 selectively down-regulates basal IL-8 expression in pig intestine: *in vivo* and *in vitro* studies. *Food Chem. Toxicol.* (In press).

Caloni, F., Spotti, M., Pompa, G., Zucco, F., Stammati, A., and De Angelis, I., 2002. Evaluation of Fumonisin B(1) and its metabolites absorption and toxicity on intestinal cells line Caco-2. *Toxicon.* 40: 1181-1188.

Casado, J.M., Theumer, M., Masih, D.T., Chulze, S., and Rubinstein, H.R., 2001. Experimental subchronic mycotoxicoses in mice: individual and combined effects of dietary exposure to fumonisins and aflatoxin B1. *Food Chem. Toxicol.* 39: 579-586.

Cawood, M.E., Gelderblom, W.C., Alberts, J.F., and Snyman, S.D., 1994. Interaction of 14C-labelled fumonisin B mycotoxins with primary rat hepatocyte cultures. *Food Chem. Toxicol.* 32: 627-632.

De Angelis, I., Frigge, G., Raimondi, F., Stammati, A., Zucco, F., and Caloni, F. 2005. Absorption of fumonisin B1 and aminopentol on an in vitro model of intestinal epithelium; the role of P-glycoprotein. *Toxicon.* 45:285-291.

Dilkin, P., Zorzete, P., Mallmann, C.A., Gomes, J.D., Utiyama, C.E., Oetting, L.L., and Correa, B., 2003. Toxicological effects of chronic low doses of aflatoxin B(1) and fumonisin B(1)-containing *Fusarium moniliforme* culture material in weaned piglets. *Food Chem. Toxicol.*, 41: 1345-1353.

Doherty, M.M., and Charman, W.N., 2002. The mucosa of the small intestine: how clinically relevant as an organ of drug metabolism? *Clin. Pharmacokinet.* 41: 235-253.

Enongene, E.N., Sharma, R.P., Bhandari, N., Voss, K.A., and Riley, R.T., 2000. Disruption of sphingolipid metabolism in small intestines, liver and kidney of mice dosed subcutaneously with fumonisin B(1). *Food Chem. Toxicol.* 38: 793-799.

Gelderblom, W.C., Abel, S., Smuts, C.M., Marnewick, J., Marasas, W.F., Lemmer, E.R., and Ramljak, D., 2001. Fumonisin-induced hepatocarcinogenesis: mechanisms related to cancer initiation and promotion. *Environ. Health Perspect.* 109(Suppl 2): 291-300.

Humpf, H.U., Schmelz, E.M., Meredith, F.I., Vesper, H., Vales, T.R., Wang, E., Menaldino, D.S., Liotta, D.C., and Merrill, A.H., Jr., 1998. Acylation of naturally occurring and synthetic 1-deoxysphinganines by ceramide synthase. Formation of N-palmitoyl-aminopentol produces a toxic metabolite of hydrolyzed fumonisin, AP1, and a new category of ceramide synthase inhibitor. *J. Biol. Chem.* 273: 19060-19064.

Krishna, D.R., and Klotz, U., 1994. Extrahepatic metabolism of drugs in humans. *Clin. Pharmacokinet.* 26: 144-160.

Mahfoud, R., Maresca, M., Santelli, M., Pfohl-Leszkowicz, A., Puigserver, A., and Fantini, J., 2002. pH-dependent interaction of fumonisin B1 with cholesterol: physicochemical and molecular modeling studies at the air-water interface. *J. Agric. Food Chem.* 50: 327-331.

Marasas, W.F., 1995. Fumonisins: their implications for human and animal health. *Nat. Toxins* 3: 193-198.

Marasas, W.F., Riley, R.T., Hendricks, K.A., Stevens, V.L., Sadler, T.W., Gelineau-van Waes, J., Missmer, S.A., Cabrera, J., Torres, O., Gelderblom, W.C., Allegood, J., Martinez, C., Maddox, J., Miller, J.D., Starr, L., Sullards, M.C., Roman, A.V., Voss, K. A., Wang, E., and Merrill, A.H., Jr., 2004. Fumonisins disrupt sphingolipid metabolism, folate transport, and neural tube development in embryo culture and in vivo: a potential risk factor for human neural tube defects among populations consuming fumonisin-contaminated maize. *J. Nutr.* 134: 711-716.

Martinez-Larranaga, M.R., Anadon, A., Diaz, M.J., Fernandez-Cruz, M.L., Martinez, M.A., Frejo, M.T., Martinez, M., Fernandez, R., Anton, R.M., Morales, M.E., and Tafur, M., 1999. Toxicokinetics and oral bioavailability of fumonisin B1. *Vet. Hum. Toxicol.* 41: 357-362.

Merrill, A.H., Jr., Sullards, M.C., Wang, E., Voss, K.A., and Riley, R.T., 2001. Sphingolipid metabolism: roles in signal transduction and disruption by fumonisins. *Environ. Health Perspect.* 109(Suppl 2): 283-289.

Norred, W.P., Plattner, R.D., and Chamberlain, W.J., 1993. Distribution and excretion of [14C]fumonisin B1 in male Sprague-Dawley rats. *Nat. Toxins* 1: 341-346.

Oswald, I.P., Desautels, C., Laffitte, J., Fournout, S., Peres, S.Y., Odin, M., Le Bars, P., Le Bars, J., and Fairbrother, J.M., 2003. Mycotoxin fumonisin B1 increases intestinal colonization by pathogenic *Escherichia coli* in pigs. *Appl. Environ. Microbiol.* 69: 5870-5874.

Oswald, I.P., Marin, D.E., Bouhet, S., Pinton, P., Taranu, I., and Accensi, F., 2005. Immunotoxicological risk of mycotoxin for domestic animals in Europe. *Food Addit. Contam.* 22: 354-360.

Oswald, I.P., 2006. Role of intestinal epithelial cells in the innate immune response of the pig intestine. *Vet. Res.* 37: 359-368.

Prelusky, D.B., Savard, M.E., and Trenholm, H.L., 1995. Pilot study on the plasma pharmacokinetics of fumonisin B1 in cows following a single dose by oral gavage or intravenous administration. *Nat. Toxins* 3: 389-394.

Prelusky, D B., Trenholm, H.L., Rotter, B.A., Miller, J.D., Savard, M.E., Yeung, J.M., and Scott, P.M., 1996. Biological fate of fumonisin B1 in food-producing animals. *Adv. Exp. Med. Biol.* 392: 265-278.

Rheeder, J.P., Marasas, W.F.O., Thiel, P.G., W.S.E., Shepard, G.S., and van Schalwyk, D.J., 1992. *Fusarium moniliforme* and fumonisins in corn in relation to human esophageal cancer in Transkei. *Phytopathology* 82: 353-357.

Riley, R.T., Hinton, D.M., Chamberlain, W.J., Bacon, C.W., Wang, E., Merrill, A.H., Jr., and Voss, K.A., 1994. Dietary fumonisin B1 induces disruption of sphingolipid metabolism in Sprague-Dawley rats: a new mechanism of nephrotoxicity. *J. Nutr.* 124: 594-603.

Rotter, B.A., Thompson, B.K., Prelusky, D.B., Trenholm, H.L., Stewart, B., Miller, J.D., and Savard, M.E., 1996. Response of growing swine to dietary exposure to pure fumonisin B1 during an eight-week period: growth and clinical parameters. *Nat. Toxins* 4: 42-50.

Schmelz, E.M., Dombrink-Kurtzman, M.A., Roberts, P.C., Kozutsumi, Y., Kawasaki, T., and Merrill, A.H. Jr., 1998. Induction of apoptosis by fumonisin B1 in HT29 cells is mediated by the accumulation of endogenous free sphingoid bases. *Toxicol. Appl. Pharmacol.* 148: 252-260.

Shephard, G.S., Thiel, P.G., Sydenham, E.W., and Alberts, J.F., 1994a. Biliary excretion of the mycotoxin fumonisin B1 in rats. *Food Chem. Toxicol.* 32: 489-491.

Shephard, G.S., Thiel, P.G., Sydenham, E.W., Vleggaar, R., and Alberts, J.F., 1994b. Determination of the mycotoxin fumonisin B1 and identification of its partially hydrolysed metabolites in the faeces of non-human primates. *Food Chem. Toxicol.* 32: 23-29.

Shephard, G.S., Thiel, P.G., Sydenham, E.W., and Savard, M.E., 1995. Fate of a single dose of 14C-labelled fumonisin B1 in vervet monkeys. *Nat. Toxins* 3: 145-150.

Shier, W. T., 2000. The fumonisin paradox: a review of research on oral bioavailability of fumonisin B_1, a mycotoxin produced by *Fusarium moniliforme. J. Toxicol. - Toxin reviews* 19: 161-187.

Spotti, M., Pompa, G., and Caloni, F., 2001. Fumonisin B1 metabolism by bovine liver microsomes. *Vet. Res. Commun.* 25: 511-516.

Stadnyk, A.W., 2002. Intestinal epithelial cells as a source of inflammatory cytokines and chemokines. *Can. J. Gastroenterol.* 16: 241-246.

Stevens, V.L., and Tang, J., 1997. Fumonisin B1-induced sphingolipid depletion inhibits vitamin uptake via the glycosylphosphatidylinositol-anchored folate receptor. *J. Biol. Chem.* 272: 18020-18025.

Theumer, M.G., Lopez, A.G., Masih, D.T., Chulze, S.N., and Rubinstein, H.R., 2002. Immunobiological effects of fumonisin B1 in experimental subchronic mycotoxicoses in rats. *Clin. Diagn. Lab. Immunol.* 9: 149-155.

Vudathala, D.K., Prelusky, D.B., Ayroud, M., Trenholm, H.L., and Miller, J.D., 1994. Pharmacokinetic fate and pathological effects of 14C-fumonisin B1 in laying hens. *Nat. Toxins* 2: 81-88.

Wang, E., Ross, P.F., Wilson, T.M., Riley, R.T., and Merrill, A.H. Jr., 1992. Increases in serum sphingosine and sphinganine and decreases in complex sphingolipids in ponies given feed containing fumonisins, mycotoxins produced by *Fusarium moniliforme. J. Nutr.* 122: 1706-1716.

WHO, 2000. *Fumonisin* B_1 (Environmental Health Criteria 219), Geneva, pp. 153.

In vivo and *in vitro* effects of fumonisin B_1 on cytokine production: implication for the immune response to vaccination

I. Taranu[1,2], D.E. Marin[1,2], S. Bouhet[1], F. Pascale[3], J.-D. Bailly[4] and I.P. Oswald[1]

[1]Laboratoire de Pharmacologie-Toxicologie, INRA, 180 Chemin de Tournefeuille, 31931 Toulouse Cedex 9, France; ioswald@toulouse.inra.fr
[2]IBNA, Institute of Biology and Animal Nutrition, Balotesti, Romania
[3]Pasteur Institute, Bucarest, Romania
[4]Ecole Nationale Vétérinaire, Toulouse, France

Abstract

Fumonisin B_1 (FB_1) belongs to the recently discovered fumonisin toxins which are produced by *Fusarium verticillioides*, fungi that commonly contaminate maize. Recent surveys on fumonisins in food and feed throughout the world, including the USA and most European countries, have raised concerns about the extent of FB_1 contamination of maize and implications for animal health and feed safety. At high concentrations, FB_1 causes a variety of species-specific toxicological effects in domestic and laboratory animals. Disruption of sphingolipid metabolism is an important aspect of the toxicity of FB_1 and it has implications on the immune response. The synthesis of both inflammatory and regulatory cytokines is affected by FB_1 treatment. Among inflammatory cytokines, particular attention has been brought to tumor necrosis factor α (TNF-α). FB_1 increases the synthesis of this cytokine which plays an important role in the toxicity of the mycotoxin. The expression of other inflammatory cytokines such as IL-12 p40, IL−1β , IL-8 and IL-18 is also increased during FB_1 intoxication. The induction of inflammatory cytokines is correlated with an increase of the synthesis of Th1 type cytokines (such as IFN-γ) and a concomitant decrease of the synthesis of Th2 type cytokines (such as IL-4). The alteration of the Th1/Th2 cytokine balance is associated with a decrease of the antibody response with implication for the immune response to vaccination.

Keywords: cytokine, fumonisin, inflammation, immune response

Introduction

Mycotoxins, present in food or feed, are able to alter immune function. Mycotoxin-induced immunomodulation may be manifested as depressed T or B lymphocyte activity, suppressed antibody production and impaired macrophage/ neutrophil-effector functions. The sensitivity of the immune system to mycotoxin arises from the vulnerability of these continually proliferating and differentiating cells (Oswald *et al.*, 2005). Mycotoxins are also able to impact the secretory function of immune cells and to alter the cytokine production. Cytokines are small peptide molecules, which are important mediators of the immune system. They play a central role in communication between cells implicated in inflammatory responses as well as in the regulation of specific immunity.

Among mycotoxins, our laboratory is particularly interested in the effect of fumonisin on the immune system. Fumonisins are a group of structurally related mycotoxins produced mainly by *Fusarium verticillioides* (*F. moniliforme*), which is a common contaminant of maize worldwide (IPCS, 2000). Fumonisin B_1 (FB_1) is the most abundant fumonisin in naturally contaminated foods and feeds. This mycotoxin induces species-specific and organ-specific toxicity, including leukoencephalomalacia in horses, pulmonary edema in swine and liver and kidney cancer in rats and mice. FB_1 has also been implicated as a contributing factor in human esophageal cancers (IPCS, 2000, IARC, 2002).

In this review, we will summarise data concerning the effects of FB_1 on cytokine production. In the first part, we will detail the effects of this toxin on the synthesis of inflammatory cytokines. Then, we will focus on the effects of FB_1 on the production of regulatory cytokines. We will particularly emphasise the Th1/Th2 cytokine balance as it orients immune response towards humoral or cellular immunity. Thus this balance impacts the outcome of many infectious diseases and the development of vaccinal immunity (Abbas *et al.*, 1996).

Effect of FB_1 on pro-inflammatory cytokine production

The inflammatory reaction is a non-specific response built to eliminate foreign material and to repair any damaged tissues. It is a complex set of interactions between cells and soluble factors among which the cytokines play a central role. An adequate response involves a balance between the pro- and anti-inflammatory cytokines to avoid persistent tissue damage (Nathan, 2002).

The effect of FB_1 on the production of inflammatory cytokines has been mainly documented for the production of tumor necrosis factor α (TNF-α) and its involvement in apoptosis. However, the influence of this mycotoxin on the synthesis of other inflammatory cytokines has also been described in the literature.

Effect of FB_1 on the production of TNF-α

TNF-α expression during FB_1 intoxication

An increased synthesis of TNF-α has often been associated with FB_1 treatment. Several studies have demonstrated an increase of this cytokine in the liver of BALB/c mice administrated subcutaneously or orally with FB_1 (Bhandari and Sharma, 2002; Bhandari *et al.*, 2002b). However, stimulation of TNF-α synthesis by FB_1 is not liver-specific as it also occurred in the lung, the heart and kidneys of treated animals (He *et al.* 2002; Bhandari *et al.*, 2002b). TNFα expression is sex dependent; its induction upon FB_1 treatment is more pronounced in males than in females (Bhandari *et al.,* 2001). A time dependent increase of TNF-like cytotoxic activity has also been observed in pigs that were fed a diet contaminated with *F. moniliforme* culture material (Guzman *et al.*, 1997). *In vitro,* elevated levels of TNF-α have been observed in porcine kidney epithelial (LLC-PK1) and murine macrophage (J774) cell lines (He *et al.*, 2001; Dugyala *et al.*, 1998).

Role of TNF-α in FB_1 toxicity

There is considerable evidence that TNF signaling pathways play an important role in fumonisin-induced toxicity. Fumonisin cytotoxicity was effectively prevented by inhibition of caspases, which are mediators of TNF signaling (Ciacci-Zanella and Jones, 1999). In addition, acute hematological toxic effects of FB_1 in mice were reversed by anti-TNF antibodies (Dugyala *et al.*, 1998). The hepatotoxicity of FB_1 was similarly reduced in mice lacking either the P75 tumor necrosis factor receptor (TNFR2) (Sharma *et al.*, 2000a) or in mice lacking the P55 TNF receptor (TNFR1) (Sharma *et al.*, 2001). However, transgenic mice over-expressing the human TNF gene were more resistant than wild-type controls to FB_1-induced hepatotoxicity (Sharma *et al.*, 2000b).

The biochemical basis of FB_1 toxicity has been identified as deregulation of sphingolipid metabolism which was correlated with apoptosis or cell-cycle arrest in several different cell lines (Riley *et al.*, 1998; Bouhet *et al.*, 2004) as well as in organs such as the liver and kidneys (Bhandari and Sharma, 2002a). TNF-α is an initiator of the intracellular events leading to apoptosis through a cell-surface mediated mechanism. As genes that inhibit FB_1-induced apoptosis also inhibit the TNF signal transduction pathway, it is therefore probable that FB_1-induced apoptosis occurred via the TNF pathway (Jones *et al.*, 2001; Gopee *et al.*, 2003). However, recent data indicate that FB_1 could induce apoptosis in the absence of TNF-α. Indeed in TNF-knock-out mice, FB_1 activated the CD95 (Fas) signaling pathway as demonstrated by the increased expression of CD95L and TNF-related apoptosis-inducing ligand (TRAIL) (Sharma *et al.*, 2003 a).

Effect of FB_1 on the synthesis of other inflammatory cytokines

Besides TNF, FB_1 also increases the expression of other inflammatory cytokines. Such observations have been made for IL-12 p40, IL-1β, IL-1α and IL-1 receptor antagonists in the liver of treated mice (Bhandari *et al.*, 2002a; Bhandari and Sharma, 2002) and for IL-8 and IL-18 in the lungs of treated piglets (Halloy *et al.*, 2005).

However, FB_1 does not cause an increase of inflammatory cytokines in every studied organ. Indeed, results obtained in our laboratory showed that FB_1 induced a decrease of IL-8 and IL-18 mRNA expression in the intestine of treated animals (Figure 1). *In vitro* treatment of a porcine intestinal epithelial cell line with the toxin confirmed the decreased production of these two inflammatory cytokines (Bouhet *et al.*, 2006).

Effect of FB_1 on the production of regulatory cytokines

Cellular immune responses generated after antigen stimulation of lymphocyte populations can be characterised by the distinct cytokines subsets that are produced. $CD4^+$ T cell clones are broadly divided into two subsets, Th1 and Th2, based on the cytokines they secrete: Th1 (IL-2, IFN-γ and TNF producing) or Th2 (IL-4, IL-5, IL-6, IL-10 and IL-13 producing). The production of cytokines belonging to one or the other subset drives the immune response towards humoral or cellular immunity. This influences the outcome of many infectious diseases and the development of vaccinal immunity (Abbas *et al.*, 1996; Romagnani, 1999).

Effect of FB_1 on the Th1 cytokine response

Several studies have determined the effects of FB_1 on the Th1 cytokine response. These studies have especially analysed the influence of this mycotoxin on the expression of interferon-γ (IFN-γ) in relation with the synthesis of TNF-α (Bhandari *et al.*, 2001, 2002a; Sharma *et al.*, 2003b). A few other publications have described the effects of FB_1 on IL-2 synthesis (Johnson and Sharma, 2001; Tonetti *et al.*, 1999).

Effect of FB1 on IFN-γ

Subcutaneous treatment with 2.25 mg FB_1/kg of body weight for 5 days specifically increased the expression of mRNA encoding for IFN-γ in the liver of male mice; the expression of this cytokine being unchanged in the spleen or kidneys (Bhandari *et al.*, 2001, 2002a). The expression of this cytokine seems to differ according to the sex of the mice: in male BALB/c mice, the toxin caused an elevated expression of IFN-γ in the liver whereas in females it showed a down-modulation of IFN-γ, suggesting gender differences in FB_1 toxicity pathways (Bhandari *et al.*, 2001). The role of

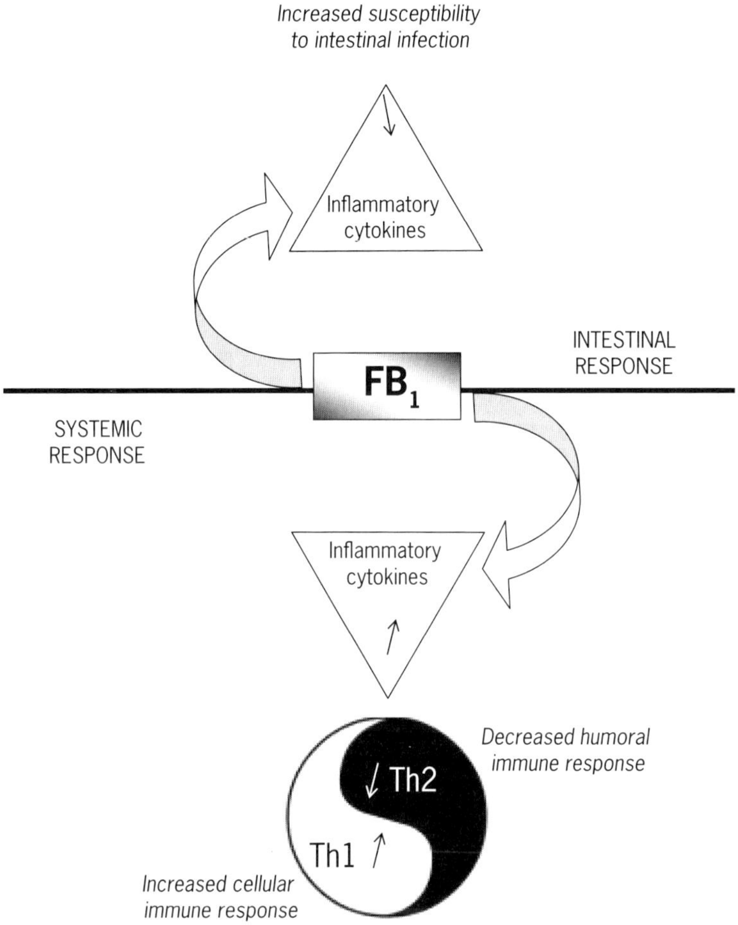

Figure 1. Effect of fumonisin B_1 on the cytokine synthesis.

IFN-γ in the FB_1-hepatotoxicity was further investigated using IFN-γ-knockout male mice (Sharma *et al.,* 2003b). Using the subcutaneous injection model, it was demonstrated that IFN-deficient mice presented a reduced hepatotoxicity associated with a reduction of hepatic enzymes in blood (Sharma *et al.*, 2003b). In mice, an elevated expression of IFN-γ was also detected after 4h of an oral administration of FB_1 (Bhandari *et al.*, 2002b) suggesting that FB_1 effects are not dependent on the route of administration. We have demonstrated in pigs, that *in vitro* exposure of peripheral blood lymphocytes to FB_1 induced a dose-dependent secretion of IFN-γ in the culture supernatant and significantly increased the mRNA expression of this cytokine in porcine lymphocytes. Similarly, a short (7 days) exposure of weanling piglets to the toxin increased the expression of IFN-γ mRNA in the mesenteric lymph nodes and in the spleen of treated animals (Taranu *et al.*, 2005).

Interaction between TNF-α and IFN-γ in hepatotoxicity

Many studies have shown a close correlation between TNF-α and IFN-γ synthesis (Boehm *et al.*, 1997). The sequential expression of IFN-γ and TNF-α in the liver was investigated in mice receiving a single high dose of the toxin. IFN-γ expression peaked at 4 hours and TNF-α at 8 hours after FB_1 oral treatment, suggesting a possible role of IFN-γ in the further amplification of the synthesis and toxicity of TNF-α (Bhandari *et al.*, 2002b). A model for hepatotoxicity induced by FB_1 has been proposed (Bhandari *et al.*, 2002a, b). The early production of low doses of TNF-α and IL-12 in the liver by liver Kupffer cells may act on NK and/or Th1 cells to produce IFN-γ. This IFN-γ further stimulates macrophages to produce TNF-α and IL-12; the vicious cycle continues leading finally, to necrosis of hepatic tissue. However, IFN-γ seems to be more important in the early toxic phase, while the TNF-α appears to be the lethal effector molecule acting at the level of the sites already sensitised by the IFN-γ. These results are supported by the decreased liver synthesis of TNF-α and IL-12 and reduced hepatotoxicity observed in IFN-γ-knockout mice treated with FB_1 (Sharma *et al.*, 2003b).

Effect of FB_1 on IL-2 synthesis

In vivo data indicate that splenocytes from female mice exposed to FB_1 present a reduced expression of IL-2 mRNA (Johnson and Sharma, 2001). This alteration was associated with a relative increase of the T lymphocyte population in the spleen of female mice but was not observed in male mice. *In vitro* experiments also demonstrated that treatment of lymphocytes with this mycotoxin decreased the production of IL-2. Pre-treatment of murine splenic primary T cell and T cell hybridomas, with 100 μM FB_1 impaired TCR-induced IL-2 production (Tonetti *et al.*, 1999). Similarly, we observed that incubation of porcine peripheral mononuclear cells with 100 to 300 μM FB_1 decrease the production of IL-2 in the culture supernatant (unpublished data). By contrast, IL-2 production was not affected in the thymoma cell line EL4, cultured with low FB_1 concentrations (14 μM) (Marin *et al.*, 1996).

Effect of FB_1 on Th2 cytokines

Because of the balance between Th1 and Th2 cytokines (Abbas *et al.*, 1996), an increase in Th1 cytokine levels should be associated with a decrease in Th2 cytokine levels. Indeed, *in vitro* treatment of pig lymphocytes with FB_1 induced a dose-dependent decrease of IL-4 concentration in the cell culture supernatant. Similarly, piglets fed with purified FB_1 or contaminated culture material demonstrated a lower expression of IL-4 in their spleen and mesenteric lymph nodes (Taranu *et al.*, 2005). Decreased synthesis of IL-10, another Th2 cytokine has also been observed in rats fed for 12 weeks with a diet contaminated with 100 mg FB_1/kg (Theumer *et al.*, 2002). By contrast, subcutaneous injection of FB_1 did not alter the expression of Th2 cytokines (IL-4, IL-6 and IL-10) in the liver of treated mice (Bhandari *et al.*, 2002a). It remains to be determined if this difference is due to the target species or to the route of administration.

The effect of FB_1 on the Th1/Th2 cytokine balance may be related to the effect of this toxin on sphingolipid metabolism and the accumulation of free sphingoid bases (Riley *et al.*, 1998). Free sphingoid bases (sphinganine and sphingosine) are known to inhibit the growth of several cell types including lymphocytes (Desai *et al.*, 2002). Th2 lymphocytes are sensitive to the action of exogenous sphingosine than Th1 lymphocytes (Tokura *et al.*, 1996). Thus, it is possible that FB_1 selectively acts on Th2 cells and decreases their synthesis of Th2 cytokines. FB_1 also modulates glycolipid synthesis (Desai *et al.*, 2002). It was first thought that FB_1 decreased the synthesis of all complex sphigoglycolipids. However, more recent data indicate that the picture is more complicated and that FB_1 has a different action on gangliosides and globosides (Meivar-Levy and Futerman, 1999). Gangliosides are known to modulate cytokine production. For example GD1b, GT1b and GQ1b enhanced Th1 cytokine production and suppressed Th2 cytokine production in human T cells (Kanda and Watanabe, 2000). Further studies are needed to determine the role of the modulation of glycolipid synthesis by FB_1 on cytokine production by lymphocytes.

Implication of the FB_1-induced cytokine modulation on vaccinal immune response

Th2 cytokines have an important role in the proliferation of the B lymphocytes and their differentiation into plasmocytes that secrete antibodies (Abbas *et al.*, 1996). Additionally, these cytokines control the isotype commutation, and especially the IgE synthesis (Vercelli *et al.*, 1989). In agreement with the decrease of Th2 type cytokines observed in FB_1-treated animals, several experiments have shown that FB_1 decreases the humoral mediated immune response.

For example, chickens inoculated with the Newcastle disease virus receiving feed contaminated with 200 mg FB_1/kg exhibited lower antibody titer during the secondary immune response than control animals (Li *et al.*, 2000). Similarly, lower antibody titers to *Brucella abortus* were observed during both the primary and the secondary responses in chickens fed a diet containing *F. verticillioides* maize cultures (Marijanovic *et al.*, 1991). In rodents, FB_1 induced a significant decrease of the primary (IgM) response to sheep red blood cells (Tryphonas *et al.*, 1997) as well as a decrease of plaque forming cells to this antigen (Martinova and Merill, 1995).

In our laboratory, we showed that the ingestion of low doses of FB_1 decreased the specific antibody response mounted during a vaccinal immune response in pigs. Indeed, a 28 days exposure to feed contaminated with 8 mg FB_1/kg significantly decreased the specific antibody response to *Mycoplasma*. By contrast, this regimen did not modify the serum concentration of total immunoglobulins (IgG, IgA and IgM). The decrease in specific antibody synthesis was related to alteration of cytokine production as peripheral blood mononuclear cells from these animals showed a decreased expression of IL-4 mRNA upon *in vitro* stimulation with mitogens (Taranu *et al.*, 2005). The effect of FB_1 on specific antibody production was gender specific and more pronounced in males than in females (Marin *et al.*, 2006).

Therefore, the presence of low levels of FB_1 in the feed can lead to a breakdown in vaccine immunity which may lead to disease, even in properly vaccinated flocks. These reactions are of considerable consequence in animal production as well as on effective vaccination programs for disease prevention (Oswald *et al.*, 2005).

Acknowledgments

Drs. I. Taranu, S. Bouhet and D. Marin were the recipients of fellowships from INRA, from the Ministère de l'Éducation Nationale, de la Recherche et de la Technologie (Paris, France) and from a French-Romanian project «Réseau Formation-Recherche», granted by the «Ministère de l'Éducation Nationale, de la Recherche et de la Technologie», respectively. This work was supported in part by the Région Midi-Pyrénées, France (DAER-Rech/99008345) and by the Transversalité INRA (Mycotoxines-P00263). Thanks are also due to Neil Ledger for his help with the English text.

References

Abbas, A.K., Murphy, K.M., and Sher, A., 1996. Functional diversity of helper T lymphocytes. *Nature* 383: 787-793.

Bhandari, N., He, Q., and Sharma, R.P., 2001. Gender-related differences in subacute fumonisin B_1 hepatotoxicity in BALB/c mice. *Toxicology* 165: 195-204.

Bhandari, N., and Sharma, R.P., 2002. Fumonisin B(1)-induced alterations in cytokine expression and apoptosis signaling genes in mouse liver and kidney after an acute exposure. *Toxicology* 172: 81-92.

Bhandari, N., Brown C.C., and Sharma R.P., 2002a. Fumonisin B_1-induced localized activation of cytokine network in mouse liver. *Food Chem. Toxicol.* 40: 1483-1491.

Bhandari, N., Enongene, E.N., Riley, R.T., Meredith, F.I., and Sharma, R.P., 2002b. Temporal expression of fumonisin B(1)-induced tumor necrosis factor-alpha and interferon gamma in mice. *Comp Biochem. Physiol. C Toxicol. Pharmacol.* 131: 113-122.

Boehm, U., Klamp, T., Groot, M., and Howard, J.C., 1997. Cellular responses to interferon-gamma. *Annu. Rev. Immunol.* 15: 749-795.

Bouhet, S., Hourcade, E., Loiseau, N., Fikry, A., Martinez, S., Roselli, M., Galtier, P., Mengheri, E., and Oswald, I.P., 2004. The mycotoxin fumonisin B1 alters the proliferation and the barrier function of porcine intestinal epithelial cells. *Toxicol. Sci.* 77: 165-171.

Bouhet, S., Le Dorze, E., Pérès, S.Y., Fairbrother, J.M., and Oswald I.P., 2006. Mycotoxin fumonisin B_1 selectively down-regulates basal IL-8 expression in pig intestine: *in vivo* and *in vitro* studies. *Food Chem. Toxicol.* (In press).

Ciacci-Zanella, J.R., and Jones, C., 1999. Fumonisin B1, a mycotoxin contaminant of cereal grains, an inducer of apoptosis via the tumor necrosis factor pathway and caspase activation. *Food Chem. Toxicol.* 37: 703-712.

Desai, K., Sullards, M.C., Allegood, J., Wang, E., Schmelz, E. M., Hartl, M., Humpf, H.U., Liotta, D.C., Peng, Q., and Merrill, A.H. Jr., 2002. Fumonisins and fumonisin analogs as inhibitors of ceramide synthase and inducers of apoptosis. *Biochim. Biophys. Acta* 1585: 188-192.

Dugyala, R.R, Sharma, R.P., Tsunoda, M., and Riley, R.T., 1998. Tumor necrosis factor-α as a contributor in fumonisin B1 toxicity. *J. Pharmacol. Exp. Ther.* 285: 317-324.

Gopee, N.V., He, Q., and Sharma, R.P., 2003. Fumonisin B1-induced apoptosis is associated with delayed inhibition of protein kinase C, nuclear factor-kappaB and tumor necrosis factor α in LLC-PK_1 cells. *Chem. Biol. Interact.* 146: 131-145.

Guzman, R.E., Bailey, K., Casteel, S.W., Turk, J, and Rottinghaus, G., 1997. Dietary *Fusarium moniliforme* culture material induces in vitro tumor necrosis factor-alpha like activity in the sera of swine. *Immunopharmacol. Immunotoxicol.* 19: 279-289.

IARC, 2002. Some Traditional Herbal Medicines, some Mycotoxins, Naphthalene and Styrene. In: *IARC Monographs on the Evaluation of Carcinogenic Risks to Humans*, vol. 82, IARC Press, Lyon, France, pp.75-366.

International Programme on Chemical Safety (IPCS). 2000. Fumonisin B_1. *Environmental Health Criteria*, 219: 43-50.

Halloy, D.J., Gustin, P.G., Bouhet, S., and Oswald, I.P. 2005. Oral exposure to culture material extract containing fumonisins predisposes swine to the development of pneumonitis caused by *Pasteurella multocida* type A. *Toxicology* 213: 34-44.

He, Q., Riley, R.T., and Sharma, R.P., 2001. Fumonisin-induced tumor necrosis factor- α in a porcine kidney cell line is independent of sphingoid base accumulation induced by ceramide synthase inhibition. *Toxicol. Appl. Pharmacol.* 174: 69-77.

He, Q., Bhandari, N., and Sharma, R.P., 2002. Fumonisin B(1) alters sphingolipid metabolism and tumor necrosis factor alpha expression in heart and lung of mice. *Life Sci.* 71: 2015-2023.

Jones, C., Ciacci-Zanella, J.R., Zhang, Y., Henderson, G., and Dickman, M., 2001. Analysis of fumonisin B_1 induced apoptosis. *Environ. Health Perspect.* 109: 315-320.

Johnson, V.J., and Sharma, R.P., 2001. Gender-dependent immunosuppression following subacute exposure to fumonisin B1. *Int. Immunopharmacol.* 1: 2023-2034.

Johnson, V.J, He, Q., Kim, S.H., Kanti, A. and Sharma, R.P., 2003. Increased susceptibility of renal epithelial cells to TNF alpha-induced apoptosis following treatment with fumonisin B_1. *Chem. Biol. Interact.* 145: 297-309.

Kanda, N., and Watanabe S., 2000. Ganglioside GD1a enhances immunoglobulin production by human peripheral blood mononuclear cells. *Exp. Hematol.* 28: 672-679.

Li, Y.C., Ledoux, D.R., Bermudez, A.J., Fritsche, K.L., and Rottinghaus, G.E., 2000. The individual and combined effects of fumonisin B_1 and moniliformin on performance and selected immune parameters in turkey poults. *Poult. Sci.* 79: 871-888.

Marin, M.L., Murtha, J., Dong, W., and Pestka, J.J., 1996. Effects of mycotoxins on cytokine production and proliferation in EL-4 thymoma cells. *J. Toxicol. Environ. Health* 48: 379-396.

Marin, D.E., Taranu, I., Pascale, F., Lionide, A., Burlacu, R., Bailly, J.D., and Oswald, I.P. 2006. Gender-related differences in the immune response of weaning piglets exposed to low dose of fumonisin extract. *Br. J. Nut.* 95: 1185-1192.

Marijanovic, D.R., Holt, P., Norred, W.P., Bacon, C.W., Voss, K.A., Stancel, P.C., and Ragland, W.L., 1991. Immunosuppressive effects of *Fusarium moniliforme* corn cultures in chickens. *Poult. Sci.* 70: 1895-1901.

Martinova, E.A. and Merrill, A.H. Jr., 1995. Fumonisin B_1 alters sphingolipid metabolism and immune function in BALB/c mice: immunological responses to fumonisin B_1. *Mycopathologia* 130: 163-170.

Meivar-Levy, I., and Futerman, A.H., 1999. Up-regulation of neutral glycoshingolipid synthesis upon long term inhibition of ceramide synthesis by fumonisin B_1. *J. Biol. Chem.* 274: 4607-4612.

Nathan, C., 2002. Points of control in inflammation. *Nature* 420: 846-852.

Oswald, I.P., Marin, D.E., Bouhet, S., Pinton, P., Taranu, I., and Accensi, F., 2005. Immunotoxicological risk of mycotoxin for domestic animals in Europe. *Food Addit. Contam.* 22: 354-360.

Riley, R.T., Voss, K.A., Norred, W.P., Sharma, R.P., Wang, E. and Merrill, F.I., 1998. Fumonisins: mechanism of mycotoxicity. *Rev. Med. Vet.* 149: 617-626.

Romagnani, S., 1999. Th1/Th2 cells. *Inflamm. Bowel. Dis.* 5: 285-294.

Sharma, R.P., Bhandari, N., Riley, R.T., Voss, K.A., and Meredith, F.I., 2000a. Tolerance to fumonisin toxicity in a mouse strain lacking the P45 tumor necrosis factor receptor. *Toxicology* 143: 183-194.

Sharma, R.P., Bhandari, N., Tsunoda, M., Riley, R.T., and Voss, K.A., 2000b. Fumonisin hepatotoxicity is reduced in mice carrying the human tumour necrosis factor alpha transgene. *Arch. Toxicol.* 74: 238-248.

Sharma, R.P., Bhandari, N., He, Q, Riley, R.T., and Voss, K.A., 2001. Decreased fumonisin hepatotoxicity in mice with a targeted deletion of tumor necrosis factor receptor 1. *Toxicology* 159: 69-79.

Sharma, R.P., He, Q., Johnson, V.J., and Voss, K.A., 2003 a. Increased expression of CD95-ligand and other apoptotic signaling factors by fumonisin B_1, a hepatotoxic mycotoxin, in livers of mice lacking tumor necrosis factor α. *Cytokine* 24: 226-236.

Sharma R.P., He, Q., and Johnson, V.J., 2003 b. Deletion of IFN-gamma reduces fumonisin-induced hepatotoxicity in mice via alterations in inflammatory cytokines and apoptotic factors. *J. Interferon Cytokine Res.* 23: 13-23.

Taranu, I., Marin, D.E., Bouhet, S., Pascale F., Bailly, J.D., Miller, J.D., Pinton P., and Oswald, I.P. 2005. Mycotoxin, Fumonisin B_1, alters the cytokine profile and decreases the vaccinal antibody titer in pigs. *Toxicol. Sci.* 84: 301–307.

Theumer, M.G., Lopez, A.G., Masih, D.T., Chulze, S.N., and Rubinstein, H.R., 2002. Immunobiological effects of fumonisin B_1 in experimental subchronic mycotoxicoses in rats. *Clin. Diagn. Lab. Immunol.* 9: 149-155.

Tonnetti, L., Veri, M.C., Bonvini, E., and D'Adamio, L., 1999. A role for neutral sphingomyelinase-mediated ceramide production in T cell receptor-induced apoptosis and mitogen-activated protein kinase-mediated signal transduction. *J. Exp. Med.* 189: 1581-1589.

Tryphonas, H., Bondy, G., Miller, J.D., Lacroix, F., Hodgen, M., Mcguire, P., Fernie, S., Miller, D., and Hayward, S., 1997. Effects of fumonisin B1 on the immune system of Sprague-Dawley rats following a 14-day oral (gavage) exposure. *Fund. Appl. Toxicol.* 39: 53-59.

Vercelli, D., Jabara, H.H., Arai, K., and Geha, R.S., 1989. Induction of human IgE synthesis requires interleukin 4 and T/B cell interactions involving the T cell receptor/CD3 complex and MHC class II antigens. *J. Exp. Med.* 169: 1295-1307.

The presence of a DNA damaging putative mycotoxin in maize from Santa Catarina, Brazil: An area of high esophageal cancer

V.M. Scussel[1], L.H. Couch[2] and P.C. Howard[2]

[1]Mycotoxicology Laboratory, Food Science and Technology Department, Centre of Agricultural Sciences, Federal University of Santa Catarina, PO.Box 476., Florianópolis-SC, Brazil; Vildes@cca.ufsc.br; nandarobert@hotmail.com
[2]National Center for Toxicological Research, U.S. Food & Drug Administration, Jefferson, Arkansas, USA

Abstract

Consumption of maize contaminated with *Fusarium* sp and FB_1 has been associated with human esophageal cancer (EC) in the Transkei, South Africa. The southern Brazilian states of Santa Catarina, Parana and Rio Grande do Sul also have a high incidence of EC. Fumonisin B_1 (FB_1) a metabolite of the common maize contaminant *F. verticilioides* induces liver and kidney cancers in experimental mice and rats. However, the specific role of the fumonisins in the etiology of human disease, including EC, remains to be determined. Recent studies using maize from the Transkei have shown the presence of a putative genotoxin. Maize for human consumption from SC state, previously shown to be predominantly infected with *F. verticillioides* was analysed for the presence of the putative genotoxin by incubation of a methanolic extract with DNA in the presence of rat liver S9. Of the 76 samples, only maize samples from the western region of the state produced the *Fusarium*-specific DNA adduct. A similar adduct was present from extracts of maize infected by *F. verticillioides* MRC826 and RRC415 from South Africa and Mississippi, respectively. These results suggest the *Fusarium* mycotoxin responsible for the DNA adduct is not widespread in the samples collected from Santa Catarina in the fall of 2000.

Keywords: fumonisins, maize, DNA adduct, *Fusarium verticillioides*

Introduction

The discovery of fumonisins (FBs) in 1988 by Gelderblom, Marasas and colleagues (Gelderblom *et al.*, 1988) has provided considerable insight into the possible etiology of esophageal cancer and the reasons for regiospecificity in areas such as the Transkei of South Africa. Their work also provided the impetus for many investigations into the toxicity of FBs, the extent of food contamination with FBs, and the possible role of combinations of mycotoxins (*e.g.* aflatoxins and FBs) in the etiology of human disease.

The presence of FBs in Brazilian maize and maize products has been the subject of several investigations, with approximately 90% of the samples having FBs at low levels of contamination (Sydenham *et al.*, 1992, Hirooka *et al.*, 1996, Machinski *et al.*, 1998, Ono *et al.*, 1998; Camargo *et al.*, 2000, and Westhuizen *et al.*, 2003). Hirooka *et al.* (1996) detected significant levels of fumonisins FB_1+FB_2 (mean 22.79 μg/g) in maize from Parana state that was destined for human consumption. Sydenham *et al.* (1992) detected a mean of 50.5 μg/g (FB_1+FB_2 + FB_3) in animal feed of the same state. The states of Parana, Santa Catarina and Rio Grande do Sul are in the southern region of Brazil, and have a regionally high incidence of esophageal cancer (Scaff and Scussel, 1999). A survey of 1999-2000 mechanically-harvested and silo-stored rural maize for human consumption from the north, west and southern parts of Santa Catarina (Van der Westhuizen *et al.*, 2003) showed that the FB producing fungus *Fusarium verticillioides* was the predominant fungus in the samples. The levels of fumonisins detected ranged from 0.022 to 18.7 μg/g (FB_1+FB_2+FB_3; mean = 2.66 μg/g).

Bever *et al.* (2000) reported that the incubation of a methanolic extract of *F. verticillioides* culture with DNA, in the presence of microsomal protein, resulted in the formation of a DNA adduct. The formed DNA adduct did not correlate with the presence of fusarin C in the extracts, and was not formed by purified fusarin C. As further evidence that fusarin C was probably not responsible for the DNA adduct, the authors demonstrated that the *Fusarium* genotoxin was stable under UV light. The DNA-adduct forming *Fusarium* genotoxin was produced by maize-based cultures of *F. verticillioides* (MRC 826, MRC 1896, MRC1897, RRC415), *F. sacchari* (MRC 1898), and *F. graminearum* (MRC1901), but not *F. graminearum* (MRC1900). The *Fusarium* genotoxin DNA adducts have also been shown to be present in >80% of DNA samples from human esophageal tumor tissue (Howard *et al.*, 2001) and beer (cwala) (Voss *et al.*, 2004) brewed as in South Africa. The aim of the present work was to evaluate the extent of contamination of maize samples from Brazil with the putative *Fusarium* genotoxin.

Materials and methods

Maize samples (n=90) from the 1999 – 2000 harvest season from Santa Catarina were shipped via overnight delivery to the National Center for Toxicological Research, Jefferson, AR, USA. Each sample was ground to approximately 250 microns using a commercial coffee grinder and stored at -20 °C. Samples of maize infected with *F. verticillioides* MRC826 and RRC415 were generated as described by Bever *et al.* (2000).

The *Fusarium* genotoxin content was determined indirectly by DNA adduct formation. A portion (2.5-4 g) of the ground maize sample was extracted with 20 ml methanol. The extract was dried *in vacuo*, dissolved in 0.12 ml dimethyl sulfoxide, and 35 μl incubated overnight at 37 °C with a 0.5 ml solution of 2 mg/ml salmon testes DNA (Sigma, St. Louis, MO) and 2 mg/ml rat liver S9 (MP Biomedicals/ CAPPEL, Irvine, CA). The incubation was terminated by the addition of an equal volume of water and 6 ml -20 °C ethanol. The DNA adduct was isolated and detected using a DNA ^{32}P-postlabeling method as described by Bever *et al.* (2000). Briefly, the ^{32}P-postlabeled sample was spotted on polyethyleneimine (PEI)-cellulose thin layer chromatography plates (Scientific Absorbants, Atlanta, GA) and the plate developed in multiple dimensions in the following solvents: dimension 1 (D1), 0.9 M sodium phosphate, pH 6.8, onto a wick; D2, 1.8 M lithium formate, 8.5 M urea, pH 3.5; D3, 0.8 M lithium chloride, 0.5 M Tris-base, 8.5 M urea, pH 8.9; D4, 0.9 M sodium phosphate, pH 6.8, onto a wick; D5, 1.8 M lithium formate, 8.5 M urea, pH 3.5, onto a wick. The level of ^{32}P-DNA adduct was quantified using a Molecular Dynamics Storm 860 Phosphorimager (Molecular Dynamics, Sunnyvale, CA). The relative level of DNA adducts is presented as cpm ^{32}P-DNA adduct per μg DNA.

Results and discussion

Methanolic extracts of 90 maize samples were incubated with DNA and rat liver S9 protein and ^{32}P-postlabeled to detect DNA adducts. In addition, extracts from uninfected maize and maize infected with *F. verticillioides* MRC 826 and RRC 415 were routinely included in the analysis for comparison. Typical results of the incubation of methanolic extracts of uninfected maize and maize contaminated with *F. verticillioides* MRC 826 with DNA and rat liver S9 proteins are shown in Figure 1. There were no visible DNA adducts formed by methanolic extracts of control maize (Figure 1A). On the other hand, two DNA adducts (arrows) from extracts from *F. verticillioides* MRC 826 are present near the abscissa following development in direction 4 (D4) (Figure 1B). Solvent 5 (D5) moved the adducts to the right (Figure 1C) a migration pattern similar to that reported by Bever *et al.*, (2000). FBs incubated with DNA alone or with the addition of S9 proteins do not form these DNA adducts.

Figure 2 shows characteristic *Fusarium* genotoxin DNA adducts from extracts of samples from Palmitos (Figure 2A) and Campus Novos (Figure 2B). Figure 3 shows quantitative results of the analysis of the 90 maize samples for the presence of the *Fusarium* genotoxin. The majority of the samples (n=68) had non-detectable to marginally present DNA adducts (0-1,000 cpm/μg DNA). Twenty of the samples had marginally detectable DNA adducts (1,000 to 30,000 cpm/μg DNA) whereas

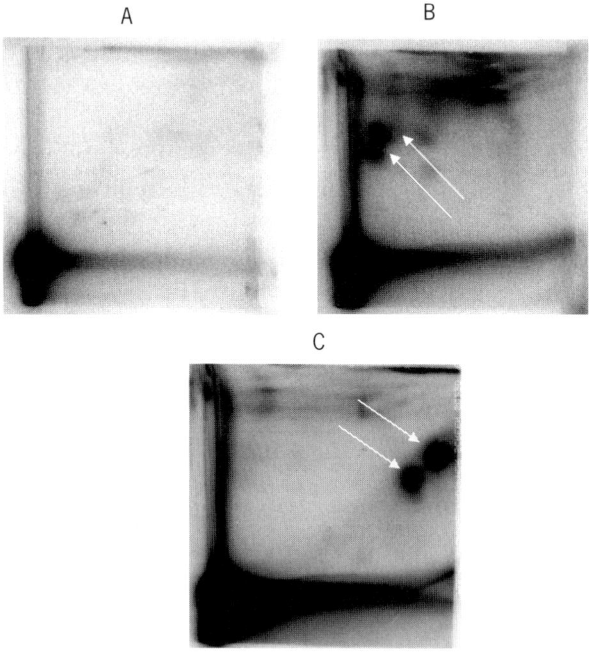

Figure 1. ^{32}P-postlabeling separation of the Fusarium *genotoxin generated DNA adducts (arrows) from methanolic extracts of control maize (A) or maize infected with* F. verticillioides *MRC 826 (B, C). Panel B is post dimension 4 (D4) and Panel C is post dimension 5 (D5).*

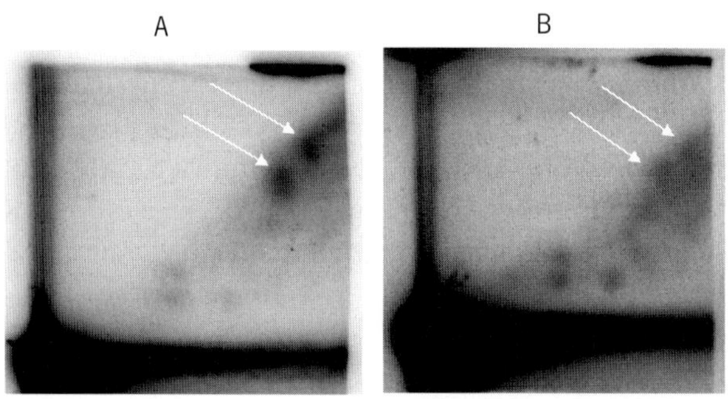

Figure 2. ^{32}P-postlabeling separation (post D5) of the Fusarium *genotoxin generated DNA adducts (arrows) from methanolic extracts of Palmitos (A) and Campos Novos (B).*

the samples from Palmitos and Campos Novos had significantly high levels of DNA adducts (>30,000 cpm/µg DNA). Of the 16 samples from Campos Novos, six had no detectable adducts, nine had marginally detectable adducts, and one had significantly

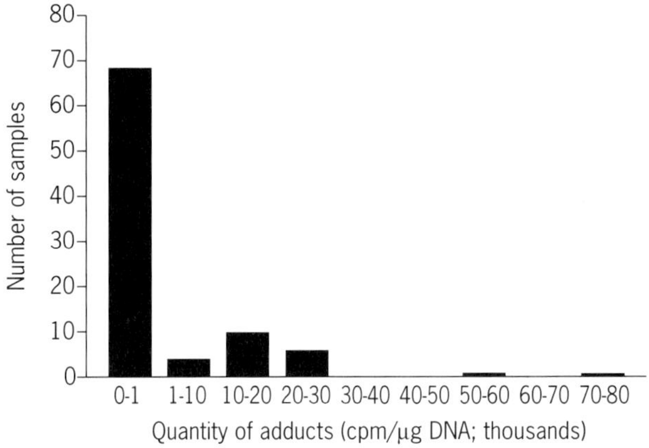

Figure 3. Histogram of the detected Fusarium *genotoxin DNA adducts following incubation of methanolic extracts of maize with DNA and rat liver S9.*

high levels of adducts. Of the two samples from Palmitos, one had marginal level of adducts and one significantly high level of adducts. Van der Westhuizen *et al.*, (2003) had earlier reported the total fumonisin content of the Palmitos and Campus Novos samples to be 1.66 and 4.93 μg/g, respectively. It may be postulated that the FBs and the *Fusarium* genotoxin co-exist in *Fusarium* contaminated maize.

Conclusions

The FBs levels in Santa Catarina maize were similar to the levels determined in other high EC regions of the world (Van der Westhuizen *et al.*, 2003). The *Fusarium* mycotoxin was present in two of the maize samples from Brazil (Palmitos and Campos Novos), marginally present in 20 of the samples, and considered not present in 68 of the samples, as evidenced by the formation of the DNA adducts. These DNA adducts are the same as those produced following incubation of methanolic extracts of *F. verticillioides* MRC 826 isolated from South Africa, and *F. verticillioides* RRC 415 isolated from maize in Mississippi, USA, with DNA in the presence of rat liver microsomal S9. This DNA adduct has also been detected in tumor tissue taken from esophageal cancer patients in South Africa. The observations that the *Fusarium* mycotoxin is produced by fungi isolated from several parts of the world suggest that a widely distributed fungus is responsible for its production. Isolation and identification of the *Fusarium* genotoxin and DNA adduct are needed to understand why there are two distinct DNA adducts detected on the ^{32}P-postlabeling plates (*e.g.* Figures 1 and 2) and the extent of the public health importance of this mycotoxin.

Acknowledgments

We thank "CIDASC - Companhia Integrada de Desenvolvimento Agricola de Santa Catarina" for the collection of the maize samples.

References

Bever, R.J., Couch, L.H., Sutherland, J.H., Williams, A.J., Beger, R.D., Churchwell, M.I., Doerge, D.R., and Howard, P.C., 2000. DNA adduct formation by *Fusarium* culture extracts: lack of role of fusarin C. *Chemico.-Biol. Interact*ions 128: 141-157.

Camargos, S.M., Soares, L.M.V., Sawaki, E., Bolonhezi, D., Castro, J.L., and Bortolleto, N., 2000. Fumonisin in corn cultivars in São Paulo state. *Brazilian Journal of Microbiology* 31: 226-229.

Gelderblom, W.C.A., Jaskiewicz, K., Marasas, W.F.O., Thiel, P.G., Horak, R.M., Vleggaar, R., and Kriek, N.P.J., 1988. Fumonisins – Novel mycotoxins with cancer-promoting activity produced by *Fusarium moniliforme. Applied Environmental Microbiology* 54: 1806-1811.

Hirooka, E.Y., Yamaguchi, M.M., Aoyama, S., Sugiura, Y., and Ueno, Y., 1996. Natural occurrence of fumonisins in Brazilian corn kernels. *Food Additives and Contaminants* 13: 173-183.

Howard, P.C., Couch, L.H., Gelderblom, W.C.A., Marasas, W.F.O., Mugwanya, D., and Stepien, A., 2001. Human esophageal tumor tissue DNA adducts that are derived from *Fusarium* mycotoxins. *The Toxicologist* 55: 419.

Machinski, M., Jr., and Soares, L.M.V., 1998. Fumonisins B_1 and B_2 in Brazilian corn-based food products. *Food Additives and Contaminants* 17: 875-879.

Ono, E.Y.S., Hashimoto, E.H., Funo, F.Y., Ono, C.J., Ono, M.A., Oda, P., and Hirooka, E.Y., 1998. Microbiota Fúngica e Fumonisinas em Milho do Estado do Paraná. In: *IX Encontro Nacional de Micotoxinas*, Florianópolis, Santa Catarina, Brasil. Proceedings. pp. 121.

Scaff, R.M.C., and Scussel, V.M., 1999. Esophageal cancer in the Southern region of Brazil – Cases from Santa Catarina State. *Proceeding of the International Symposium of Mycoytoxicology´99, Chiba, Japan. Mycotoxin Supplement*, pp. 226-230.

Shephard, G.S., 2001. Liquid chromatographic method for FBs in corn. *Methods Molecular Biology* 157: 147-158.

Sydenham, E.W., Marasas, W.F.O., Shephard, G.S., Thiel, P.G., and Hirooka, E.Y., 1992. Fumonisin concentration in Brazilian feeds associated with field outbreaks of confirmed and suspected animal mycotoxicoses. *Journal of Agricultural and Food Chemistry* 40: 994-997.

Van der Westhuizen, L., Shephard, G.S., Scussel, V.M., Costa, L.L.F., Vismer, H.F., Rheeder, J.P., and Marasas, W.F.O., 2003. Fumonisin contamination and *Fusarium* incidence in corn from SC, Brazil. *Journal of Agriculture & Food Chemistry* 51: 5574-5578.

Voss, K.A., Couch, L.H., Howard, P.C., Keller, N.P., Mabathoana, L., and Bacon, C.W., 2004. Toxicological evaluation of *Joala*, a home-brewed beverage, prepared from corn contaminated with *Fusarium verticillioides* culture material. *The Toxicologist* 58: 1022.

Degradation of patulin in rats and humans quantified by stable isotope dilution assays: A review

M. Rychlik

Chair for Food Chemistry, Technical University Munich, Lichtenbergstr. 4, D-85748 Garching, Germany; michael.rychlik@ch.tum.de

Abstract

Up to now, little is known about absorption and metabolism of the mycotoxin patulin. Therefore, in a first series of experiments the absorption of patulin in the tissue of perfused rat stomachs and its vascular appearance after luminal application was quantified by a stable isotope dilution assay. The new method involves the use of [$^{13}C_2$]-patulin as the internal standard, trimethylsilylation of patulin isotopomers and detection of patulin derivatives by GC-MS. After application of juices, to which patulin at doses of 350 and 3.5 mg/l, respectively, had been added, the mycotoxin appeared almost instantly in the perfusate. Additionally, the luminal disappearance of the mycotoxin could be attributed in part to its addition to intracellular glutathione (GSH). Regarding the GSH content in the tissue, a decrease of 87% compared to that of control stomachs was observed for the high PAT dose, whereas in case of the low patulin dose no significant GSH decrease occurred. In further studies, the absorption and degradation of patulin in humans was investigated. No patulin was found in the serum of a volunteer, whose blood was drawn shortly after consumption of a juice containing 50 µg/l patulin. Subsequent *in-vitro* assays showed that after addition of patulin to human whole blood, only 6.1% of the mycotoxin was detected after 2 min at 37 °C. It was concluded that, even high naturally occurring concentrations of patulin in foods are quickly degraded in the gastro intestinal tract before reaching other tissues. Possible reaction products were investigated. In model experiments of patulin with blood, glutathione adducts of the mycotoxin were identified in the blood. However, local toxic effects on the tissues of stomach and intestine still have to be considered, in particular as the majority of reaction products of PAT in organisms are still unknown.

Keywords: glutathione adducts, patulin, perfused rat stomach, stable isotope dilution assay

Introduction

H-furo[3,2-*c*]pyran-2(6*H*)-one, (PAT) is produced by different fungi in rotting fruits, and, therefore, occurs very frequently in commercially available apple juices (Rychlik and Schieberle, 1999). Although it has been investigated since the 1940s, the toxicology of PAT has still remained a controversial topic. In prior feeding trials, its acute and chronic toxicity (Becci *et al.*, 1981) and teratogenicity (Sugyanto *et al.*, 1993) were shown. Moreover, in mutagenicity assays the mycotoxin induced single-strand DNA breaks (Lee and Röschenthaler, 1986) as well as micronuclei (Pfeiffer *et al.*, 1998) were demonstrated indicating that patulin could be a potent carcinogen. Orally given, PAT produced no tumors in Sprague-Dawley rats and Swiss mice (Osswald *et al.*, 1978). Based on a long-term investigation in rats (Becci *et al.*, 1981), the WHO has set a provisionally tolerable weekly intake of 7 μg/kg body weight (WHO, 1985). Thus, the content of patulin in foods has been restricted to 50 μg/kg in many countries. The EU even has set a limit of 10 μg/kg in childrens' food (EC, 2003). However, these allowances are still under debate. A major obstacle to deciding whether mutagenic concentrations can reach any target cells has been the lack of sensitive and accurate analytical methods to obtain detailed information about absorption, toxicokinetic behavior and metabolism of PAT in humans.

Standard methods in patulin analysis

The most frequently used method to quantify PAT is high performance liquid chromatography (HPLC) coupled to UV-detection. However, this method is known to be susceptible to matrix interferences and, therefore, cannot be employed for the analysis of PAT in clinical matrices such as blood, as the necessary clean-up procedures cause substantial losses of the mycotoxin. On the other hand, GC procedures did not gain much interest because of incomplete derivatisation and the lack of a suitable internal standard.

Stable isotope dilution assays

Recently we reported on the excellent accuracy of stable isotope dilution assays (SIDA) for quantification of pantothenic acid (Rychlik, 2000), the vitamin of the folate group (Freisleben *et al.*, 2003), and the mycotoxin ochratoxin A (Lindenmeier *et al.*, 2004) by employing isotopomers of the analytes as internal standards. This enabled an optimal compensation for losses of the analytes in all analytical steps.

Therefore, we synthesised [$^{13}C_2$]-PAT in a six-step preparation (Rychlik and Schieberle, 1998) from L-(+)-Arabinose. As displayed in Figure 1, unequivocal

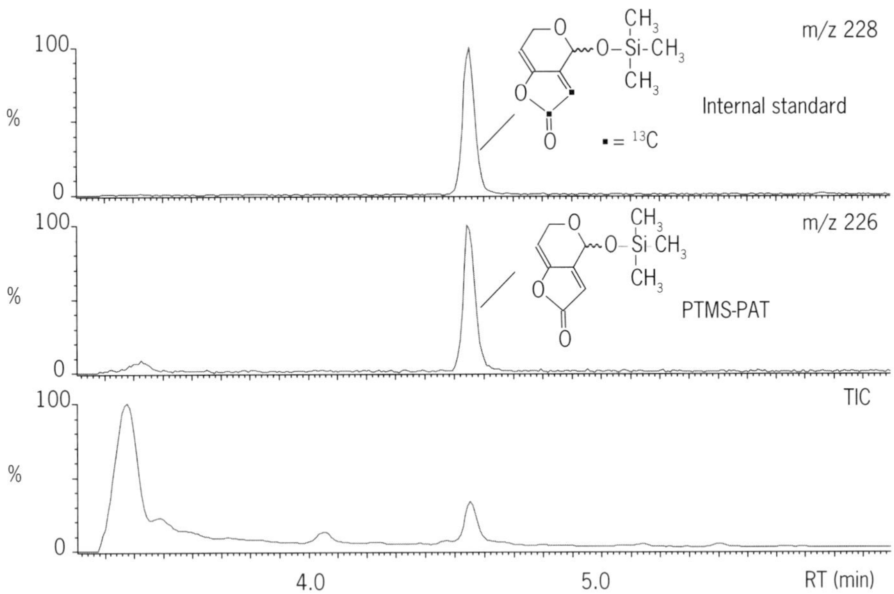

Figure 1. Gas chromatography / mass traces of a gastric tissue extract after luminal application of an apple juice containing 350 mg/l patulin. The PAT amount of 10.1 μg/g is detected in trace m/z =226 (as TMS-patulin), whereas the internal standard [$^{13}C_2$]-patulin is monitored in the trace m/z =228 (as TMS-[$^{13}C_2$]-patulin). TIC total ion current, RT retention time.

quantification was achieved by employing [$^{13}C_2$]-PAT as the internal standard and detecting the trimethylsilyl derivatives of the isotopomers by GC-MS, which proved to be extremely specific and a hundred fold more sensitive than HPLC-UV.

Absorption of patulin from the rat stomach

The stomach is the first organ to come into prolonged contact with PAT after its ingestion. Furthermore, this organ is known to enable absorption of toxic compounds which could result in damage of gastric tissue or affect the function of more remote organs after entering the circulation. Therefore, we used the model of the isolated perfused rat stomach (Schusdziarra *et al.*, 1983) to study the gastric absorption of luminally applied PAT. For studying the gastric absorption of PAT we applied commercial apple juice in concentrations of 3.5 and 350 mg/l, which corresponded to absolute doses of 35 μg and 3.5 mg, respectively (Figure 2). The PAT concentrations in juices, perfusate fractions and tissues were quantified by SIDA.

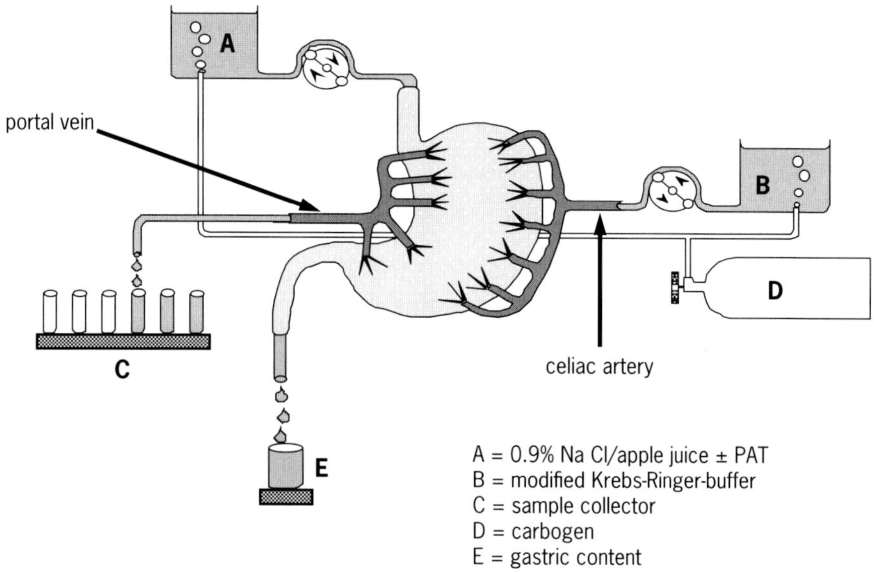

Figure 2. Scheme of the isolated perfused rat stomach.

At the high PAT dose, the mycotoxin appeared vascularly shortly after its luminal application and its concentration rose to 2.8 ± 0.6 μg/ml in the perfusate at t = 30 min. At the low concentration, PAT appeared likewise directly after its application, but increased only to a concentration of 1.9 ± 0.6 ng/ml (Rychlik *et al.*, 2004).

The virtually absorbed amount of PAT during the perfusion assays was estimated from its decrease in the luminally applied juice and is detailed in Table 1. In all trials, a mean decrease of nearly 30% of the applied PAT amount was measured. Of the amounts "absorbed" from the high dose PAT trials, about 17% and 3% were recovered in the sum of all perfusate fractions and in the stomach tissue, respectively. For the low PAT dosage, the recoveries in the perfusate and the gastric tissue only amounted to 2% and 0.1%, respectively. This behaviour points to the existence of a degrading mechanism, which is very effective in lower doses, whereas in high doses only relatively smaller amounts of the mycotoxin were degraded. This raised the question how the mycotoxin had disappeared. Since PAT bears an α, β, γ, δ-unsaturated carbonyl moiety, it is likely to react with cellular nucleophiles. One of the most abundant nucleophiles is the tripeptide glutathione (GSH), which has been shown to form adducts with PAT spontaneously (Fliege und Metzler, 2000).

Table 1. Absorption balances of the perfusion trials with apple juices (10 ml) containing different amounts of patulin (PAT).

Dose	**Low PAT concentrations 35 μg PAT (n=4)**	**High PAT concentrations 3.5 mg PAT (n=4)**	**Controls No PAT[a] (n=2)**
Decrease of luminally applied patulin = virtual "absorption"	8600 ± 1840 ng	899 ± 122 μg	260 ng
(percentage of the applied PAT amount)	(25 ± 6%)	(26 ± 4%)	(> 96%)
Sum of vascularly recovered PAT	184 ± 75 ng	153.6 ± 31.4 μg	n. d.
(percentage of the "absorbed" PAT amount)	(2.0 ± 0.5%)	(17 ± 1%)	
Recovered amount of PAT in gastric tissue	6 ± 3 ng	27.4 ± 11.7 μg	n. d.
(percentage of the "absorbed" PAT amount))	(0.06 ± 0.03%)	(3.1 ± 0.8%)	
Sum of recovered PAT after the perfusions	190 ± 74 ng	181 ± 43 μg	n. d.
(percentage of the "absorbed" PAT amount)	(2.1 ± 0.5%)	(20.1 ± 1.9%)	
Remaining GSH content in gastric tissue	57 ± 9[b] nmol/mg protein	7.4 ± 0.9 nmol/mg protein	58 ± 5[c] nmol/mg protein

n.d. not detectable.

[a]commercial apple juices containing endogenously 27.4 μg/l PAT.

[b]Value very significantly different from HP ($p<0.001$) according to the t-test.

[c]Value not significantly different from LP ($p>0.05$) and very significantly different from HP ($p<0.001$) according to the t-test.

We, therefore, determined the residual GSH content in the stomach tissue after the trials. Compared to the GSH content of 58 nmol/mg protein in the controls, no significant decrease was observed in the stomachs treated with the low PAT concentration. By contrast, GSH decreased to 7.4 nmol/mg protein in the high PAT trials. A reaction of PAT with GSH had obviously occurred during our high PAT dose experiments since PAT degradation was correlated with a decrease in GSH content in the gastric tissue.

PAT quantification in blood serum

To clarify whether PAT was detectable in blood after it's oral uptake, an apple juice (1 l) adjusted to a PAT content of 49.9 µg/l was drunk by a volunteer and blood samples were drawn before and one hour after consumption. In three experiments within two weeks with 4 day breaks between the trials, no PAT was detectable by SIDA in the volunteer's serum before and after consumption (Rychlik, 2003).

Reaction of PAT with whole blood

We hypothesised that GSH in blood, at the high content of 1.07 mmol/l, rapidly degrades PAT. To verify this hypothesis, we added PAT to blood to evaluate its detoxification capacity. As presumed, only mean residues of 14.0% and 6.1% could be detected at one and two min, respectively after addition of 100µg PAT to the blood.

Detection of adducts by LC-MS

To get a more detailed insight into the underlying reaction of PAT degradation in blood, the mycotoxin was first reacted with GSH in a buffer solution at physiological pH, as Fliege and Metzler (2000) suggested that addition to GSH is the most important reaction for degradation of PAT in blood. The resulting mixture was analysed for GSH-PAT adducts. Using an LC-MS-MS system, we were able to tentatively identify several adducts reported by Fliege and Metzler (2000) by their molecular mass, MS-MS decomposition behaviour and their UV-spectra.

After adjusting the MS conditions for detection of most adducts, we added PAT to freshly drawn blood and deprotonised the mixture with metaphosphoric acid. The resulting LC-MS chromatogram is displayed in Figure 3 which clearly shows the detection of the respective adducts that were already found in the model experiments.

Conclusions from the toxicity studies

In our study on rat stomachs PAT was quickly absorbed from the gastrointestinal tract and was partly degraded when diffusing through the mucosa wall. Moreover, no traces of PAT were detectable in the serum of the volunteer after drinking a PAT dose of 49.9 µg, although we used the highly sensitive SIDA. Furthermore, we demonstrated that PAT disappears rapidly when added to whole blood. It can, therefore, be concluded that PAT could only exert local toxicity and cannot evoke systemic effects in other parts of the organism. However, even at low concentrations,

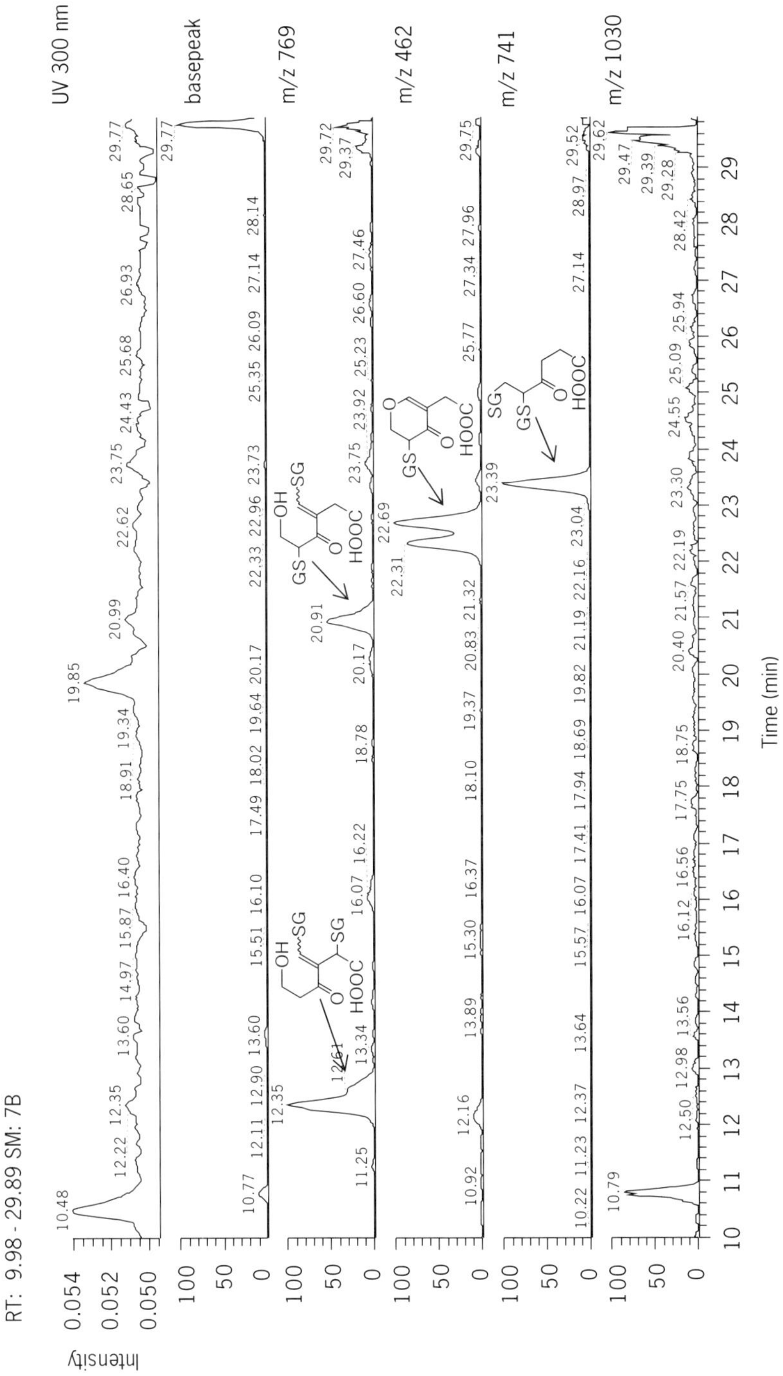

Figure 3. LC-MS chromatogram of patulin-glutathione adducts in whole blood.

a fraction of PAT passed through the gastric mucosa and appeared in the vascular system. In addition, as an altered intestinal permeability may result in an inflammatory response (Mahfoud *et al.*, 2002), gastric inflammation cannot be excluded.

Furthermore, methods for quantification of GSH-PAT adducts have to be developed to clarify their role in PAT degradation. As proteins have also been shown to react with PAT (Fliege and Metzler, 1999), their reaction products have to be detected and quantified in blood as well. Moreover, PAT could also have been bound to proteins or been rapidly degraded to PAT metabolites, which might be toxic. Therefore, in order to protect the consumer´s health, the legal limit of PAT concentrations stipulated by many countries is justified. The mycotoxin's content should be reduced to a technologically possible level.

References

Becci, P.J., Hess, F.G., Johnson, W.D., Gallo, M.A., Babish, J.G., Dailey, R.E., and Parent, R.A., 1981. Long-term carcinogenity and toxicity studies of patulin in the rat. *J. Appl. Toxicol.* 1: 256-261.

European Commission, 2003. EC No. 1425/2003.

Fliege, R., and Metzler, M., 1999. The mycotoxin patulin induces intra- and intermolecular protein crosslinks in vitro involving cysteine, lysine, and histidine side chains, and α-amino groups. *Chem.-Biol. Interactions* 123: 85-103.

Fliege, R., and Metzler, M., 2000. Electrophilic properties of patulin. N-Acetylcysteine and glutathione adducts. *Chem. Res. Toxicol.* 13: 373-381.

Freisleben, A., Schieberle, P., and Rychlik, M., 2003. Specific and sensitive quantification of folate vitamers in foods by stable isotope dilution assays using high-performance liquid chromatography-tandem mass spectrometry. *Anal. Bioanal. Chem.* 376: 149-156.

Lee, K S., and Röschenthaler, R.J., 1986. DNA-damaging activity of patulin in Escherichia coli. *Appl. Environm. Microbiol.* 52: 1046-1054.

Lindenmeier, M., Schieberle, P., and Rychlik, M., 2004. Quantification of ochratoxin A in foods by a stable isotope dilution assay using high-performance liquid chromatography-tandem mass spectrometry. *J. Chromatogr. A* 1023: 57-66.

Mahfoud, R., Maresca, M., Garmy, N., and Fantini, J., 2002. The mycotoxin patulin alters the barrier function of the intestinal epithelium: mechanism of action of the toxin and protective effects of glutathione. *Toxicol. Appl. Pharmacol.* 181: 209-218.

Osswald, H., Frank, H. K., Komitowski, D., and Winter, H., 1978. Long-term testing of patulin administered orally to Sprague-Dawley rats and Swiss mice. *Food Cosm. Toxicol.* 16: 243-247.

Pfeiffer, E., Gross, K., and Metzler, M., 1998. Aneuploidogenic and clastogenic potential of the mycotoxins citrinin and patulin. *Carcinogenesis* 19: 1313-1318.

Rychlik, M., 2000. Quantification of free and bound pantothenic acid in foods and blood plasma by a stable isotope dilution assay. *J. Agric. Food Chem.* 48: 1175-1181.

Rychlik, M., 2003. Rapid degradation of the mycotoxin patulin in man quantified by stable isotope dilution assays. *Food Addit. Contam.* 20: 829-837.

Rychlik, M., and Schieberle, P., 1998. Synthesis of 13C-labeled patulin [4-hydroxy-4H-furo[3,2-c]pyran-2(6H)-one] to be used as internal standard in a stable isotope dilution assay. *J. Agric. Food Chem.* 46: 5163-5169.

Rychlik, M., and Schieberle, P., 1999. Quantification of the mycotoxin patulin by a stable isotope dilution assay. *J. Agric. Food Chem.* 47: 3749-3755.

Rychlik, M., Kircher, F., Schusdziarra, V., and Lippl, F., 2004. Absorption of the mycotoxin patulin from the rat stomach. *Food Chem. Toxicol.* 42: 729-735.

Schusdziarra, V., Bender, H., and Pfeiffer, E.F., 1983. Release of bombesin-like immunoreactivity from the isolated perfused rat stomach. *Regul. Pept.* 7: 21-29.

Sugiyanto, J., Inouye, M., Oda, S.-I., Takagishi, Y., and Yamamura, H., 1993. Teratogenicity of patulin, a mycotoxin, in mice. *Environm. Med.* 37: 43-46.

World Health Organisation, International Agency for Research on Cancer (1985) Patulin. In: *IARC Monographs on the evaluation of the Carcinogenic Risc of Chemicals to Humans.* IARC, Geneva, Switzerland, pp. 83-98.

Metabolism of mycotoxins in mammalian cells: Advantages of a stable isotopic enrichment strategy

S. Yen[1], M. Pean[2], O. Puel[3], N. Loiseau[3], F. André[1] and M. Delaforge[1]

[1]CNRS URA 2096, DBJC, CEA-Saclay, Gif/Yvette cedex, France; marcel.delaforge@cea.fr
[2]CEA Cadarache, DEVM/GRAP, St Paul les Durance, France
[3]INRA, UR66 Laboratoire Pharmacologie-Toxicologie, Toulouse, France

Abstract

Pharmacological studies of exogenous compounds often encounter problems: these compounds are in such infinitesimal amounts in their biological matrices that they require particular detection methods such as use of radio-labelled compounds and HPLC-MS^n techniques. The use of stable isotopes is an interesting alternative for *in vitro* and *in vivo* studies including human. *Fusarium graminearum* was grown on ^{13}C wheat seeds to produce ^{13}C-labeled zearalenone which was used for metabolism studies. The metabolic profile of the ^{13}C-labeled zearalenone was similar to that of commercial zearalenone. The isotopic pattern made metabolite detection easier, and identification could be made from both MS and MS^n spectra including those resulting from cleavage or conjugation. The compound could also be used as an internal standard with high reliability in order to quantify zearalenone in contaminated food samples.

Keywords: zearalenone, *Fusarium graminearum*, mycotoxin quantification, metabolism, isotopic enrichment, ^{13}C ^{15}N uniform labelling culture, xenobiotics, pharmacology, mass spectrometry

Introduction

Pharmacological studies of foreign compounds often encounter detection problems due to the fact that the studied compound is mostly present in very low amounts in the biological matrix. Detection is classically done using radio-labelling and high performance liquid chromatography with online mass spectrometry product ion scan mass spectrometry (HPLC-MS^n) techniques. In this respect, the use of stable isotopes is an interesting alternative for *in vitro* and *in vivo* studies including man (Korach-André *et al.*, 2004).

Isotopic enrichment with ^{13}C and ^{15}N is a powerful tool in metabolic studies (Ralph *et al.*, 1998; Chabannes *et al.*, 2001). In order to produce ^{13}C, ^{15}N-enriched mycotoxins of natural origin, host plants could be grown in the presence of 10% $^{13}CO_2$, as the only source of carbon, and enriched ^{15}N nitrogen to facilitate the uniform incorporation of ^{15}N and 10% ^{13}C by the whole plant during photosynthesis. The plant is then inoculated with an appropriate toxigenic fungal species.

Materials and methods

Production of double isotope enriched plants with ^{13}C and ^{15}N

To perform uniform enrichment with ^{13}C, plants were grown in hermetically closed chambers for a complete growth cycle *i.e.* several weeks or months. Seeds were sown in pots containing vermiculite or perlite. Vermiculite or perlite are used to limit soil respiration which releases carbon dioxide with about 1% natural ^{13}C leading to a decrease of ^{13}C-enrichment level. The pots were placed in air-tight chambers (750 l) in which gas composition and environmental parameters (photoperiod, photon flux rate, night and day temperatures and relative humidity) were precisely regulated. CO_2 concentration in the chamber was continuously monitored with an infra-red analyzer. During light periods, CO_2 concentration was maintained at 360 μl/l by automatic injections of concentrated CO_2 enriched at 10% ^{13}C to compensate for photosynthetic assimilation. The plants were automatically watered with nutrient solution containing ^{15}N (10%) enriched nitrate and ammonium salts.

Uniformly ^{13}C ^{15}N-labeled zearalenone production

Zearalenone production was carried out using autoclaved ^{12}C or ^{13}C-enriched durum wheat (Antares strain) as the only nutrient source for the fungi (Llorens *et al.*, 2004). Wheat grains were moistened with sterile distilled water for 4 days at 4 °C, before sterilisation (20 min, 120 °C). Approximately 30 g of grain was distributed into a 140 mm diameter Petri dish. Each Petri dish was inoculated with a 2×10^5 conidia suspension of *Fusarium culmorum* G5G7 and incubated at 25 °C for 30 days.

Zearalenone analysis

The fungal cultures were extracted with 100 ml acetonitrile+water (70:30). Zearalenone analysis was carried out using high performance liquid chromatography (HPLC). The HPLC unit (Kontron instruments, Milan, Italy) was fitted with a 250 × 4 mm Kromasil 5 μm C18 column and a diode-array detector (DAD). Standard solutions were prepared from commercial zearalenone (Sigma, St. Louis, Mo.). The solvent system was composed of 33 mM acetic acid as solvent A and methanol as solvent B. The toxin was eluted by varying the gradient from 20% methanol in acetic acid for 15 minutes, followed by a step up to 30% and by a linear gradient to 75% methanol within 10 minutes. After an isocratic elution for 20 minutes, the gradient was decreased to 20% methanol within 10 minutes. Zearalenone was purified by semi-preparative high-pressure chromatography on a 250 x 7.5 mm Kromasil C18 column using 2 ml acetonitrile+water linear gradient (20 to 80% acetonitrile in 30 minutes). Fractions collected every 0.5 minute were dried under vacuum and 15 μg of pure ^{13}C-enriched zearalenone were recovered from 1 ml grain extract mixture. This led to a total yield of extraction and purification of 0.67 mg zearalenone/g wheat, this amount being similar to that produced using classical liquid nutritive media.

Use of labelled zearalenone as internal standard

The purified ^{13}C-enriched zearalenone can be used as an internal standard in order to determine more accurately the amount of zearalenone in a given sample. The experiment consisted of adding a known amount of the ^{13}C-enriched zearalenone internal standard before extraction and evaluating the $^{12}C/^{13}C$ ratio by mass spectrometry. 20 μM ^{13}C-enriched zearalenone and various amounts (0 to 20 μM) of commercial zearalenone were added to a rat liver microsomal protein suspension.

The rat liver microsomal suspensions (1 mg protein/ml) were incubated in presence of 50 μM of zearalenone and an NADPH generating system (1 mM NADP, 10 mM Glucose-6-Phosphate, 2.5 UI Glucose-6-phosphate dehydrogenase) in 0.1 M phosphate buffer (pH 7.4) at 37 °C. Incubations were stopped after 30 min by adding the same volume of acetonitrile. The final mixture was cooled and centrifuged at 10,000 rpm for 10 minutes before HPLC-MS^n analysis.

Mass spectrometry analysis

The HPLC-MS^n instrument used was an LCQ DUO Ion Trap coupled with an HPLC system from Thermo-Finnigan, Courtaboeuf, Les Ulis, France fitted with a 150 x 2.1 mm Kromasil 5 C_{18} reverse-phase column. 20 μl of a 50% solution of buffer in CH_3CN was directly injected into the LC system. Gradient chromatography (run for 33 minutes) was performed with a water+acetonitrile mixture (eluent A: 90/10 water: acetonitrile and eluent B: 10/90 water:acetonitrile) at a flow rate of 250 μl/min. The program started with 100% of eluent A, then eluent B was increased from 0 to 80% within 25 minutes and held constant during the next 3 minutes before returning to the initial conditions in 2 minutes. LC-MS measurements were performed using electrospray ionisation mode (ESI). ESI was performed at room temperature in negative mode, the voltage was maintained at 4.5 kV and the capillary temperature at 250 °C. The MS analyzer and collision energy set at 35% were optimised using zearalenone.

Results and discussion

Under negative ionisation mode, the mass spectrum of the U ^{13}C-zearalenone extract exhibited a molecular isotopic cluster centered at m/z 318.8, having a 1.8 mean mass shift (Figure 1B) compared to authentic zearalenone (Figure 1A). This shift originated from the 10% ^{13}C content of the 18 carbon atoms. The isotopic cluster follows a polynomial repartition (solid line in Figure 1C) which is in good agreement with the theoretical calculated profile (polynomial fit in dotted line in Figure 1C).

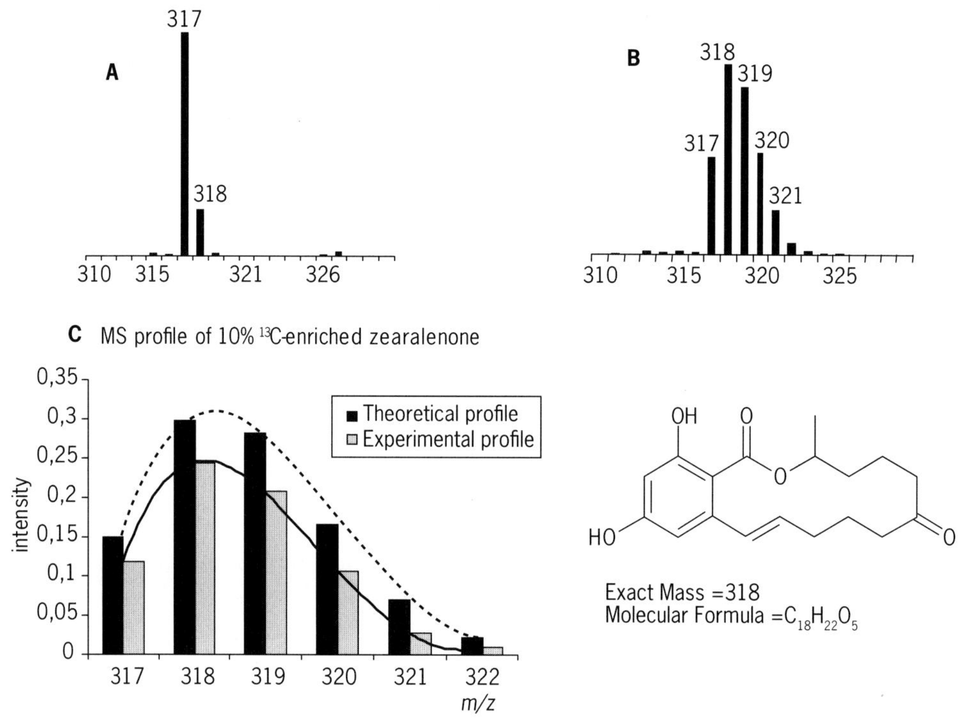

Figure 1. Mass spectrum in negative ionisation mode of (A) authentic zearalenone, and (B) purified U ^{13}C-zearalenone. (C) comparison of the theoretical and experimental MS spectrum of 10% ^{13}C-enriched zearalenone.

Use of labelled zearalenone as internal standard

After acetonitrile extraction, mass spectra were recorded. When the ^{12}C zearalenone m/z 317 peak intensity was plotted as a function of zearalenone concentration, the result of a linear regression show a R^2 correlation coefficient of 0.96. This correlation could be significantly improved by plotting the 317/320 relative intensity ratio as a function of zearalenone concentration. In this case the R^2 value was higher than 0.996 (Figure 2). The peak at m/z 320 arising from three ^{13}C carbons from the ^{13}C-enriched zearalenone was used for normalisation since at this mass, there was no isotopic contribution coming from natural abundance zearalenone (see Figure 1) and the signal intensity is still high enough relative to the m/z 317.

It can be concluded that ^{13}C-enriched zearalenone can be used as an internal standard upon addition into complex matrices to perform MS quantification whatever the level of ^{12}C zearalenone concentration. In this experiment, concentration of ^{13}C internal standard was 20 μM and amounts of commercial zearalenone ranged from 0 up to 20 μM. These conditions can be considered as typical from maize extracts.

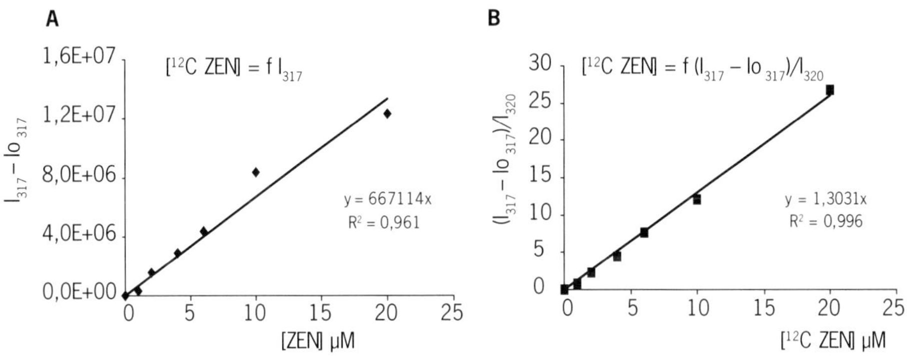

Figure 2. Calibration curves obtained from m/z 317 peak intensities plotted versus analyte concentration. Results of a linear regression (A) of m/z 317 peak intensity (I_{317}) of authentic zearalenone. The peaks intensity has been corrected from the m/z 317 signal contribution of the ^{13}C internal standard ($_{I0\,317}$), (B) m/z 317/320 peak intensity ratio as a function of authentic zearalenone concentration; the m/z 317 peak intensity is corrected from the signal contribution of ^{13}C zearalenone internal standard ($(I_{317} - I_{0\,317})/I_{320}$).

Use of ^{13}C-enriched zearalenone in a metabolism study

^{13}C-Enriched Zearalenone preparations are presently used for *in vitro* experiments such as structural elucidation and metabolic transformation. To illustrate the use of this technique as a new possibility for structural elucidation, two incubations were carried out: one performed with ^{13}C-enriched zearalenone and the other with commercial zearalenone. They were performed at 37 °C using phenobarbital-treated rat microsomes and stopped at 30 minutes using the same volume of cold acetonitrile.

After centrifugation, the supernatant was analysed by HPLC-MS^n on a Thermo-Electron ion trap equipped with an ESI source operating in negative mode. The total ion currents were similar for both types of incubations with formation of more polar peaks which were absent in the blank incubations performed without NADPH generating system. These observations indicate that metabolism occurred in the reaction medium whether zearalenone was isotopically enriched or not (Figure 3).

Metabolites identification

The isotopic pattern makes metabolite detection easier, and identification can be made from both MS and MS^n spectrum even those resulting from cleavage or conjugation.

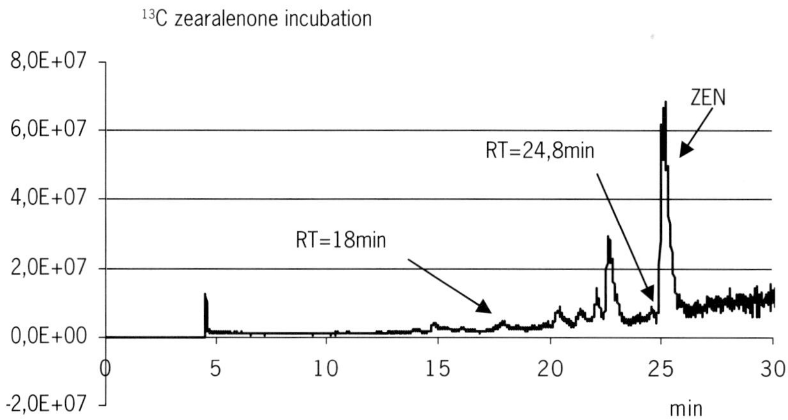

Figure 3. Total Ion Chromatogram traces after incubation with 10% ^{13}C-enriched zearalenone.

The metabolite eluted at R_t = 24.8 min exhibited an isotopic cluster centered at m/z 320.7 (Figure 4). This value corresponds to a 2 unit mass increase in comparison to the mass spectrum of zearalenone. This metabolite exhibited a retention time identical to that of zearalanone, derived from natural zearalenone by reduction of the aliphatic double bond. Its MS^2 fragmentation leads to a loss of a fragment having 44 mass units, corresponding to a carboxyl moiety.

The compound eluted at 18 min (Figure 3) displayed a UV spectrum similar to that of zearalenone and could be postulated to be a metabolite of zearalenone. However, the absence of an isotopic cluster signature means that this compound was not a metabolite from zearalenone.

Conclusion

The technique developed in this study has several advantages. The stable isotope uniformly enriched compound is fully identical to a natural ^{12}C except for a molecular weight difference lower than 1%. Indeed, it had similar extraction recovery (more than 95% from C_{18} Solid Phase Extraction cartridge), similar chromatographic

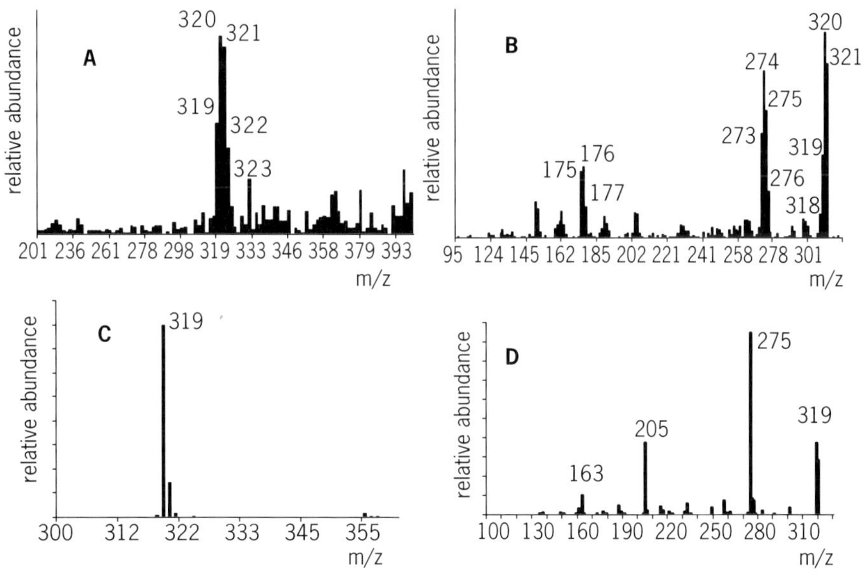

Figure 4. Mass Spectra of metabolite eluted at Rt = 24.8 min; MS (A) and MS^2 (B) from ^{13}C zearalenone microsomal incubation compared to the MS (C) and MS^2 (D) of authentic zearalanone.

retention times, and similar ionisation response in mass spectrometry. It can be used for multiple purposes such as structural elucidation, since during fragmentation the isotope cluster signature can be preserved, and allows distinction between oxygen or carbon losses. Physicochemical properties matching with the natural compound allow the use of the enriched compound as an internal standard in complex matrices where recovery yields are low. With this technique, compounds can be followed in *in vitro* model systems. Pharmacological studies can be performed with a mycotoxin isolated from the natural environment. Contrary to deuterium enrichment known for its isotopic effects and exchanges (Stokvis *et al.*, 2005), ^{13}C-enrichment is relatively safer and can be used for human drug or steroid pharmacokinetics (Setchell *et al.*, 2003; Baume *et al.*, 2006), although it seems that isotopic effects may also occur (Miron *et al.*,2006). This technique can also be applied to large variety of compounds of natural origin. A large number of ^{13}C mycotoxins, such as deoxynivalenol, nivalenol, mycophenolic acid or veruculogen have already been prepared in our laboratory. Currently solid cultures of *Claviceps purpurea* on cereals are being tested for production of ergot derivatives.

From an isotopic point of view, this method is safer than use of radio-labelled compounds which can induce cellular damage. Besides, it does not influence the mode of action of the compound, so that the measured activity is totally reliable. The use of stable isotopes is thus an interesting alternative for *in vitro* and *in vivo* studies including human.

References

Baume, N., Saudan, C., Desmarchelier, A., Strahm, E., Sottas, P.E., Bagutti, C., Cauderay, M., Schumacher, Y.O., Mangin, P., and Saugy, M., 2006.Use of isotope ratio mass spectrometry to detect doping with oral testosterone undecanoate: inter-individual variability of 13C/12C ratio. *Steroids* 71: 364-370.

Chabannes, M., Barakate, A., Lapierre, C., Marita, J.-M., Ralph, J., Péan, M., Danoun, S., Halpin, C., Grima-Pettenati, J., and Boudet, A.M., 2001. Strong decrease in lignin content without significant alteration of plant development is induced by simultaneous down-regulation of cinnamoyl CoA reductase (CCR) and cinnamyl alcohol dehydrogenase (CAD) in tobacco plants. *Plant Journal* 28: 257-270.

Korach-André, M., Roth, H., Barnoud, D., Péan, M., Perronet, F., and Leverve, X., 2004. Glucose appearance in the peripheral circulation and liver glucose output in men after a large 13C starch meal. *American Journal of Clinical Nutrition* 80: 881-886.

Llorens, A., Mateo, R., Hinojo, M.J., Logrieco, A., and Jimenez, M., 2004. Influence of the interactions among ecological variables in the characterization of zearalenone producing isolates of Fusarium spp. *Systematic and Applied Microbiology* 27: 253-260.

Miron, M.L.L., Herrera, M.L.G., Ramirez, P.N., and Hobson, K. A., 2006. Effect of diet quality on carbon and nitrogen turnover and isotopic discrimination in blood of a New World nectarivorous bat. *Journal of Experimental Biology* 209: 541-548.

Ralph, J., Hatfield, R.D., Piquemal, J., Yahiaoui, N., Péan, M., Lapierre, C., and Boudet, A.M., 1998. NMR characterization of altered lignins extracted from tobacco plants down-regulated for lignification enzymes cinnamyl-alcohol dehydrogenase and cinnamoyl-CoA reductase. *Proceedings of the National Academy of Sciences, U S A* 95: 12803-12808.

Setchell, K.D., Faughnan, M.S., Avades, T., Zimmer-Nechemias, L., Brown, N.M., Wolfe, B.E., Brashear, W.T., Desai, P., Oldfield, M.F., Botting,N.P., and Cassidy, A., 2003. Comparing the pharmacokinetics of daidzein and genistein with the use of 13C-labeled tracers in premenopausal women. *American Journal of Clinical Nutrition* 77: 411-419.

Stokvis, E., Rosing, H., and Beijnen, J.H., 2005. Stable isotopically labelled internal standards in quantitative bioanalysis using liquid chromatography/mass spectrometry: necessity or not? *Rapid Communication in Mass Spectrometry* 19: 401-407.

Section V:

Occurrence of mycotoxins

Study of aflatoxin exposure in a Brazilian population using an aflatoxin – albumin biomarker

V.M. Scussel[1], P. Haas[1], Y.Y. Gong[2], C.P. Turner[2] and C.P. Wild[2]

[1]Mycotoxicology Laboratory - Labmico, Food Science and Technology Department, P.O. Box 476, Federal University of Santa Catarina, Florianopolis, SC, Brazil
[2]Molecular Epidemiology Unit, Centre for Epidemiology and Biostatistics, Leeds Institute of Genetics, Health and Therapeutics, University of Leeds, United Kingdom

Abstract

The evaluation of exposure to aflatoxins (AF) by measurement of the level of contamination in food is hampered due to the heterogeneous distribution of AF in food. Therefore, an alternative is to estimate the exposure using specific biological markers (biomarkers) based on an understanding of the metabolism of the compound. For AF, these include aflatoxin-N^7-guanine in the urine, or AFB_1-albumin (AF-alb) in the blood. This study assessed the level of exposure to AF in Brazilian individuals using a biomarker approach, *i.e.* the AF-alb adducts. Blood samples were collected from urban residents (n=50; aged 18-52) in June 1999, at the Blood Center of Antonio Carlos de Camargo Hospital, Sao Paulo, Brazil. AF-alb adduct levels were determined, by ELISA following serum albumin extraction and digestion. AF-alb adducts were detected in 31/50 (62%) samples [range 0-57.3 pg AFB_1-lys adducts/mg of blood albumin (pg/mg)]. The mean level of positives was 14.9 pg/mg and males had the two highest levels measured (57.1 and 57.3 pg/mg). There was no correlation with age or profession. This is the first study of Brazilian, or indeed South American, individuals that has determined exposure to AF at the individual level using a biomarker approach. These levels are similar to those observed in the Philippines. These data warrant further investigation of both the sources and consequences of exposure to this potent toxin in Brazil.

Keywords: aflatoxin exposure, AFB_1-albumin, adduct, biomarker

Introduction

Aflatoxins (AF), are highly potent carcinogens and immunotoxins that can contaminate stored foods, such as maize and peanuts, due to fungal growth. However, the evaluation of exposure to aflatoxin B_1 (AFB_1) at the individual level by measurement of the amount of contamination in food is hampered due to the heterogeneous distribution of aflatoxins in food. Additionally dietary intake of a given food can be highly variable and unreliably reported (IARC, 1993).

Genetic variability in AF metabolism may also influence the level of toxicity at the individual level. Therefore, an alternative is to estimate the exposure using specific biological markers (biomarkers) based on an understanding of the metabolism of the compound. For AFB_1, these include the aflatoxin-N^7-guanine adduct in urine, or the aflatoxin-albumin (AF-alb) adduct in blood (Wild and Hall, 1996; Wild *et al.*, 1995). This study was aimed at assessing the level of exposure to AF, at the individual level, in Brazilians using AF-alb adducts as a biomarkers.

Materials and methods

Sample collection

Blood samples were collected from 50 volunteer blood donors, aged 18 to 52 years, at the Blood Center of Antonio Candido Camargo Hospital, Sao Paulo, Brazil. Collection of the blood samples from the donors was timed to coincide with a June Festival (when peanut and maize products are frequently consumed). The sampled subjects included 29 males and 21 females, and each donor provided 9 ml of blood.

Analysis of AF-alb adducts

ELISA was performed on hydrolysed albumin according to the method of Chapot and Wild (1991). Briefly, albumin was isolated from serum and two mg digested with pronase. Aflatoxin residues were isolated on a Sep-pak® C_{18} cartridge. The AF-residues were dried, dissolved in phosphate buffered saline (PBS) and assayed using the ELISA. AFB_1-Lysine (AFB_1-Lys) adducts were used to generate the ELISA standard curve and the results were expressed as pg AFB_1-Lys adduct equivalent/mg albumin (pg/mg). AF-albumin adduct data was natural log transformed, normalised and then back transformed for calculation of geometric means and 95% confidence intervals.

Results and discussion

The ELISA procedure limit of detection (LOD) for AF-alb adducts was 3 pg/mg. AF-alb adducts were detected in 31/50 (62%) samples (Figure 1). Adducts were detected more frequently in females (15 of 21, 71.4% compared to males 16 of 29, 55.2%) but the two highest levels measured were in males (57.1 and 57.3 pg/mg) (Table 1 and 2). The geometric mean for females (6.3 pg/mg; 95% CI: 3.8; 10.0, N=21) was not significantly (p=0.05) higher than that for males (4.1 pg/mg, 95% CI 2.6; 6.2, N=29).

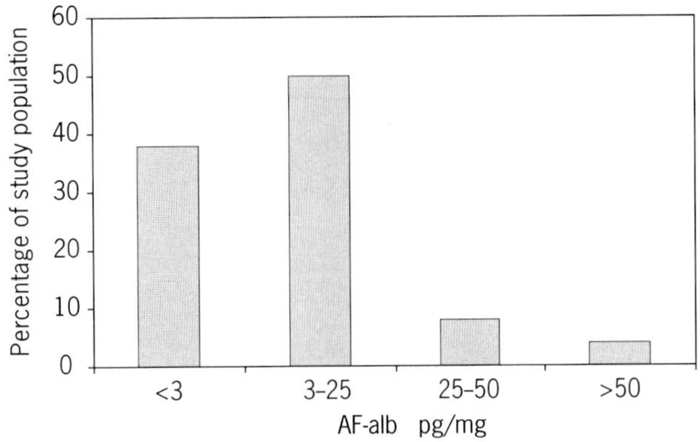

Figure 1. Variation in aflatoxin-albumin adduct among all study subjects Brazil.

The geometric mean for all positive samples was 10.1 pg/mg (95% CI: 7.4; 13.8 pg/mg). There was no correlation between AF-alb and age or profession.

This is the first study of Brazilians, or indeed South Americans, that has determined exposure to AF at the individual level using a biomarker. AF exposure was detectable in >60% of the study group. The level of adducts was greater than observed in Western Europe (low risk) but not as high as in West Africa (high risk), (Wild *et al.*, 1993, 2000; Turner *et al.*, 2002) where generally >95 of individuals have detectable AF-albumin adducts (Figure 2). These intermediate levels of exposure are similar to those observed in Thailand and Bangladesh (Vinitketkumnuen *et al.*, 1997; Dawlatana *et al.*, 2002). These data warrant further investigation of both the sources and consequences of exposure to this potent toxin in Brazil.

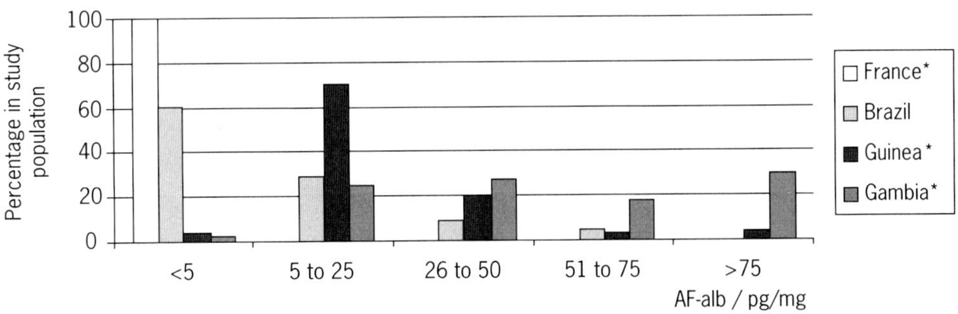

*Figure 2. Variation in aflatoxin-albumin adduct levels in Brazil compared to other countries (marked with *). Additional data from Wild* et al., *2000, and Turner* et al., *2002.*

Table 1. Levels of aflatoxin exposure through AFB_1-Lys adduct in male residents from São Paulo City, Brazil.

Sample (Nº)	Age	Activity	AFB_1-Lys Adduct (pg/mg)
1	18	Student	< 3
2	33	Taxi driver	< 3
3	47	Porter	7.8
4	26	Distributor	< 3
5	31	Mailman	17.1
6	34	Mechanic	< 3
7	24	Unemployed	3.6
8	26	Salesperson	14.6
9	18	Student	4
10	33	System programmer	4
11	18	Student	< 3
12	26	Mechanic	< 3
13	31	Porter	22.7
14	30	Mechanic	4.1
15	32	Tourism agent	3.3
16	34	System analyst	17.8
17	29	Mechanic	19.1
18	26	News dealer	< 3
19	42	Office boy	< 3
20	18	NI[a]	< 3
21	37	Public employee	< 3
22	30	Taxi driver	< 3
23	26	Nurse	57.1
24	23	Office boy	< 3
25	28	Student	< 3
26	31	Military	3.7
27	43	Sales person	4.6
28	21	Teacher	57.3
29	52	Office boy	3.9

[a] – not informed.

Table 2. Levels of aflatoxin exposure through AFB_1-Lys adduct in female residents from São Paulo City, Brazil.

Sample (N°)	Age	Activity	AFB_1-Lys Adduct (pg/mg)
30	29	Housekeeper	7
31	28	System analyst	3.9
32	42	Dealer	31
33	26	Operator	5.9
34	24	Taxi driver	3.8
35	53	Hair stylist	5.3
36	54	Housekeeper	4.7
37	47	Retired	14.8
38	58	Housekeeper	< 3
39	22	Student	29.7
40	32	Housekeeper	20.1
41	28	Housekeeper	27.2
42	39	Teacher	< 3
43	38	Maid	< 3
44	30	Sales person	< 3
45	28	Secretary	< 3
46	43	Social worker	< 3
47	27	Housekeeper	19.2
48	46	Salesperson	7.8
49	39	Secretary	10.7
50	56	Psychologist	27.4

Acknowledgements

This work was supported by a grant (ES06052) from the National Institute of Environmental Health Sciences (NIEHS), USA.

References

Chapot, B., and Wild, C.P., 1991. Elisa for quantification of aflatoxin-albumin adducts and their application to human exposure assessment. In: Warhol, M., Van Velzen, D., and Bullock, G.R. (Eds.), *Techniques in Diagnostic Pathology 2*, pp. 135-155.

Dawlatana, M., Coker, R.D., Nagler, M.J., Wild, C.P., Hassan, M.S., and Blunden, G., 2002. The occurrence of mycotoxins in key commodities in Bangladesh: surveillance results from 1993 to 1995. *Journal of Natural Toxins* 11: 379-386.

IARC, 1993. *Some naturally occurring substances, food items and constituents, heterocyclic aromatic amines and mycotoxins.* IARC monographs on the evaluation carcinogenic risks to human, ed. IARC v.56,, Lyon, Chapter: mycotoxins, pp. 245–552.

Turner, P.C., Sylla A., Diallo M.S., Castagnaro M., HallL A.J., and Wild C.P., 2002. The role of chemical carcinogens and hepatitis viruses in the aetiopathogenesis of hepatocellular carcinoma: a basis for primary prevention in Guinea Conakry, West Africa. *J. Gastroenterol. Hepatol.* 17: S441-S448.

Vinitketkumnuen, U., Chewonarin, T., Kongtawelert, P., Lertjanyarak, A., Peerakhom, S., and Wild, C.P., 1997. Aflatoxin exposure is higher in vegetarians than in nonvegetarians in Thailand. *Natural Toxins* 5: 168-171.

Wild, C.P., and Hall, A.J., 1996. Epidemiology of mycotoxin-related diseases. In: *The Mycota, 6, Human and animal relationships*, Howard/Miller, Springer-Verlang, Berlin, Heidelberg, pp.213-227.

Wild, C.P., Fortuin, M., Donato, F., Whittle, H.C., Hall, A.J., Wolf, C.R., and Montesano, R., 1993. Aflatoxin, liver enzymes, and hepatites B virus infection in Gambia children. *Cancer Epidemiology and Biomarkers Prev.* 2: 555-561.

Wild, C.P., Hasegawa, R., Barraud, L., Chutimataewin, S., Chapot, B., Ito, N.E., and Montesano, R., 1995. Aflatoxin-albumin adducts: a basis for comparative carcinogenesis between animals and humans. *Cancer Epidemiology Biomarkers Prev*. 5: 179-189.

Wild, C.P., Yin F., Turner P.C., Chemin, I., Chapot, B., Mendy, M., Whittle, H., and Hall, A.J., 2000. Environmental and Genetic Determinants of Aflatoxin-Albumin Adducts in The Gambia. *Int. Journal of Cancer* 86: 1-7.

The occurrence of aflatoxin M_1 in fresh cow milk retailed in Dar es Salaam, Tanzania

E.M. Urio, A. Juma, S. Mwanyika, N.V. Mlingi, G.T. Ndunguru and G.D. Ndossi

Tanzania Food and Nutrition Centre (TFNC), Department of Food Science and Nutrition, P.O. Box 977, Dar es Salaam, Tanzania; tfnc@muchs.ac.tz or elisamuri@yahoo.com

Abstract

Aflatoxins are toxic substances produced by *Aspergillus flavus* and *Aspergillus parasiticus* during their growth on foods or feeds. Aflatoxin M_1 (AFM_1) occurs in cow's milk as a result of dairy cows consuming aflatoxin B_1 (AFB_1) contaminated feeds. The number of dairy cattle in Dar es Salaam has been on the increase over the years as part of individual efforts to supplement income. In addition to grass, dairy cattle are fed feeds that are likely to contain AFB_1 because of susceptibility to fungal contamination. Little is known about the occurrence of aflatoxins in fresh milk retailed in Dar es Salaam. The aim of this investigation was to determine the occurrence of AFM_1 in local milk supplies retailed in Dar es Salaam. Thirty seven samples of raw cow's milk were randomly collected from milk stalls and retail shops in Dar es Salaam and analysed for the presence of AFM_I. The samples were analysed using high-performance liquid chromatography. Thirty four of the 37 milk samples (92%) had AFM_1 levels between 0.005 and 0.855 μg/kg with a mean concentration of 0.246 μg/kg. Sixty eight percent contained low levels of AFM_1 ranging from 0.005 to 0.336 μg/kg while twenty four percent contained high levels ranging from 0.545 to 0.855 μg/kg. Eight percent contained no AFM_1 or undetected amounts. Dairy cattle in Dar es Salaam are fed on feed contaminated with AFB_I and therefore dairy and dairy products consumers are at risk of aflatoxin M_1 exposure. Dairy cattle keepers and local feed producers need to be educated on the potential hazard of mycotoxins and available preventive measures that are based on principles of Hazard Analysis Critical Control Points (HACCP). Also surveillance on aflatoxin contamination in animal feeds should be strengthened.

Keywords: Aflatoxin B_1, Aflatoxin M_1, cow's milk, animal feed, high-performance liquid, chromatography (HPLC), Dar es Salaam

Introduction

Aflatoxin B_1 (AFB_1) is a mycotoxin produced by strains of the fungus *Aspergillus flavus* and *Aspergillus parasiticus* while growing on various food commodities and animal feeds. These fungi can grow on food crops while on the farm, during harvest or during storage. AFB_1 has been reported to have carcinogenic, mutagenic and teratogenic effects in several animal species including humans (Sabbioni and Sepai, 1998). The International Agency for Research on Cancer (IARC) has classified AFB_1 into category 1A of active carcinogenic compounds (IARC, 1993). Aflatoxin M_1 (AFM_1) is produced as a metabolite of AFB_1 (Stoloff, 1980). It is secreted in the milk after the feeding of AFB_1 containing feed to the lactating cow (Saitanu, 1997). AFM_1 is considered a potential carcinogenic compound for humans and it is included in the class 2B of the carcinogenic compounds (IARC, 1993). Since AFM_1 is relatively stable during pasteurisation, processing and storage (Stoloff *et al.*, 1975), the occurrence of AFM_1 in retail milk is an obvious health hazard of concern.

For the past two decades in Dar es Salaam City, the number of dairy cattle has been on the increase as part of individual efforts to complement income. In addition to green grass, dairy cattle in Dar es Salaam are fed on feed that is likely to be contaminated by AFB_1. The feed is produced locally by small-scale producers using maize bran and cottonseed cake, which are susceptible to fungal contamination. It is therefore, not clearly known whether milk produced in Dar es Salaam is free from aflatoxins. The aim of this study was to determine whether aflatoxins are present in local milk supplies in Dar es Salaam to such an extent that it constitutes a potential public health hazard.

Materials and methods

Equipment

The HPLC system used consisted of a pump: Waters 600 multisolvent delivery system and controller with a flow-rate 0.8 ml/min, autoinjector: Shimadzu model SIL – 10 AD VP model # 7120 with 20 µl loop (Tokyo, Japan). A C_{18} SU 150 x 3.0 mm column (Phenomenex, Torrance CA) fitted with a Waters symmetry C_{18}, 5µm guard column. The detector was Shimadzu Fluorescence monitor model # RF – 10AXL, and a Shimadzu chromatopac model – C- R3A (Tokyo, Japan) integrator.

Reagents

All solvents for mobile phase were HPLC grade and consisted of acetonitrile+water (25:75). Other chemicals used included dichloromethane, chloroform, acetone, n-Hexane and trifluoroacetic acid which were all analytical grade.

Preparation of standards

Aflatoxin M_1 (10 μg) was dissolved in 10 ml benzene+acetonitrile (9:1) to get a stock solution. 5 ml of stock solution was diluted with 10 ml benzene+acetonitrile (9:1) to get intermediate standard solution. Working standard solutions in ng/ml were prepared from the intermediate standard to obtain a calibration curve.

Sampling, sample preparation and analysis

Thirty-seven fresh milk samples (about 500 ml each) were collected randomly from milk stalls and retail shops in Dar es Salaam. These stalls were located in the suburbs of Dar es Salaam city, namely: Mwenge, Ukonga, Tabata, Ubungo and Kiwalani. During the collection, samples were kept in labelled and sealed plastic containers and placed in a cool box. Promptly upon arrival at the laboratory, all samples were stored in a deep freezer at – 20 °C until analysis.

Samples were prepared according to a VICAM aflatoxin M_1 Instruction Manual (1998). Fifty (50 ml) millilitres of each sample was centrifuged at 1,540 × g for five minutes to separate out fat. After separating the fat layer, the method of Bijl and Peteghem (1985) was followed. The centrifuged sample was then passed through the pre-washed column (Sep pak C_{18} cartridge (Waters Corporation, Milford, MA) with 5 ml acetonitrile and 5 ml water at 4 – 5 ml/min. After passing the sample, the column was rinsed with 5 ml water and 20 ml acetonitrile: water (1:9) and the eluate was discarded. After rinsing, the column was eluted with 2 ml acetonitrile and the eluate evaporated under nitrogen at 40 °C to near dryness in a heating block. The residue was dissolved in 1 ml dichloromethane.

The Sep-Pak® silica C_{18} cartridge was pre-washed with 5 ml dichloromethane prior to the application of the sample residue dissolved in dichloromethane. After passing the residue through the column, the cartridge was rinsed twice with 1 ml dichloromethane followed by washing with 5 ml diethyl ether. After washing the cartridge, aflatoxin M_1 was eluted with 8 ml chloroform+ acetone (4:1) and collected in a vial. The eluate was then evaporated to near dryness under nitrogen at 40 °C in a heating block. The residue was dissolved in 1 ml mobile phase Acetonitrile+water (25:75) and then subjected to

HPLC analysis. Twenty microliters (20 µl) of the standard and the sample were injected into the HPLC unit and the concentration of AFM_1 was calculated as follows:

$$\frac{\text{(ng std in 20 µl)}}{\text{Peak area of std}} \times \text{Peak area of sample} \times \frac{\text{Dilution factor}}{\text{Aliquot taken}}$$

Confirmation of identity by derivatisation

The sample extract dissolved in mobile phase and AFM_1 working standard were evaporated to dryness. Hexane (200 µl) and trifluoroacetic acid (50 µl) were added to each vial containing residues of the sample extract and AFM_1 standard. The vials were covered with stoppers and heated to 45 °C in a heating block for 20 min. Residues were diluted with the mobile phase and then subjected to HPLC analysis where the appearance of the aflatoxin M_1 derivative confirmed $AFM_{1.}$

Method performance (recovery)

Recovery experiments were performed by adding 5 µl of AFM_1 standard (194.153 ng/ml) to 50 ml of a blank milk sample. The level at which standards were added to the test portion was about 2µg/ml. Recovery was found to be 82.89%.

Results and discussion

Thirty four of the 37 milk samples (92%) showed AFM_1 levels between 0.005 and 0.855 µg/kg. Twenty five (68%) of the 37 milk samples had AFM_1 levels between 0.005 and 0.336 µg/kg while the remaining nine samples (24%) showed levels ranging from 0.545 to 0.855 µg/kg. Three samples (8%) contained no AFM_1 or undetected amount (Table 1). The method limit of detection was 0.005 µg/kg.

These results suggest that the majority of dairy cattle in Dar es Salaam were fed aflatoxin B_1 contaminated feed. The study also revealed that 24% of the milk samples

Table 1. Levels of aflatoxin M_1 in fresh cow milk.

Number of milk samples	Range of aflatoxin M_1 in µg/kg	Percent
25	0.005 – 0.336	68
9	0.545 – 0.855	24
3	0	8

had higher levels of aflatoxin M_1 compared to the allowable limit of 0.5 μg/kg. These results are comparable to a similar study conducted in Dar es Salaam by Tanzania Food and Nutrition Centre (TFNC) almost ten years ago (TFNC Report No.1570). The study showed that 24% of milk samples had aflatoxin M_1 levels above the permissible limit. Aflatoxin M_1 levels higher than the recommended limit are a significant food safety concern.

Conclusions

Milk samples collected from Dar es Salaam contained aflatoxin M_1 indicating that they originated from herds of dairy cattle that had consumed feed contaminated with aflatoxin B_1. The presence of aflatoxin M_1 in 92% of the milk samples clearly indicates that a large segment of the population in Dar es Salaam City was ingesting low levels of aflatoxin. The aflatoxin contamination problem revealed by this study is of serious food safety concern and therefore two important points are suggested. First, the surveillance system for aflatoxin contamination in foods and feeds should be strengthened. Secondly, extension services should be extended to dairy cattle keepers and animal feeds processors with the specific aim of raising their awareness of the potential hazard of mycotoxin contamination in animal feeds and the available preventive measures that are based on principles of Hazard Analysis Critical Control Points (HACCP). Lastly, more studies are suggested in order to establish the actual magnitude of aflatoxin contamination in milk and milk products.

References

Bijl, J., and Van Peteghem, C., 1985. Rapid extraction and sample clean up for the fluorescence densitometric determination of aflatoxin M1 in milk and milk powder. *Anal. Chem. Acta* 170: 149–152.

IARC, 1993. International Agency for Research on Cancer. *Monograph on the evaluation of carcinogenic risks to humans*. Vol. 56. IARC, Lyon, pp. 445–466.

Sabbioni, G., and Separi, O., 1998. Determination of human exposure to aflatoxin. In:

Mycotoxins in Agriculture and Food Safety, Sinha, K.K., and Bhatnagar, D. (Ed.), Marcel Dekker Inc., New York, pp. 183–226.

TFNC, 1993. Tanzania Food and Nutrition Centre. *Toxicological and microbiological quality of raw, fresh cow milk from Dar es Salaam region*. TFNC Report No. 1570.

Saitanu, K., 1997. Incidence of aflatoxin M_1 in Thai milk products. *Journal of Food Protection* 60: 1010-1012.

Stoloff, L,, Trucksess, M., Hardin, N., Francis, O.J., Hayes, J.R., Polan, C.E., and Campbell, T.C., 1975. Stability of aflatoxin M_1 in milk. *Journal Dairy Science* 58: 1789–1793.

Stoloff, L., 1980. Aflatoxin M_1 in perspective. *Journal Food Protection* 43: 226–230.

VICAM Science and Technology, 1998. *Instruction Manual for aflatoxin M_1 analysis*. VICAM, L.P., Watertown, MA 02474 USA, 313 pp.

Exposure to fumonisins through *Kenkey* – a Ghanaian fermented maize product

K.A. Kpodo[1], G.S. Ayernor[2], G.S. Shephard[3] and M. Jakobsen[4]

[1]Food Research Institute, P. O. Box M.20, Accra, Ghana; kpodofri@ghana.com
[2]Department of Nutrition and Food Science, University of Ghana, Legon, Accra, Ghana
[3]PROMEC Unit, Medical Research Council, P. O. Box 19070, Tygerberg, 7505 South Africa
[4]Department of Dairy and Food Science, The Royal Veterinary and Agricultural University, Rolighedsvej 30, DK 1958 Frederiksberg C, Denmark

Abstract

Fumonisins are a group of mycotoxins produced mainly by *Fusarium verticillioides* [= *F. moniliforme*], *F. proliferatum*, and several other *Fusarium* species that are widespread in maize throughout the world. The most popular way of consuming maize, a dietary staple in Ghana, is in the form of *Kenkey*, a ready-to-eat food. For the preparation of *Kenkey*, maize kernels are steeped, milled, fermented, moulded, wrapped in maize husks and boiled for 3 hours. There is however a growing concern that a high dietary intake of maize-based foods, notably *Kenkey*, may expose consumers in Ghana to fumonisins. The study measured fumonisin levels in maize and *Kenkey* from 75 maize processing sites in Ghana. The maize and *Kenkey* samples were analysed for fumonisins using HPLC after extraction and clean-up on either strong anion exchange (SAX) or reversed-phase (C_{18}) cartridges respectively. More than 90% of the maize samples contained fumonisins ranging from 11 to 2,534 μg/kg whilst over 70% of the *Kenkey* samples also contained fumonisins at levels ranging from 15 to 1,035 μg/kg confirming the direct intake of fumonisins. Further experiments to determine the fate of fumonisins during the preparation of *Kenkey* using maize naturally contaminated with fumonisins at different levels revealed significant reductions during the 48-hour steeping stage. Reduction levels were dependent on the initial fumonisin content of the maize kernels. Fumonisin B_1 was reduced by between 59 and 86%. The initial fumonisin B_1 content of the maize kernel samples ranged from 1,206 to 78 μg/kg. Reductions in fumonisin levels ranged from 16 to 31% for fumonisin B_2 and between 57 and 64% for fumonisin B_3. No significant changes in fumonisin levels were recorded during the three-day fermentation and the three-hour boiling processes. This study has established the carry-over of fumonisins from maize into *Kenkey* and the possible risk posed to consumers.

Keywords: mycotoxins, fumonisins, *Kenkey*, Ghana, maize, fermentation

Introduction

Fumonisins are toxic metabolites of *Fusarium verticillioides* (= *F. moniliforme*) and *F. proliferatum*. *F. verticillioides* has been reported as the most prevalent fungus in maize growing areas around the world (Nelson *et al.*, 1981; Marasas *et al.*, 1984) but much more widespread in tropical and subtropical regions.

In Ghana, maize is the principal cereal produced and is consumed as a staple crop in most parts of the country and contributes the major part of the total calories in the diet of Ghanaians especially in the coastal areas. Mycological studies have revealed *F. verticillioides* as the dominant *Fusarium* species on Ghanaian maize kernels, and also confirmed their ability to produce substantial amounts of fumonisins on maize substrates (Kpodo *et al.*, 2000).

Fumonisins have been associated with several diseases in animals such as leukoencephalomalacia in horses (Kellerman *et al.*, 1990), pulmonary oedema in swine (Harrison *et al.*, 1990), and have also been found to be hepatocarcinogenic in rats (Gelderblom *et al.*, 1991). In humans, fumonisins have been associated with the high incidence of human esophageal cancer in the Transkei region of the Eastern Cape Province of South Africa (Rheeder *et al.*, 1992).

In Ghana as in most other West African countries, a large proportion of the maize is consumed in the fermented form as *Kenkey*, produced through traditional spontaneous and uncontrolled fermentations. Most studies have centered on the fumonisin content of maize kernels and not the ready-to-eat form of the foods made from maize. Speculation about fumonisins in *Kenkey* has therefore been high and, recently, there has been a growing concern that a high dietary intake of maize-based foods, notably *Kenkey*, may be exposing consumers to fumonisins. This study was, therefore, undertaken to ascertain the presence and levels of fumonisins in maize kernels and *Kenkey* in Ghana and also determine the fate of fumonisins during the *Kenkey*-making process using maize naturally contaminated with fumonisins.

Materials and methods

Samples and chemicals

Seventy-five maize kernel samples and an equal number of *Kenkey* samples were purchased from 75 different maize processing sites situated in five agro-ecological zones of Ghana. Fumonisin-free maize kernels used for control experiments were obtained from Brogaarden Company Ltd. in Copenhagen, Denmark. Fumonisins B_1, B_2, and B_3 were isolated from *Fusarium verticillioides* MRC 826 culture material at PROMEC Unit, Tygerberg, South Africa according to the method of Cawood *et al.* (1991). Reagents for HPLC separations were of the HPLC grade (Merck Chemicals, Darmstadt, Germany). All other chemicals used were of analytical grade.

Preparation of Kenkey

Three samples of maize kernels with varying amounts of fumonisins were purchased from three maize processing sites in Accra, Ghana and designated as "low-fumonisin", "medium-fumonisin" and "high-fumonisin" maize samples. Each sample was used to prepare *Kenkey* as follows: 1,200 g of maize kernels were steeped in 2 litres water for 48 hours at ambient temperature (28 °C). The steeped kernels were milled into flour and dough prepared by adding 400 ml water to the flour and mixed. The dough was left to ferment spontaneously for three days after which a 500 g portion was cooked into a smooth paste (*Aflata)* with 500 ml water. The *aflata* was kneaded with 1,000 g of un-cooked dough till free of lumps after which portions were moulded into balls, wrapped in maize husk and boiled for 3 hours.

Samples were taken at each processing stage and analysed for moisture, and fumonisins B_1, B_2, and B_3. Data obtained was subjected to Analysis of Variance and differences in fumonisin levels were analysed using the Duncan's Multiple Range Test.

Determination of fumonisins in maize, fermented maize dough, and Kenkey

Maize samples were analysed for fumonisins according to the method of Sydenham *et al.* (1992). For the analysis of fermented maize dough and *Kenkey*, the developed method described by Kpodo (2001) was used. Briefly, 25 g of fermented dough or *Kenkey* was extracted by homogenising for 5 minutes with 100 ml methanol+water (3:1, v/v) after which the slurry was centrifuged and filtered. Thirty millilitres water were added to 10 ml filtrate and the pH adjusted to 3.0 with 0.1 M HCl and centrifuged. The 40 ml supernatant was cleaned-up on a C_{18} solid phase extraction cartridge and fumonisins eluted with methanol after washing with methanol+water (1:4, v/v). The eluate was evaporated to dryness at 60 °C under a stream of nitrogen and fumonisins determined by HPLC. Fumonisins, after pre-column derivatisation with *o*-phthaldialdehyde (OPA) mercaptoethanol reagent, were resolved on an Ultracarb 5-μm ODS column maintained at ambient temperature.

The method was validated with respect to precision and accuracy for both fermented maize dough and *Kenkey*. For precision, six separate determinations each on naturally contaminated fermented dough and *Kenkey* were conducted. Accuracy of the procedure was assessed through recovery studies conducted on fermented dough and *Kenkey* samples spiked at two levels. Six replicate determinations were conducted at each spiking level.

Results and discussions

Results of the spiking experiments

At the spiking levels of 1.0, 0.4, and 0.2 μg/g, mean recoveries for fumonisins B_1, B_2, and B_3 from *Kenkey* were 89, 78, and 75%, respectively. Recoveries from fermented dough spiked at similar levels were 74, 79 and 74%, respectively. Recoveries at lower spiking levels of 0.5, 0.2, and 0.1 μg/g were 90, 80, and 84% for *Kenkey* and 77, 77 and 75% for fermented maize dough. Six replicate determinations of fumonisins in *Kenkey* gave a reproducibility of 10.5, 11.7 and 17.8% for fumonisins B_1, B_2, and B_3. Similar studies conducted on fermented maize dough gave values of 11.0, 11.4 and 17.2%.

Fumonisins in commercial Kenkey samples

Sixty-eight (90.7%) of the 75 maize kernel samples collected from the processing sites contained fumonisins at various levels ranging from 11 to 2,534 μg/kg. Table 1 shows the distribution of maize samples at different fumonisin levels. For *Kenkey*, 55 (73.3%) of the 75 samples contained fumonisins (FB_1+FB_2+FB_3) at levels ranging from 15 to 1,035 μg/kg (Table 2). The fact that fumonisins were detected in *Kenkey*, which is a ready-to-eat food confirms speculation about the presence of fumonisins in *Kenkey*. The Joint FAO/WHO Expert Committee on Food Additives (JECFA) has allocated a group provisional maximum tolerable daily intake (PMTDI) of 2 μg/kg body weight to fumonisins B_1, B_2, and B_3 alone or in combination (WHO, 2002). In the light of these levels, the data presented in Table 2, and the fact that *Kenkey* is widely consumed sometimes on a daily basis in some Ghanaian communities where an average man can consume about 1 kg of *Kenkey* (moisture content ≈ 70%) in one meal, is cause for concern. From the data presented, no significant variations were observed in fumonisin contents of maize and *Kenkey* samples within the different agro-ecological zones of Ghana.

Table 1. Distribution of maize samples at different fumonisin levels (µg/kg).

Zone	Total fumonisin level (µg/kg)			
	<100	100–500	501–1000	>1000
Coastal Scrub	6	5	0	4
Tropical Rain Forest	11	3	1	0
Forest	9	5	1	0
Transitional	3	9	2	1
Guinea Savannah	3	4	4	4
Total No. of samples	32	26	8	9

Table 2. Distribution of Kenkey samples at different fumonisin levels (µg/kg).

Zone	Total fumonisin level (µg/kg)			
	<100	100–500	501–1000	>1000
Coastal Scrub	6	6	3	0
Tropical Rain Forest	12	3	0	0
Forest	13	2	0	0
Transitional	5	10	0	0
Guinea Savannah	6	4	4	1
Total No. of samples	42	25	7	1

Effect of processing on fumonisins

To study the fate of fumonisins during the *Kenkey* making process and possible carry-over of the toxins from contaminated maize into the end product, three batches of maize with varying fumonisin contents designated as "low", "medium" and "high" fumonisin maize kernels were used. "Low" fumonisin maize contained only fumonisin B_1 at a level of 68 µg/kg. Fumonisin contents of "medium" and "high" fumonisin maize kernels expressed on dry weight basis together with changes in fumonisin levels during processing into *Kenkey* are shown in Tables 3 and 4.

Table 3. Effect of steeping, fermentation and cooking on "medium" contamination levels of fumonisins in maize.

Treatment	Fumonisin levels (μg/kg)[1, 2]		
	FB_1	FB_2	FB_3
Maize kernels	503 ± 9.9[a]	160 ± 9.2[a]	88 ± 7.1[a]
Steeped kernels	165 ± 12.7[b]	135 ± 8.5[a]	32 ± 5.7[b]
Dough (0 hour)	104 ± 10.6[c]	92 ± 12.7[b]	44 ± 10.6[bc]
Dough (24 hour)	100 ± 13.4[c]	92 ± 10.6[b]	59 ± 11.3[c]
Dough (48 hour)	106 ± 12.0[c]	77 ± 8.5[b]	28 ± 5.7[b]
Dough (72 hour)	110 ± 15.6[c]	78 ± 12.7[b]	32 ± 14.1[b]
Kenkey	100 ± 14.8[c]	80 ± 9.9[b]	ND[d]

[1]Means within column sharing the same letter are not significantly different ($p>0.05$).
[2]Values are means ± standard deviations of one determination each on duplicate samples.
ND = None detected (<10 μg/kg).

Table 4. Effect of steeping, fermentation and cooking on "high" contamination levels of fumonisins in maize.

Treatment	Fumonisin levels (μg/kg)[1, 2]		
	FB_1	FB_2	FB_3
Maize kernels	1206 ± 40.3[a]	315 ± 17.0[a]	150 ± 19.1[a]
Steeped kernels	498 ± 27.6[b]	216 ± 12.0[b]	64 ± 9.2[b]
Dough (0 hour)	304 ± 13.4[c]	132 ± 16.3[c]	62 ± 17.7[b]
Dough (24 hour)	298 ± 23.3[c]	134 ± 21.9[c]	76 ± 19.8[b]
Dough (48 hour)	270 ± 17.0[c]	102 ± 15.6[cd]	66 ± 13.4[b]
Dough (72 hour)	313 ± 12.7[c]	106 ± 12.0[cd]	54 ± 18.4[b]
Kenkey	395 ± 25.4[d]	86 ± 16.3[d]	68 ± 16.3[b]

[1]Means within column sharing the same letter are not significantly different ($p<0.05$).
[2]Values are means ± standard deviations of one determination each on duplicate samples.

The study revealed significant reductions in fumonisin levels in the maize kernels during the 48-hour steeping stage with partial recovery in the steep water (data not shown). The percentage reduction was dependent on the initial fumonisin content of the maize kernels. Regarding the "low" fumonisin maize with 68 μg/kg of FB_1, fumonisin was no longer detected in the maize kernels after the 48 hr steeping period. Reductions in FB_1 levels in individual samples ranged from 59 to 86%, and from 16 to 31% for FB_2 and 57 to 64% for fumonisin B_3, respectively. These reductions reflect the solubility of fumonisins in aqueous solutions and confirm similar observations by Canela *et al.* (1996). In contrast, no significant changes in fumonisin levels were recorded during the three-day spontaneous fermentation process as practised in Ghana or the three-hour boiling process. Previous studies on the fermentation of naturally contaminated maize by yeasts for ethanol production have also indicated that fumonisins are stable during these fermentation processes (Bothast *et al.*, 1992).

Conclusion

Through this work, the presence of fumonisins in *Kenkey* is reported for the first time in Ghana. It is the conclusion of this study that despite significant reductions in fumonisin levels during the 48 hour steeping process, there is still a carry-over into the end product *Kenkey* thus exposing a section of the consuming Ghanaian populace to the health hazards of fumonisins.

Acknowledgments

The authors express their appreciation to the Danish International Development Assistance (DANIDA) for funding this study.

References

Bothast, R.J., Bennett, G.A., Vancauwenberge, J.E., and Richard, J.L., 1992. Fate of fumonisin B_1 in naturally contaminated corn during ethanol fermentation. *Applied and Environmental Microbiology* 58: 233-236.

Canela, R., Pujol, S., Sala, N., and Sanchis, V., 1996. Fate of fumonisins B_1, and B_2 in steeped corn kernels. *Food Additives and Contaminants* 13: 511–517.

Cawood, M.E., Gelderblom, W.B.C., Vleggaar, R., Behrend, Y., Thiel, P.G., and Marasas, W.F.O., 1991. Isolation of the fumonisin mycotoxins: a quantitative approach. *Journal of Agricultural and Food Chemistry* 39: 1958–1962.

Gelderblom, W.C.A., Kriek, N.P.J., Marasas, W.F.O., and Thiel, P.G., 1991. Toxicity and carcinogenicity of the *Fusarium moniliforme* metabolite, fumonisin B_1, in rats. *Carcinogenesis* 12: 1247-1251.

Harrison, L.R., Colvin, B.M., Greene, J.T., Newman, L.E., and Cole, J.R. Jr., 1990. Pulmonary edema and hydroxythorax in swine produced by fumonisin B_1, a toxic metabolite of *Fusarium moniliforme*. *Journal of Veterinary Diagnosis and Investigation* 2: 217-221.

Kellerman, T.S., Marasas, W.F.O., Thiel, P.G., Gelderblom, W.C.A., Cawood, M.E., and Coetzer, J.A.W., 1990. Leukoencephalomalacia in two horses induced by oral dosing of fumonisin B_1. *Onderstepoort Journal of Veterinary Research* 57: 269-275.

Kpodo, K.A., 2001. *Fusarium species and fumonisins in corn and fermented corn products in Ghana*. PhD Thesis. University of Ghana, Legon.

Kpodo, K., Thrane, U., and Hald, B., 2000. *Fusaria* and fumonisins in corn from Ghana and their co-occurrence with aflatoxins. *International Journal of Food Microbiology* 61: 147–157.

Marasas, W.F.O., Nelson, P.E., and Toussoun, T.A., 1984. *Toxigenic Fusarium species. Identity and mycotoxicology*. The Pennsylvania State University Press, University Park, Pennsylvania.

Nelson, P.E., Toussoun, T.A., and Cook, R.J., 1981. *Fusarium: Diseases, Biology and Taxonomy*. The Pennsylvania State University Press, University Park, Pennsylvania and London.

Rheeder, J.P., Marasas, W.F.O., Thiel, P.G., Sydenham, E.W., Shephard, G.S., and Van Schalkwyk, D.J., 1992. *Fusarium moniliforme* and fumonisins in corn in relation to human esophageal cancer in Transkei. *Phytopathology* 82: 353-357.

Sydenham, E.W., Shephard, G.S., and Thiel, P.G., 1992. Liquid chromatographic determination of fumonisins B_1, B_2, and B_3 in foods and feeds. *Journal of AOAC International* 75: 313–318.

World Health Organization, 2002. *Evaluation of certain mycotoxins in food*. WHO Technical Report Series 906. 56th Report of the Joint FAO/WHO Expert Committee on Food Additives, WHO, Geneva.

Analysis of ochratoxin A in serum, urine and food consumed by inhabitants from an area with Balkan endemic nephropathy: A one month follow-up study

A. Pfohl-Leszkowicz[1], T. Vrabcheva[2], T. Petkova-Bocharova[3], L. Garren[4], F. Grosso[5], I. Nikolov[3], S. Dragacci[5], I.N. Chernozemsky[3] and M. Castegnaro[1]

[1]Unit of Toxicology ml Food Safety, Institut National Polytechnique de Toulouse, UMR CNRS N° 5503, Avenue de l'Agrobiopole, BP 107, F-31326 Auzeville-Tolosane, Toulouse, France; leszkowicz@ensat.fr
[2]National Center of Hygiene, Mycotoxicological Laboratory, 15 Dimitar Nestorov Street,1431 Sofia, Bulgaria
[3]Institute of Oncology, 6 Plovdivsko Pole, Sofia Darvenitza, Bulgaria
[4]Unit of Gene-Environment Interactions, International Agency for Research on Cancer, 150 cours Albert Thomas, 69372 Lyon Cedex 08, France
[5]Unité des Toxines Microbiennes, Agence Française de Sécurité Sanitaire des Aliments, Maison Alfort, France

Abstract

A series of publications from Bulgaria, Yugoslavia and Romania described a kidney disease occurring in geographically limited areas of these three Balkan countries. Several studies implicated mycotoxins, and particularly ochratoxin A (OTA). In Bulgaria, the exposure of populations to OTA was supported by a very high prevalence of OTA levels exceeding 2 ng/ml in the blood of the affected population. OTA has also been found more often in the urine of people living in Balkan endemic nephropathy (BEN)-endemic villages than in those in non-endemic villages, and the highest concentrations were seen in patients with BEN or urinary tract tumors. The present study was designed to confirm the exposure of a population to OTA in the Vratza district in Bulgaria and follow-up the intake of OTA during one month. Sixteen healthy (20-30 years-old) volunteers were selected from two villages considered at high risk for BEN disease and whether it was correlated to serum or urinary OTA. Samples of food were collected daily whereas samples of urine and blood were collected at the beginning and end of each week for one month. The average weekly intake of OTA varied from 130 to 14,152 ng/person, corresponding to 1.9 to 206 ng/kg body weight: the highest intake being twice the tolerable weakly intake recommended by JECFA. The OTA blood concentrations were in the same range as those reported previously in this region with highest concentrations around 10 μg/l. Weekly OTA food intake did not directly correlate with concentrations in blood and in urine but seemed to modulate them.

Keywords: ochratoxin A, Balkan endemic nephropathy, toxicokinetic, food, blood, urine

Introduction

The attention of nephrologists was attracted in the 1950s by scientific publications from Bulgaria (Tanchev *et al.*, 1956), Yugoslavia (Danilovic *et al.*, 1957) and Romania (Fortza and Negoescu, 1961) describing a kidney disease found in geographically limited areas of these three Balkan countries called Balkan Endemic Nephropathy (BEN). An association between BEN and urinary tract tumors was also recognised early on and the problem of BEN thus turned out to be not only nephrological but also oncological. Many hypotheses have been advanced including the mycotoxin one which seems the most relevant and supported by several studies in Bulgaria and former Yugoslavia (Pfohl-Leszkowicz *et al.*, 2002).

An approach for evaluation of exposure to ochratoxin A is the measurement of the concentrations of the mycotoxin in human blood. The long persistence of OTA in the body (binding to serum albumin, reabsorption after bilary excretion and/or kidney excretion), reflecting both recent and chronic exposures at the level of the individual and could be very useful in epidemiological studies. Urinary OTA analysis is relatively unexplored but presents the advantage of being less invasive than blood monitoring.

In order to confirm the exposure of the population in the Vratza district in Bulgaria to OTA and to establish if there is a correlation between OTA food intake, OTA blood concentration and urinary OTA excretion, study participants were monitored for one month following the protocol described by Gilbert *et al.* (2001).

Materials and methods

Study subjects

Sixteen 20-30 years old volunteers from two villages [Gorno Peshtene (GP) and Beli Izvor (BI)] in the Vratza district of Bulgaria, an area of high incidence of BEN were selected. Eleven volunteers were from Beli Izvor and the remainder were from Gorno Peshtene. The volunteers were healthy (normal serum and urinary formulae, without any renal, hepatic, cardiovascular or gastrointestinal disorders) and did not consume alcohol or drugs.

Collection of food

Participants were asked to eat their usual food for one month. Food samples were prepared following the duplicate diet method described by Gilbert *et al.* (2001). Participants prepared duplicate portions of each daily meal. They recorded all ingredients used for the preparation of the meal. The portion of the meal reserved for analysis was stored in a freezer until collection. The food was collected each day and stored at –20 °C. Each week the food from an individual was pooled and homogenised. Some beverages were collected separately. All homogenates were stored at –80 °C until analysis. The food samples were analysed as described by Vrabcheva *et al.*, (2004).

Blood and urine collections

About 10 ml of blood was collected from each participant on days 0, 7, 14, 21 and 28. Twenty-four-hour urine samples were collected 1 day before day 0, and on days 6, 13, 20 and 27. All samples were stored at –80 °C until analysis. The samples were analysed following the method of Petkova-Bocharova *et al.* (2003). Briefly, acidic extraction of OTA was used, followed by immunoaffinity column clean-up and detection by HPLC/spectrofluorimetry (excitation, 335 nm and emission, 465 nm).

Results and discussion

Table 1 summarises the data on OTA quantities in food, blood and urine. Ochratoxin A was detected in 50% of the food samples from GP and in 78% of the food samples from BI. Concentrations ranged from 0.07 to 0.25 μg/kg in the food samples GP and from 0.07 to 2.60 μg/kg in food samples from BI. The amounts of food consumed by an individual varied from 2,000 to 7,950 g for participants from GP and from 700 to 8,000 g for those from BI. Based on the food consumption data and determined concentrations of OTA in the food, weekly consumption of OTA by participants from GP ranged from 98 to 1,550 ng/person; and from 110 to 14,152 ng/person for those from BI. For some individuals, the weekly variations in OTA intake were very large (*i.e.* GP09, BI02, BI07, BI08), while others had a steady intake.

Ochratoxin A was detected in all serum samples. Serum quantities ranged from 0.1 to 1.6 μg/l in GP and 0.1 to 10.9 μg/l in BI. The toxin was detected in 88% of urine samples from GP ranging from 0.01 to 0.33 μg/l and in 97.6% of urine samples from

Table 1. Average and SD of weekly OTA intake, blood and urine concentrations, and OTA urinary excretion.

Averages ± SD	OTA consumed ng/week	OTA concentration in serum (ng/l)	OTA concentration in urine (ng/l)	OTA excreted in urine (ng/day)
GP 01	224 ± 184	460 ± 190	22 ± 23.9	11.8 ± 11.8
GP 02	521 ± 263	440 ± 150	16 ± 8.9	9.42 ± 5.63
GP 03	198 ± 120	540 ± 360	54 ± 23.0	23.0 ± 6.6
GP 04	582 ± 360	1460 ± 190	64 ± 36.5	39.0 ± 17,2
GP 09	369 ± 389	440 ± 290	98 ± 129.9	72.7 ± 77.9
BI 02	1968 ± 1669	260 ±150	54 ± 36.5	56.0 ± 28.8
BI 03	1220 ± 280	1700 ± 610	92 ± 43.8	66.2 ± 22.6
BI 05	321 ± 170	560 ± 310	38 ± 20.6	27.4 ± 14.3
BI 06	85 ± 93	720 ± 220	74 ± 50.8	55.1 ± 31.7
BI 07	6373 ± 6656	8360 ± 1560	860 ± 656.2	1272.4 ± 888.3
BI 08	6489 ± 4607	7700 ± 2200	455 ± 237.0	618.5 ± 304.9
BI 09	442 ± 189	540 ± 90	36 ± 31.3	32.3 ± 29.5
BI 11	305 ± 188	700 ± 220	83 ± 67.0	88.5 ± 72.4
BI 12	115 ± 30**	360 ± 210	34 ± 21.9	19.9 ± 11.9

** Based on limit of quantification.

BI (range, 0.01 to 1.91 μg/l). Based urine volume, urinary OTA excretion varied from 4.3 to 212 ng/day/person in GP and from 6.4 to 2,672 ng/day/person in BI. The serum concentration of OTA and urinary excretion were fairly constant over the five analysis for some individuals, for others, concentrations varied up to 5 fold in serum and 7 fold in urine.

When the average OTA intake, serum OTA and urinary OTA excretion are compared, there seems to be an apparent correlation between the three parameters as shown in Figure 1. Indeed for participants BI-07 and BI-08, high excretion and high blood contents correlated with high OTA intake. Because participants BI-07 and BI-08 appeared to be outliers, they were removed from the data pool, in order to better study the relation between intake, distribution and excretion. The resulting correlation pattern is presented in Figure 2.

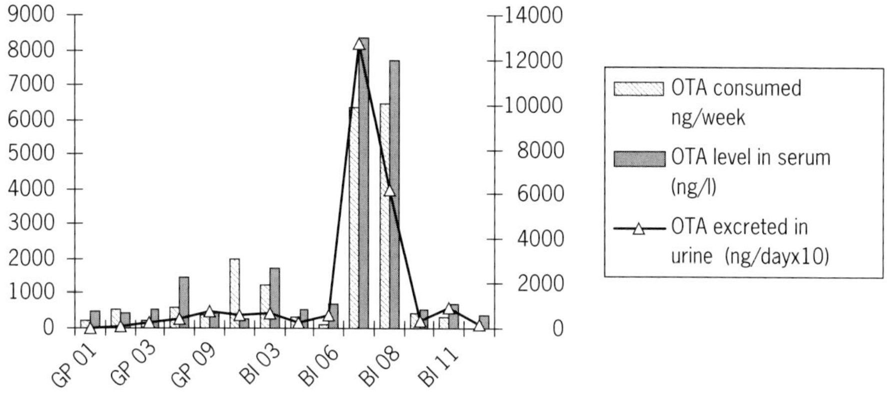

Figure 1. Overall comparison of OTA intake, serum OTA and urinary OTA excretion.

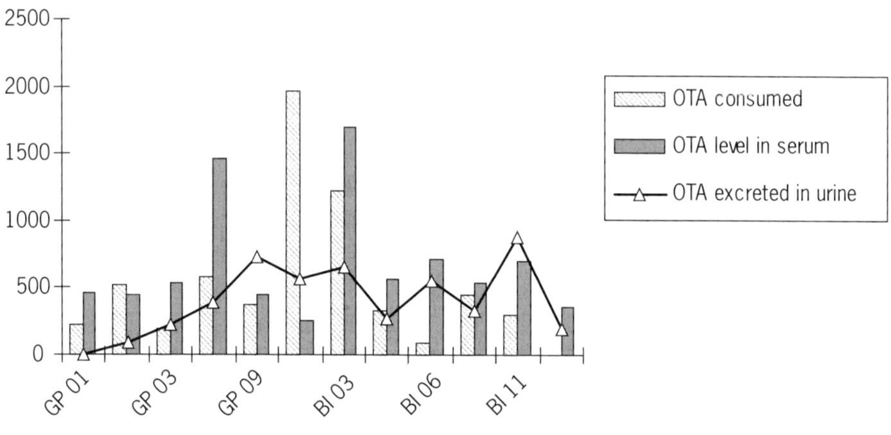

Figure 2. Comparison of OTA intake, serum OTA and urinary OTA excretion after exclusion of the two individuals with extremely high intake of OTA.

It becomes clear that high OTA intake is not reflected by high serum OTA concentration (individual BI-02). Similarly, a high serum concentration may not necessarily relate to high OTA consumption (individuals GP-04 and BI-03).

Figure 3 shows OTA profiles (OTA consumption, serum OTA and OTA excretion) of three individuals (BI-12, BI-06 and BI-02). The weekly OTA food intake of participant BI-02 was approximately 10 to 70 times higher than that of BI-06 and BI-12. However, their blood OTA levels were comparable over the one month period.

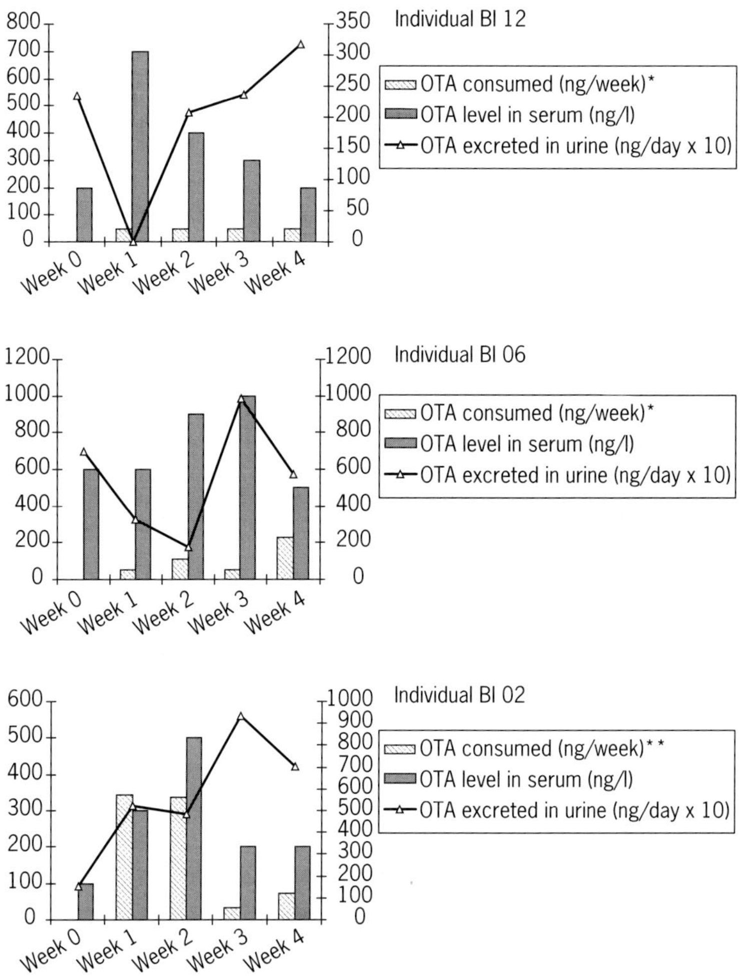

Figure 3. Comparison of OTA intake, serum OTA and urinary OTA excretion from three volunteers.
**the values in the diagram correspond to ½ the limit of quantification.*
*** the values for food OTA intake were divided by 10 to have comparable scales.*

Although BI-06 has a low OTA consumption (comparable to that of BI-12), he has a constant higher serum OTA concentration which increases slightly with decreasing urinary excretion. For all three participants, at almost constant OTA intake, a sharp increase in urinary excretion led to a drop in blood OTA levels.

The data indicate that the blood OTA level of a given individual is relatively stable over a period of one month and is regulated by urinary excretion. As long as the OTA food intake is relatively low, a variation of OTA food intake is not reflected by a variation of OTA blood concentration. This could be explained by the fact that the free fraction of toxin in plasma is less than 0.2% in all species including man (Hagelberg *et al.,* 1989) because OTA is bound to proteins and stored in tissues and not excreted. These interpretations would have been more conclusive had OTA consumption data been generated on a daily basis instead of weekly.

Conclusions

Results obtained during this study confirm previous observations that stated that, food from BEN areas can be highly contaminated by OTA and at high frequency. The highest OTA blood concentrations are around 10μg/l with 12.5% having levels of 2 μg/l or more. The high blood concentrations result in a high incidence of urinary OTA excretion, reaching up to 2 μg/l.

A comparison of the three parameters for each volunteer may help explain the OTA storage capacity for each volunteer. Ochratoxin A residence in the blood is modulated by the urinary excretion. Since the age of the volunteers (20-30) is low compared to that at which diseases are expected to develop (50-60), poor excretion may indicate early damage to kidney functions, although no "overt" sign could be discovered.

Finally it should be noted that the Provisional Tolerable Weekly Intake (PTWI) proposed by JECFA (100 ng/kg body weight) was almost attained over the one month period by two of the volunteers (BI-07, 93 ng/ kg bw and, BI-08, 91 ng/kg bw), and sometimes exceeded (206 ng/kg BW).

Acknowledgments

The authors want to thank NATO for supporting this project, the "Institut National Polytechnique de Toulouse" (INPT) for providing a fellowship to Dr T. Petkova-Bocharova to come to Toulouse, France, and Financial supports from the French Ministry of Research and Universities and the "Midi-Pyrénées Region" N° 99008345.

References

Danilović, V., Djurišić, M., Mokranjac, M., Stojimirović, B., Živijinović, J., and Stojaković, P., 1957. Chronic nephritis caused by poisoning with lead via the digestive tract (flour). *Presse Médicale* 65: 2039-2040.

Fortza, N., and Negoescu, M., 1961. Nefrita cronica azotemica endo-epidemica. *Stud. Cercet Medicine* 1: 217-221.

Gilbert, J, Brereton, P., and MacDonald, S., 2001. Assessment of dietary exposure in the UK using the duplicate diet approach and analysis of urine and plasma samples. *Food Additives and Contaminants* 18: 1088-1093.

Hagelberg, S., Hult, K., and Fuchs, R., 1989, Toxicokinetics of ochratoxin A in several species and its plasma- binding properties. *Journal of Applied Toxicolology* 9: 91-96.

Petkova-Bocharova, T., Castegnaro, M., Pfohl-Leszkowicz, A., Garren, L., Grosso, F., Nikolov, I.N., Vrabcheva, T., Dragacci, S., and Chernozemsky, I.N., 2003. Analysis of ochratoxin A in serum and urine of inhabitants from an area with Balkan endemic nephropathy: a one month follow up study *Facta Universitatis Series: Medicine and Biology* 10: 57 - 62.

Pfohl-Leszkowicz, A., Petkova-Bocharova, T., Chernozemsky, I.N., and Castegnaro, M., 2002. Balkan endemic nephropathy and the associated urinary tract tumours: review on etiological causes, potential role of mycotoxins. *Food Additive and Contaminants* 19: 282-302.

Tanchev, I., Evstatiev, Z.V., Dorosiev, G., Pencheva, Z.H., and Zvetkov, G., 1956. Studies on nephritis in the region of Vratza. *Savr. Med.* 7: 14-29 (in Bulgarian).

Vrabcheva, T., Petkova-Bocharova, T., Grosso, F., Nikolov, I.N., Chernozemsky, I.N., Castegnaro, M., and Dragacci, S., 2004. Analysis of Ochratoxin A in Foods Consumed by Inhabitants from an Area with Balkan Endemic Nephropathy: A 1 Month Follow-up Study. *Journal of Agriculture and Food Chemistry* 52: 2404-2410.

Section VI:

Management of mycotoxins

Pre-harvest forecasting of deoxynivalenol for regulatory action in wheat grain in Uruguay using readily available weather inputs

A.W. Schaafsma[1], D.C. Hooker[1], M. Pineiro[2], M. Díaz de Ackermann[3], S. Pereyra[3] and J.P. Castaño[3]

[1]*University of Guelph at Ridgetown College, Main Street East, Ridgetown, NOP 2CO Ontario, Canada; aschaafs@ridgetownc.uoguelph.ca*
[2]*Food Quality and Standards Service, Food and Nutrition Division, Food and Agriculture Organization, Rome, Italy*
[3]*Instituto Nacional de Investigación Agropecuaria, La Estanzuela, Colonia, Uruguay*

Abstract

In Uruguay, a severe *Fusarium* epidemic in wheat in 2001 resulted in concentrations of deoxynivalenol (DON) in baked goods from 1-5 mg/kg. DONcast is a prediction model used in Canada mainly to aid fungicide spraying, but was adapted to Uruguay to help target pre-harvest regulatory and marketing actions for reducing DON contamination in wheat destined for food markets. Daily temperature (T), and relative humidity (RH) were collected from four automated weather stations, and daily precipitation data were collected from 50 manual stations in the wheat growing regions in Uruguay. Temperature and RH data were interpolated for each of the manual sites. The level of DON was predicted for each day across a range of heading dates, for each weather station. Before harvest, regions were identified across Uruguay with low (<1.0 mg/kg), moderate (1.0-5.0 mg/kg), and high (>5 mg/kg) frequencies of predictions, depending on the range of heading dates within each region. DONcast was modified for unique environmental conditions in Uruguay, by accounting for RH between 3 and 10 days after heading (increases DON), precipitation and RH between 20 and 36 days after heading (increases DON), and T_{max} >32 °C between 10 and 18 days after heading (decreases DON). DON was predicted using one of two equations, depending on whether rain was recorded during heading (R^2 = 0.80) or where no rain was recorded during heading (R^2 = 0.65). Validation analysis showed that the revised model forecasted a mean of 6.1 mg/kg of DON, compared to 7.5 mg/kg measured in 20 fields, and that trends of elevated DON concentrations were identified, depending on the heading date. The results from this study suggest that DONcast is a practical tool to forecast the severity of regional DON contamination before harvesting wheat in Uruguay and other similar regions.

Keywords: deoxynivalenol, contamination, wheat, forecasting, prediction, Uruguay

Introduction

Wheat is an important crop in the southwest part of Uruguay (UY) (Figure 1). It is produced during the winter months primarily for domestic consumption through baked goods. In 2002, deoxynivalenol (DON) was found in Uruguayan market samples of wheat-based baked goods at levels exceeding current health guidelines (Miller 1994). These high levels resulted from a severe epidemic of *Fusarium graminearum* Schwabe in the 2001 Uruguayan wheat crop. The wheat crop in 2002, while not as severe as 2001, was also contaminated by DON. Recently, the Uruguayan Ministerio Salud Pública (MSP) has decreed a limit of 1 ppm in wheat flour or other products destined for human consumption. A model to forecast the level of DON at harvest could be a useful tool to help restrict the entry of DON-contaminated grain into the food supply.

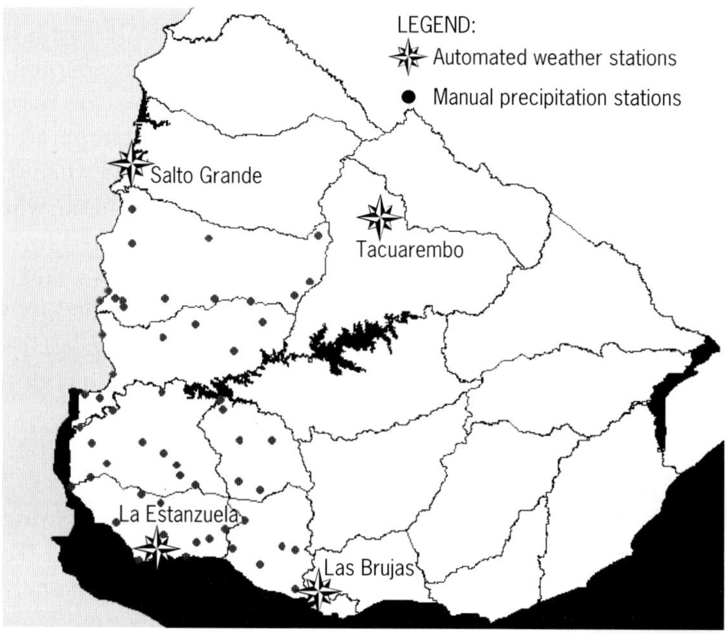

Figure 1. South-Western region (containing dots) of wheat production in Uruguay showing locations of automated weather and manual precipitation stations used to run DONcast.

The Uruguayan Government requested the Food and Agriculture Organisation of the United Nations (FAO) to help with the problem, and the Food and Nutrition Division, through its Food Quality and Standard Service, urgently approved a technical assistance project for the prevention and control of *Fusarium* (FAO TCP/URU/2801, 2002-2004 "Assistance in prevention and control of *Fusarium* and mycotoxins in grains"). The FAO was asked to review the Uruguayan chain of wheat production from farmer's field to consumer to determine how best to manage the mycotoxin contamination. The primary objective of the project was to reduce the health risk and the secondary objective was to reduce the economic impact, by developing an integrated management system from the start to the end of the wheat production and utilisation chain. The project activities included awareness raising, setting the legal basis and regulatory framework, designing an integrated system based on GAP/GMP/HACCP, training in "best practices", detection methods, and communication strategies. The outcome was a National Action Plan developed by a team comprised of international consultants, national experts, government agents and industry representatives. The DON forecasting for Uruguay was developed and validated within this project, and subsequently incorporated into the National Action Plan

DONcast is a model that has been used successfully in Ontario, Canada, as a forecast tool mainly to help with fungicide spray decisions at heading (Hooker *et al.*, 2002; Hooker and Schaafsma, 2003). It was proposed that this model might be adapted to Uruguay and become part of the integrated management strategy for DON.
The objectives for this project were: (1) to test the existing DONcast model in Uruguay using a limited number of samples from 2002 harvest, (2) to make adjustments to the model to account for discrepancies, (3) to deliver the forecast from October to November in the years 2002, 2003 during the period from wheat heading to maturity, and (4) to validate the model with additional historical DON and meteorological data collected at Estanzuela in 2000, 2001 and 2002.

Methods

Daily maximum and minimum temperatures (Tmax and Tmin) and relative humidity (RH) at 11:00:00 were collected from four automated weather stations managed by Instituto Nacional de Investigación Agropecuaria, Uruguay (INIA) (Figure 1). Daily precipitation data were collected manually by Dirección Nacional de Meteorología, Uruguay and updated every 10 days via INIA from about 50 stations in the region (Figure 1). Temperature and RH data were interpolated for each of the 50 sites where precipitation data were collected using a weighted, nearest- neighbour calculation. All weather data were sent via Internet by INIA to Ridgetown College, Canada, and were used as input variables to the 2002 version of DONcast (Hooker *et al.*, 2002). The forecast was updated approximately once per week. A predicted value for DON (mg/kg) was calculated for specific heading dates for each of the 50 weather stations. These points were then used to construct maps of varying levels using SURFER software (Golden Software Inc. http://www.goldensoftware.com). One map was created corresponding to each heading date. The date of heading was defined as 75% of the heads fully emerged from the boot. Maps were then posted on the Internet for internal use by Food and Agriculture Organisation of the United Nations (FAO), Ministerio de Ganadería, Agricultura y Pesca, Uruguay (MGAP), and INIA.

To estimate the relative severity of DON accumulation on a regional scale, MGAP and INIA were consulted to estimate the length of the heading period for each region. In each day of the heading period for each station, the DON prediction was categorised into low (<1 mg/kg), moderate (1 to 5 mg/kg), or high (>5 mg/kg), and the frequency of days were plotted on maps specific to each DON level category.

Model validation

Blocks of wheat were planted on different dates at four locations in Uruguay (Table 1) with different varieties to achieve a range of heading dates for 24 samples. Heading dates were recorded and grain samples were harvested and threshed at maturity according to Schaafsma *et al.* (2001). DON was determined for each sample at INIA according to Malone *et al.* (1998). Predicted values were then compared with observed values using simple regression.

Model modification

The model was recalculated using the same approach as Hooker *et al.* (2002), but this time using a total of 596 fields (data points), including the 2003 data from Ontario and 2002 data from Uruguay. The revised model (DONcast 2003) included an additional 30 days of weather variables after heading, which more or less corresponded to the period from heading to maturity. Relative humidity was introduced as an input variable in the new model, where 80% RH at 11:00:00 was considered a damp day. Relative humidity was estimated for all the 596 fields from nearby weather stations. Variety coefficients were applied to DON data from Uruguay to adjust for cultivars that are slightly more susceptible to DON accumulation than the average cultivars in Ontario.

Host residue on the soil surface was also considered. For Ontario, minimum or no-tillage, and maize or wheat or barley within 2 years before wheat returned the binary variable of 1 in the model, suggesting host residue was present on the soil surface in the

Table 1. Information on sample points from Uruguay in 2002 used to validate DONcast model and corresponding output for the old and revised versions of the model.

Cultivars	Location	Lat:Long	Heading date 75% spikes out	Weather Station RH	Weather Station PPT, TEMP	DON Observed (mg/kg)	Predicted DONcast v. 2002	Predicted DONcast v. 2003
I. Tijereta	Paysandú	-32,54: -57,96	12/10/02	Salto	Paysandu	3.7	4.6	9.3
E. Pelón 90	Paysandú	-32,54: -57,96	14/10/02	Salto	Paysandu	5.2	2.7	3.3
I. Churrinche	Paysandú	-32,54: -57,96	14/10/02	Salto	Paysandu	2.9	2.7	3.3
I.Tijereta (2)	Paysandú	-32,54: -57,96	19/10/02	Salto	Paysandu	3.8	1.5	6.6
I. Tijereta (3)	Paysandú (2)	-32,40: -57,97	12/10/02	Salto	Paysandu	5.3	4.6	9.3
I. Gorrión	Estanzuela	-34,33: -57,68	20/10/02	Estanzuela	Estanzuela	5.3	4.1	4.0
I. Boyero	Estanzuela	-34,33: -57,68	31/10/02	Estanzuela	Estanzuela	9.4	3.2	7.9
I. Gavilán	Estanzuela	-34,33: -57,68	22/10/02	Estanzuela	Estanzuela	10.0	2.1	5.0
I. Tijereta	Estanzuela	-34,33: -57,68	19/10/02	Estanzuela	Estanzuela	12.0	3.4	3.8
E. Pelón 90	Estanzuela	-34,33: -57,68	19/10/02	Estanzuela	Estanzuela	6.7	3.4	3.8
I.Tijereta	Young	-32,68: -57,65	23/09/02	Salto	Young	4.1	5.2	8.5
I.Tijereta	Young	-32,68: -57,65	09/10/02	Salto	Young	6.6	0.9	4.2
I. Boyero	Young	-32,68: -57,65	25/09/02	Salto	Young	11.0	4.0	9.7
I. Boyero	Young	-32,68: -57,65	04/10/02	Salto	Young	20.0	1.7	16.8
I. Boyero	Young	-32,68: -57,65	28/09/02	Salto	Young	10.0	0.3	4.6
I. Boyero	Young	-32,68: -57,65	28/09/02	Salto	Young	9.1	0.3	4.7
I. Tijereta	Young	-32,68: -57,65	13/10/02	Salto	Young	5.1	2.7	4.4
I. Tijereta	Young	-32,68: -57,65	20/10/02	Salto	Young	5.2	1.4	6.6
I. Tijereta	Estanzuela	-34,33: -57,68	26/10/02	Estanzuela	Estanzuela	3.3	0.2	1.6
I. boyero	Estanzuela	-34,33: -57,68	16/10/02	Estanzuela	Estanzuela	8.9	0.4	5.3
I. Boyero	Estanzuela	-34,33: -57,68	25/09/02	Estanzuela	Estanzuela	20.0	0.3	14.0
I. Tijereta	Estanzuela	-34,33: -57,68	30/10/02	Estanzuela	Estanzuela	2.4	0.3	3.5
I. Boyero	Estanzuela	-34,33: -57,68	04/10/02	Estanzuela	Estanzuela	6.1	0.8	3.8
I. Tijereta	Estanzuela	-34,33: -57,68	15/10/02	Estanzuela	Estanzuela	3.3	0.4	3.2
						7.5	2.1	6.1
						R-square =	0.024	0.489*

*significant at $P < 0.001$

wheat crop. For Uruguay, all fields were assumed to have high potentials for inoculum, therefore, it was assumed that all fields had sufficient residue on the soil surface.

Results and discussion

The DONcast version 2002 underestimated the amount of DON (Table 1) by about 30%; and the correlation between observed and predicted values was poor. A comparison between the weather data obtained from Uruguay and the database from Canada showed that RH was generally higher in Uruguay than in Ontario, which may be an important input-variable in the model. Furthermore, observations in both Uruguay and Ontario suggested that rain late in the season also affected the final accumulation of DON. In general, the 2002 DONcast was effective in identifying regions most seriously affected, but underestimated the magnitude of the DON problem in Uruguay.

In updated DONcast 2003, all of the variables in DONcast 2002, and their timing relative to heading, were retained in the model as important predictors of DON. The coefficients were adjusted slightly with the new data. Several other important variables were identified and included in the updated model: (1) the number of days RH> 80% at 11:00:00 between 3 and 10 days after heading (increases DON); (2) the number of days of precipitation and RH between 20 and 36 days after heading (increases DON), and (3) the number of days when Tmax >32 °C between 10 and 18 days after heading (decreases DON).

The new model (DONcast 2003) resulted in a much closer approximation of DON than DONcast 2002. For example, average DON concentrations of 6.1 mg/kg compare to 7.5 mg/kg actual DON, versus 2.1 mg/kg predicted with the earlier DONcast. The effects of RH are illustrated in Figure 2, where the models were compared with actual DON for weather data collected at Estanzuela, Uruguay, in 2002. Both outputs show similar periods of elevated DON, but the 2003 version provided better predictions of DON, for these data. The predicted versus actual data collected in 2002 from Uruguay fit nicely into the data collected in Canada (Figure 3).

The correlation between observed and expected values, for DON in field samples collected in 2003, appears less than satisfactory (Figure 4), but the means for both are relatively similar. The apparent lack of correlation appears greater at higher levels of DON. This could be a function of variation due to variety, sampling, analysis for DON, and estimation of heading date. To illustrate the sensitivity to errors in heading date, two model outputs are shown at Estanzuela for Julian day 278 (Figure 5a) and 280 (Figure 5b). The top portion of each figure shows the number of days exposed

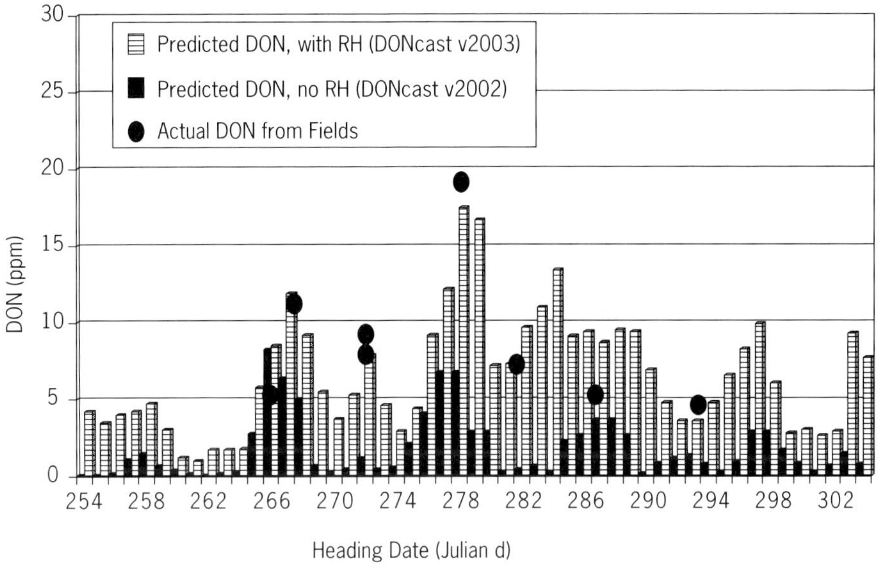

Figure 2. Predicted and actual deoxynivalenol levels (ppm = mg/kg) for a series of wheat heading dates using two versions of DONcast at La Estanzuela, UY 2002.

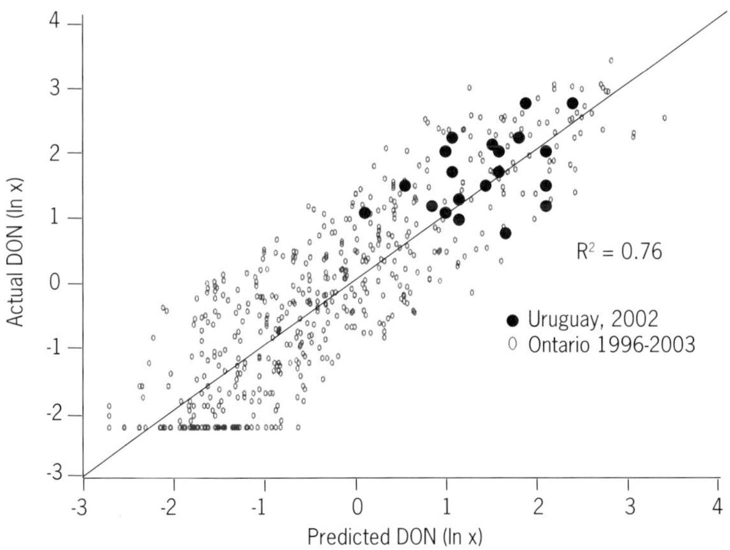

Figure 3. Scatter plot of observed versus predicted values of ln(DON) mg/kg showing how the data from Uruguay (black dots) fit into the overall data set driving DONcast version 2003.

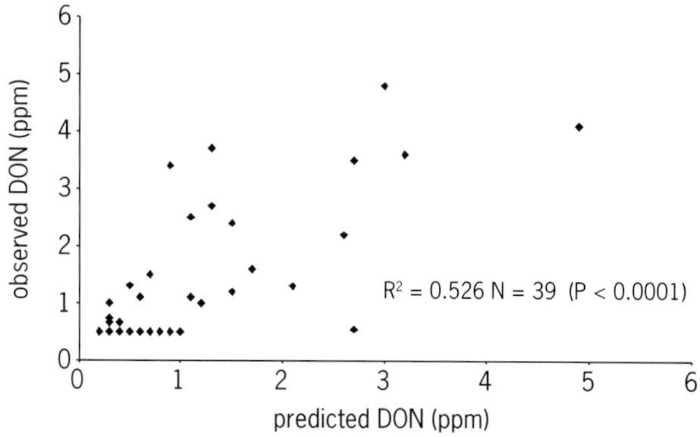

Figure 4. Scatter plot of predicted and measured DON (ppm = mg/kg) data collected from commercial scale wheat samples in Uruguay in 2003.

to critical weather factors during the three critical periods of wheat development. The bottom histogram in each figure shows the predicted level of DON at harvest. During those two days, DON predictions varied from 17.5 to 7.2 mg/kg, but more importantly, the model predicted accurately elevated DON concentrations for wheat heading in this period.

The relative severity of DON contamination in wheat fields across Uruguay was determined by plotting the frequency of wheat heading dates during the period of actual heading when DONcast predicted DON < 1 mg/kg (Figure 6a), 1-5 mg/kg (Figure 6b) and > 5 mg/kg (Figure 6c). Using these maps as a tool, regulators and grain handlers could identify regions of highly contaminated wheat, and also regions where they were likely to find wheat of good quality. It is important to note that, because of the long period of heading across regions, troublesome levels of DON could be found in some fields in all regions. Figure 6 illustrates the potential of finding good quality grain in specific regions.

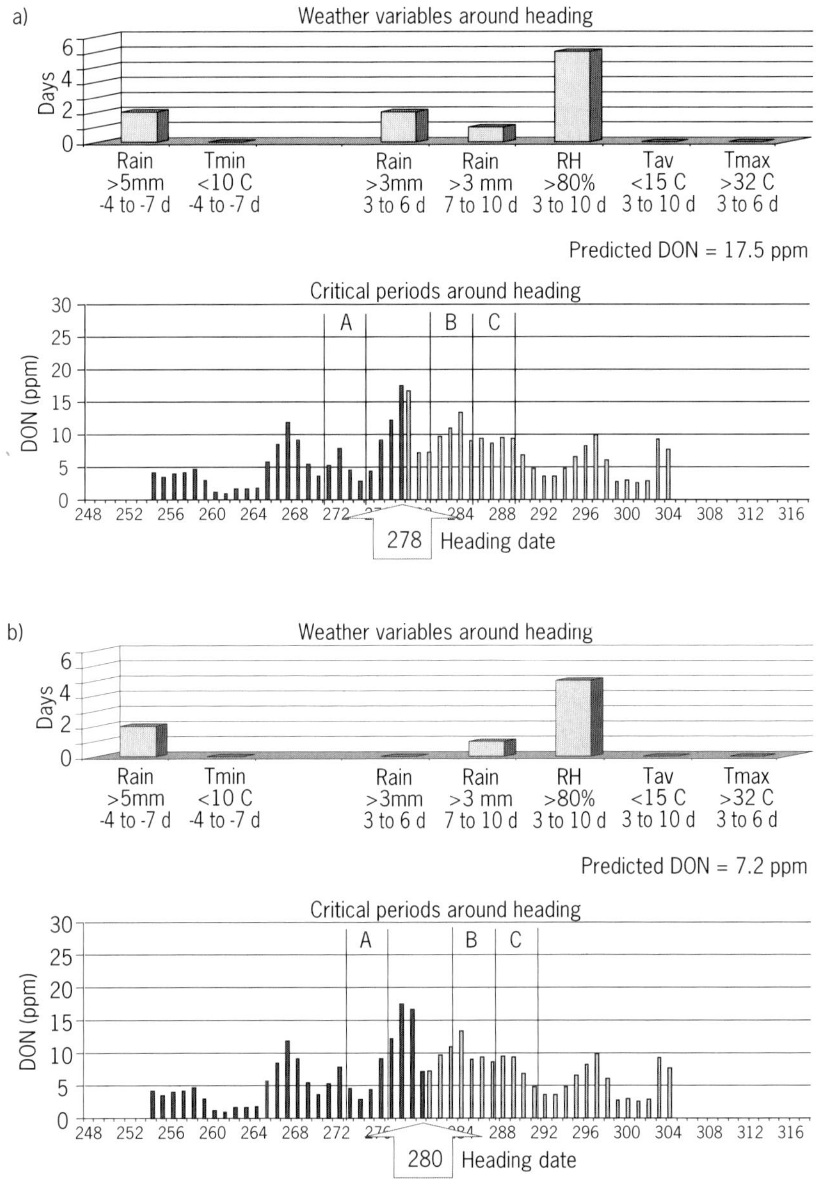

Figure 5. DONcast version 2003 inputs (top histogram show number of days meeting critical values during critical periods) and outputs [forecasted DON (mg/kg)]by heading date) for two heading dates a) Julian date 278 and b) Julian date 280 showing highly variable critical inputs and highly sensitive outputs from day to day. La Estanzuela, UY 2003.

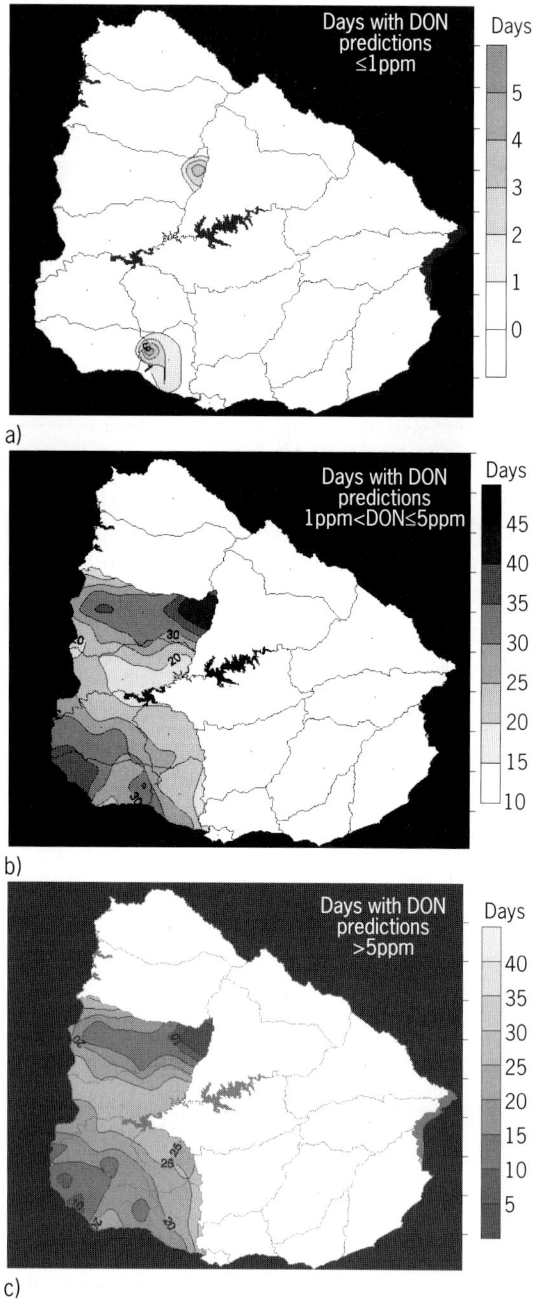

Figure 6. Frequency of wheat heading dates during the period of actual heading in Uruguay when DONcast forecasted (a) < 1 mg/kg, (b) 1- 5 mg/kg and (c) > 5 mg/kg deoxynivalenol. Uruguay 2002.

Conclusions and recommendations

Validations suggest that the DONcast version 2003 is a practical tool to forecast the relative severity of regional DON contamination before wheat harvest. This tool will be helpful to MSP and MGAP to determine marketing and regulatory actions on a regional and field-to-field basis. For example, MSP would be alerted to potential monitoring for compliance to the national decree on DON limits. MGAP can use the information to guide the industry in dealing with the post harvest handling and utilisation of contaminated grain. The pre-harvest forecast can realistically be delivered using currently accessible weather inputs with weekly updates.

DONcast may be used to identify regions of elevated DON concentrations by categorising wheat fields from low to high DON predictions, and calculating the frequency of fields within each category by region. Agronomists and grain handlers would use this information to determine specific marketing actions. It is likely more practical to use this approach, rather than integrating a wheat phenological model with DONcast, given the resources and infrastructure currently available in Uruguay.

This model has potential also to be used by wheat producers to make spray decisions at the flowering time of wheat. Because the window for fungicide application is relatively narrow around heading, the delivery of this information is much more time sensitive than predicting DON levels at harvest. Weekly updates would be insufficient for use as a tool to determine the need for fungicide application.

The wheat trade in Brazil, Uruguay and Argentina is tightly interconnected and they share the same *Fusarium* problem, with DON contamination seriously affecting the movement and pricing of grain. A regional forecast across the wheat production regions of Brazil, Uruguay, and Argentina would be useful to help manage the risk of DON contamination in food across these regions.

References

Hooker, D.C., Schaafsma, A.W., and Tamburic-Ilincic, L., 2002. Using weather variables pre- and post- heading to predict deoxynivalenol in winter wheat. *Plant Disease* 86: 611-619.

Malone, B.R., Humphrey, C.W., Romer, T.R., and Richard, J.L., 1998. One step solid-phase extraction cleanup and fluorometric analysis of deoxynivalenol in grains. *Journal AOAC International* 81: 448-451.

Miller, J.D., 1994. Epidemiology of *Fusarium* ear diseases. In: *Mycotoxins in grain: compounds other than aflatoxin.* Miller J D, Trenholm H L (Editors), Eagan Press, St. Paul, MN. pp. 19-36.

Schaafsma, A.W., Hooker, D.C., Tamburic-Ilinic, L, and Miller, J.D., 2001. Agronomic considerations for reducing deoxynivalenol content in wheat grain. *Canadian Journal of Plant Pathology* 23: 279-285.

Identification of agronomic factors that influence the level of deoxynivalenol (DON) in wheat grown in France

B. Barrier-Guillot[1], M. Delambre[2], A.Morel[1], C. Maumene[1], H. Gouet[1], F. Grosjean[1] and M. Leuillet[1]

[1]ARVALIS-Institut du végétal, Station Expérimentale, 91720 Boigneville, France; b.barrier@arvalisinstitutduvegetal.fr
[2]Union INVIVO, 83-85 Avenue de la Grande Armée, 75782 Paris Cedex 16, France

Abstract

Deoxynivalenol (DON) is the most common mycotoxin found in cereals produced in France and Europe. The study was initiated to assess the effect of different factors such as cultivation system, previous crops, varieties and fungicide application on the DON content of wheat. Between 1999 and 2002, samples were collected from 765 wheat fields and the agronomic and climatic factors pertaining to those fields recorded. The different factors involved in DON contamination were then evaluated by the segmentation method. The results showed that occurrence of DON and contamination levels were correlated with various risk factors. Conducive climate at flowering, previous cultivation of maize, no tillage and a susceptible cultivar combine to create conditions for very high DON content in a wheat crop. With a low risk previous crop, DON contamination depends on the sensitivity of cultivars against *Fusarium*. The hierarchy of the different agronomical and climatic factors has become better appreciated and could assist the wheat industry better manage potential DON contamination.

Keywords: deoxynivalenol, agronomic factors, climate, prevention, risk prediction

Introduction

The European Community intends to drastically reduce food and feedborne risks for human and animals. Consequently, it has become important for cereal producers to identify the types and content of contaminants that may be present in grains, the factors that influence their occurrence, and explore possible ways of minimising the occurrence and reducing the levels of the contaminants. This paper deals with the case of deoxynivalenol (DON) the most common mycotoxin found in cereals produced in

France and Europe. Deoxynivalenol belongs to the group called trichothecenes and is produced by some strains of *Fusarium graminearum* and *F. culmorum*; it is considered to be very stable and not substantially degraded during storage and processing, including high temperature cooking. Therefore, the best way to minimise exposure to DON and reduce the consequent risks is to prevent its formation in the field.

Materials and methods

The study was carried out from 1999 to 2002 and included a total of 765 wheat (*Triticum aestivum*) fields. It is a composite of different field trials carried out by ARVALIS-Institut du végétal and a survey of farms carried out by ARVALIS-Institut du végétal in collaboration with INVIVO cooperative union.

The field trials involved 16 different areas in France, although more than half of the trials were conducted in mainly 3 regions (Centre, Ile de France and Pays de Loire). Data included in this report is from only 28 of the field trail sites. The collaborative farm fields survey was carried out in mainly four areas in France (Bretagne, Centre, Champagne-Ardennes and Lorraine). Farm fields involved were chosen based on their agronomic characteristics: previous crop, cultivation system and wheat variety (susceptibility to *Fusarium* head blight). The survey provided 392 samples.

In either study, an assessment of the effects of different agronomic factors such as cultivation system, previous crops, wheat varieties and fungicide usage, either individually or in combination, on DON contamination of the wheat was carried out. Overall, the two studies involved a total of 765 fields (Table 1); data from all 765 fields was examined for this report.

All agronomic parameters were recorded for each site (experimental plots and farm fields) before cropping and sampling. In addition, weather variables (daily rainfall, minimum, average and maximum daily temperatures) around flowering period were estimated from nearby weather stations. The date of flowering was estimated for each site according to the date of sowing and the wheat variety based on a model established by ARVALIS-Institut du vegetal (unpublished model). Cultivar susceptibility was evaluated from an initial score of *Fusarium* susceptibility assigned at the registration of each cultivar, and the results of DON accumulation observed in previous experiments at our institute.

Three 1 kg samples were taken from each site at harvest (during emptying of the combine harvester). The 3 samples were thoroughly mixed, cleaned (laboratory cleaner mini Petkus, Rohr, Aulnay-Sous-Bois, France) to remove impurities (straw,

Table 1. Main characteristics of samples involved in the study (n=765).

French area	%	Previous crop	%	Tillage stystem	%	Varietal susceptibility	%
Centre	27.6	Maize	32.1	Ploughing	80	High	30
Bretagne	23.8	Rapeseed	23.0	No-ploughing	20	Medium	64
Champagne-Ardennes	9.5	Pea	10.6			Low	6
Ile de France	8.2	Wheat	7.5				
Lorraine	8.1	Sugar beet	6.9				
Normandie	6.2	Potato	5.5				
Pays de Loire	4.4	Sunflower	3.1				
Picardie	3.7	Ray grass	2.5				
Midi Pyrénées	3.0	Others	8.8				
Rhône-Alpes	2.2						
Others	3.3						

Different columns are independent hence blanks do not mean missing data.

glumes, small and broken grains, dusts) and reduced to 1 kg. The 1 kg sample was ground in a mill fitted with a 1 mm screen. Analysis for trichothecenes (types A and B) was carried out by gas chromatography/mass spectrometry (CEN, 2005a). The method was based on extraction with acetonitrile+water (84:16, v/v), SPE column purification (R-Biopharm® P51), derivatisation with TMSI+TMSCl (90:10, v/v) and detection by mass spectrometry. In 2002 DON was analysed by liquid chromatography/UV detection (CEN, 2005b). This method was based on extraction with polyethylene-glycol+water (5:200, v/v), immunoaffinity column purification (R-Biopharm® DON PREP) and UV detection at 218 nm. Type A trichothecenes comprise T-2 toxin, HT-2 toxin, diacetoxyscirpenol and neosolaniol and type B comprise deoxynivalenol (DON), nivalenol, fusarenon-X, 3-acetyldeoxynivalenol, 15-acetyldeoxynivalenol.

Each sample was classified according to agronomic and climatic variables inputs and DON content. A median score was calculated for each set of inputs of a variable (*e.g.* weather at flowering). Correlations between the quantitative variables and DON content were calculated. The DON content was converted into classes to become a qualitative variable. The different factors involved in DON contamination were then evaluated by the segmentation method whereby a set of qualitative variables is used to explain a specific qualitative variable. Segmentation gives rise to a tree structure which allows a hierarchical placement of the factors involved, and a ranking of the different inputs of each factor. Segmentation divides the initial population of data points at each level into two sections by placing inputs with the same DON content in the same branch of the tree structure.

Results and discussion

Type A trichothecenes were not detected in any of all the wheat samples; among type B trichothecenes, DON was detected more frequently (58%) whereas nivalenol was the least common (detected in less than 2% of trichothecene positive samples). These results contrast those recently reported by the European Commission where 21%, 12% and 14% of the wheat and wheat flour samples were positive for T-2 toxin, HT-2 toxin and nivalenol, respectively (Reports on tasks for Scientific Cooperation, SCOOP task, 2003).

Because of the lack or the very low level of occurrence of the trichothecenes other than DON in our study and the fact that the European Commission projects a regulation on *Fusarium* toxins for DON only in food and grains, we chose to analyse only DON content in 2002. Therefore, the remainder of this paper will only deal with

Table 2. Comparison of DON content (μg/kg) in wheat in from the present study to the results of the SCOOP task force for France.

	The present study	French data* (from SCOOP task, 2003)
Number of data	765	844
Median value	60	lod-190
Mean value	275	177
% ≥ 1000 μg/kg	6.9	1.3
% ≥ 1500 μg/kg	3.7	-
% ≥ 2000 μg/kg	2.5	0.1
Median value in 1999	290	-
Median value in 2000	120	190
Median value in 2001	lod**	lod
Median value in 2002	116	100

* Only data from chromatographic methods are selected.

**lod: limit of detection: 30 μg/kg.

deoxynivalenol (DON) occurrence between 1999 and 2002. The results for DON content from our study are compared to those from SCOOP Task force in Table 2. The median content *i.e.* the 50th percentile of the population was very low (60 µg/kg) and close to the limit of quantification of DON. These results are in good agreement with those reported by the SCOOP task for France for approximately the same years of 2000 to 2002. In both studies the levels were between the limit of detection and 190 µg/kg. However it is important to notice a clear year-effect. The year 2001 was characterised by very low levels of DON, with median values at the level of the limit of detection of the methods in our study as well as in the SCOOP survey, respectively. On the other hand, the years 1999, 2000 and 2002 could be characterised as the so-called "*Fusarium*-years" with higher median and mean concentrations of DON for both studies (Table 2). From our study, DON levels above 1,000 µg/kg [a possible future threshold to be adopted by the European Commission for DON regulation (European Commission, 2003)], 1,500 and 2,000 µg/kg were respectively found in 6.9, 3.7 and 2.5% of the samples tested. In the SCOOP task force report for 2003 only approximately 1% of the samples were above 1,000 µg/kg. The difference could be attributed to two major reasons. First, some of the fields in our study were intentionally selected based on the existence of the main agronomic factors known to contribute to DON accumulation. As a result, our agronomic practices population was not representative of the real agronomic practices in the country leading to an over estimation of the risk of DON contamination. Secondly, most samples collected for general studies or surveys are taken from silos or before the first processing stage of combined grains from several fields. Our samples were obtained directly from individual fields increasing the variability in our data and retaining in our data pool individual field samples with high DON content that would otherwise have been diluted due to mixing at silos or the first processing stage.

Figure 1 shows the DON content of wheat grown under three different systems (ploughing or tillage, minimum tillage and no tillage or direct drilling) on plots previously cultivated with maize and barley. The wheat was grown in 2000 and 2001, years that had different climatic conditions. The results in Figure 1 show that in addition to favourable climatic conditions, growing wheat without tillage (direct drilling) on a plot previously cultivated with maize leads to very high levels of DON in the crop. On the contrary, much lower levels of DON are detected when wheat is directly planted (direct drilling) on a plot that was previously cultivated with wheat. It is well established that crop residues play a role in the disease cycle, especially when decomposition is slow. Propagules from crop debris are the main source of initial inoculum (Xu, 2003), and populations of *Fusarium graminearum* on crop residues are affected by incorporation depth (Yi *et al.*, 2002). Furthermore, some crops are known to be serious inoculum reservoirs than others. Dill-Macky and Jones (2000),

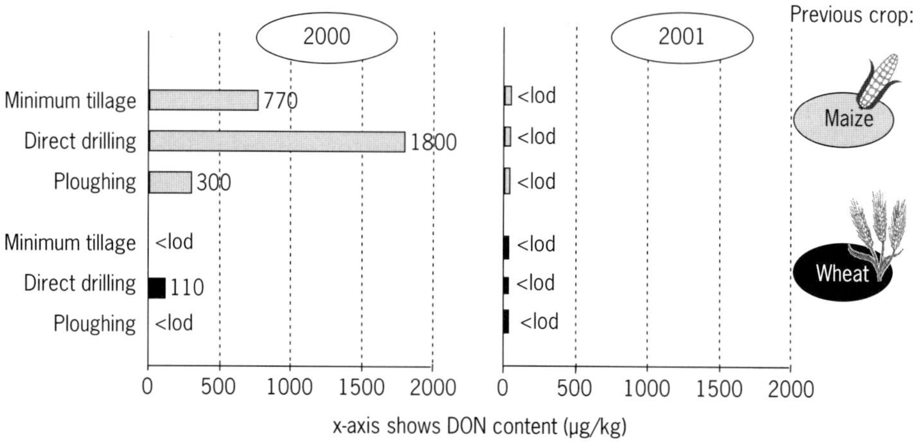

Figure 1. Effect of cultivation system and previous crop (maize or wheat) on DON content (µg/kg) in wheat grain at the Boigneville experimental station in 2000 and 2001.

Bauer, (2000) and Obst *et al.* (2000) reported that no tillage after a previous maize crop was the highest risk factor leading to high levels of DON contamination. Our results are in agreement with these previous observations.

Fusarium head blight epidemics are usually associated with abundance of inoculum. The crucial role of climatic conditions on the extent of *Fusarium* infection and subsequent DON formation is illustrated in Figure 1. The results show that regardless of the level of inoculum available, favourable weather conditions are critical for infection to occur in wheat heads, an observation also reported by Hooker *et al.*, (2002). Recent investigations on understanding *Fusarium* head blight development and mycotoxin production has been based on field observations (Obst *et al.*, 2000; Schaasma *et al.*, 2001; Hooker *et al.*, 2002; Xu, 2003), and these field-based studies have confirmed that temperature and moist conditions during heading and anthesis are the key factors.

After combining all agronomic parameters and climatic variables, a tabular description of six risk classes derived from the tree structure segmentation is shown in Table 3. Figure 2 shows a distribution of DON contamination within and between the risk classes. The level of risk increases in severity from risk class 1 to risk class 6. Our results confirm the importance of climate in the model used particularly around the flowering period. Unfortunately, the actual model developed is, for now, considered proprietary information and cannot be disclosed. Without any climatic risk, the

Table 3. Classification of risks and their component risk variables used in the model of prediction of risk of DON in wheat.

Risk class	Weather at flowering	Previous crop	Cultivation system	Variety susceptibility	Median level (µg/kg)
Risk 1	No risk	No risk	All systems	All	30 (=lod)
Risk 2	No risk	High risk	All systems	All	30 (=lod)
Risk 3	Risk	No risk	All systems	Low, medium	60
Risk 4	Risk	No risk	All systems	High	120
Risk 5	Risk	High risk	Ploughing	All	315
Risk 6	Risk	High risk	No ploughing	All	1220

Each risk is a combination of the different risk variable. Risk 1 represents the following situation: no risk for weather, no risk for previous crop, all cultivation systems, and all varieties.

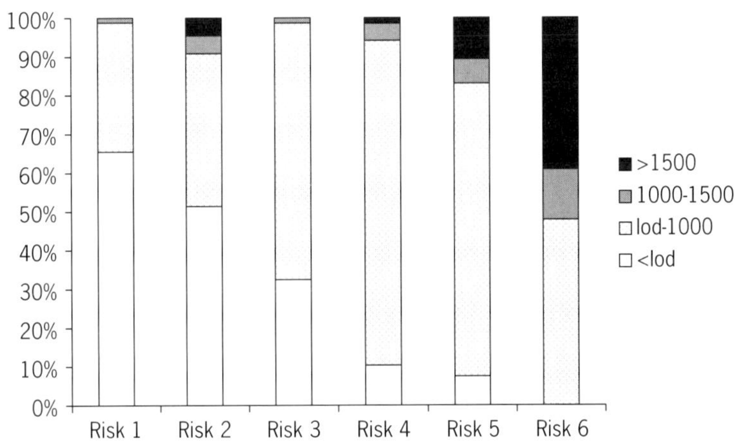

Figure 2. Allotment of DON results according to the 6 risks of the model.

model leads to very low occurrence of DON as shown in risk classes 1 and 2. Most samples (66%) that fell in risk class 1 were negative (below the limit of detection) for DON. The presence of a previous high risk crop such as maize slightly increases the risk of DON contamination as shown in risk class 2 where a few samples had DON contamination above 1,000 µg/kg (Figure 2).

The combination of risky (*Fusarium* favourable) climatic conditions, a high risk previous crop (*e.g.* maize) and a no tillage system (risk class 6) leads to the highest DON content with 52% of sample population contaminated at > 1,000 µg/kg (a possible threshold for future European regulation) and a median of 1,220 µg/kg. Variables falling into risk class six, abundant inoculum (high risk previous crop without ploughing) and favourable weather conditions around the time of flowering, occur at between 1 and 3% the wheat growing locations in France. Reducing the availability of crop residues at the soil surface by ploughing under greatly reduces DON contamination (risk class 5). Only 11% of the population of samples in this class was above 1,000 µg/kg, and the median contamination was of 335 µg/kg. Finally, with a low risk previous crop, DON contamination will depend on the sensitivity of the cultivars against *Fusarium*. Cultivars resistance against *Fusarium* head blight will moderate DON accumulation even when climatic and inoculum risks are high (risk class 3).

Conclusion

The inferences from this study need to be understood in the context of some weaknesses in the study design. Our data per crop was limited and the crops were only present in our database a year. More importantly, our database is skewed towards years and/or sites with high DON content. However, we have contributed to a better appreciation of the hierarchy of the different agronomical and climatic factors. It is clear that the presence of DON in wheat grain cannot be completely avoided with preventive measures due to the critical influence of weather conditions at the time of flowering, although further study is needed to ascertain this interaction. Overall, even models like the one derived from this study may not predict DON level in wheat grain field by field with absolute accuracy, they have a general usefulness to the wheat industry for forecasting potential DON occurrence.

References

Bauer, G., 2000. Zur Analyse der Daten des Fuarium-Monitorings Bayern. In: *Risiken durch den Ährenparasiten Fusarium graminearum – Ergebnisse eines LPB –Forschungsverbunds.* Bodenkultur und Planzenbau, Schriftenreihe der Bayerische Landesanstalt für Bodenkultur und Pflanzenbau, pp. 33-37.

CEN, 2005a. European Committee for Standardisation. *Food Products - Determination of type B trichothecenes in wheat and corn –Method by GC.* CEN TC/275/WG5 Working Document N 341.doc STD version 2.2. Nederlands Normalisatie-instituut (NEN), Delft, the Netherlands, 10 pp.

CEN, 2005b. European Committee for Standardisation. *Food Products - Determination of deoxynivalenol in wheat and corn – Immunoaffinity clean-up and HPLC with UV detection method.* CEN TC/275/WG5 Working Document N 344.doc STD version 2.2. Nederlands Normalisatie-instituut (NEN), Delft, the Netherlands, 10 pp.

Dill-Macky, R., and Jones R.K., 2000. The effect of previous crop residues and tillage on Fusarium head blight of wheat. *Plant disease* 84: 71-76.

European Commission, 2003. Consultation paper on *Fusarium*-toxins. European Commission, Health and Consumer Protection DG, October 2003, pp. 1-6.

Hooker, D.C., Schaafsma, A.W., and Tamburic-Ilincic, L., 2002. Using weather variables pre- and post-heading to predict deoxynivalenol content in winter wheat. *Plant disease* 86: 611-619.

Obst, A., Lepschy, J., Beck, R., Bauer, G, and Bechtel, A., 2000. The risk of toxins by Fusarium graminearum in wheat – interactions between weather and agronomic factors. *Mycotoxin research* 16A/1: 16-20.

Schaafsma, A.W., Tamburic-Ilincic, L., Miller, J.D., and Hooker, D.C., 2001. Agronomic considerations for reducing deoxynivalenol in wheat grain. *Canadian Journal of Plant Pathology* 23: 279-285.

SCOOP task, 2003. Collection of occurrence data of *Fusarium* toxins in food and assessment of dietary intake by the population of EU member states. Final Report SCOOP Task 3.2.10. Part A: Trichothecenes. European Commission, Health and Consumer Protection DG, October 2003, pp. 1-237. http://europa.eu.int/comm/food/fs/scoop/task3210.pdf.

Xu, X., 2003. Effects of environmental conditions on the development of *Fusarium* ear blight. *European Journal of Plant Pathology* 109: 683-689.

Yi, C., Kaul, H.-P., Klüber, E., and Aufhammer, W., 2002. Populations of *Fusarium graminearum* on crop residues as affected by incorporation depth, nitrogen and fungicide application. *Journal of Plant Diseases and Protection* 109: 252-263.

Potential bio-control of patulin producing *Penicillium expansum* in post-harvest fruits using antagonistic yeasts

A.R. Coelho[1], R.M. Levy[1], F.L. Hoffmann[2], M.H. Taniwaki[3], C. Kmelmmeier[4], F.C. Pagnocca[5] and E.Y. Hirooka[1]

[1]Dept Food and Drug Technology, State University of Londrina, P.O. Box 6001, 86051-990, Londrina, Parana, Brazil; hirooka@uel.br
[2]Dept Food Engineering and Technology, UNESP-Sao Jose do Rio Preto, Sao Paulo, Brazil
[3]Institute of Food Technology, P.O. Box 139, 13073-001, Campinas, Sao Paulo, Brazil
[4]Dept. Biochemistry, State University of Maringa, Parana, Brazil;
[5]UNESP-Rio Claro, Sao Paulo, Brazil

Abstract

The use of innocuous yeasts as bio-control agents to prevent *Penicillium expansum* contamination of apples and subsequent production of patulin can enhance refrigerated storage and has less environmental impact than conventional fungicides. The antagonistic potential of yeasts of natural origin against a toxigenic *P. expansum* strain was evaluated. In a preliminary assessment, *Pichia membranifaciens* E.C. Hansen and *Sporobolomyces roseus* Kluyver & van Niel, showed antagonistic/biodegradation properties which decreased patulin in broth from 5.9 mg/l to 2.9 mg/l. Further investigations were performed with 17 yeast isolates from various sources including maize silage, papaya, and grape fruit. *Debaryomyces hansenii* var. *hansenii* strain C1 from maize silage showed nutrient competition antagonism in its inhibition of the growth of *P. expansum* strain #2 on solid media after 120 hr incubation at 21 °C, while *D. hansenii* strain C7 displayed a antibiosis/parasitism pattern of antagonism. In nutrient broth *P. ohmeri* and *C. guilliermondii* P3. *P. ohmeri* inhibited conidia germination and mycelial growth the most after 12 hours at 21 °C. The data indicates a potential application of these yeasts in post-harvest control of spoilage/mycotoxigenic fungi in fruits.

Keywords: yeast, antagonism, patulin, *P. expansum*, fruits

Introduction

Environmental factors and mechanical damage during harvest and storage could trigger fungal deterioration in post-harvest fruits (Arras *et al.*, 1998; Hussein and Brasel, 2001). Mycotoxins are a food safety concern for international agribusiness, where Brazilian fresh fruit have been introduced as promising natural foods (WHO-IARC, 1996; FDA, 2002).

Chemical fungicides control post-harvest diseases in fruits, but progressive loss of effectiveness, resistant pathogens and the potential for agrotoxic residues (Sugar and Spotts, 1999; Castoria *et al.*, 2001) renders them less favourable. Biological control by antagonistic micro-organisms has emerged as a promising alternative that is safe for both human health and the ecosystem (Janisiewicz *et al.*, 2000; Usall *et al.*, 2001). Recent emphasis has been placed on the use of yeasts that are generally regarded as safe antagonists of low ecological impact and reliable in refrigerated storage of fruit (Usall *et al.*, 2001).

Penicillium expansum is the predominant post-harvest fungus in apples and produces patulin, which has been shown to cause gastrointestinal distress and neurotoxic effects in rodents, immunotoxic effects in mice and rabbits, and genotoxic effects in mammalian cells (Hopkins, 1993; Sharma, 1993; Wouters *et al.*, 1996; Moss and Long, 2002). Prevention of *P. expansum* infection has been reported to prolong the quality of apples, pears, grapes and other fruits (Lima *et al.*, 1998, Janisiewicz *et al.*, 2000; Fan and Tian, 2001; Castoria *et al.*, 2001). *Saccharomyces cerevisiae* has been reported to degrade patulin by a fermentative process that yields E-ascladiol (Sumbu *et al.*, 1983; Moss and Long, 2002).

In a recent study, we reported the antagonism of *Pichia membranifaciens* and *Sporobolomyces roseus* strains against *P. expansum*, as a decrease in patulin production *in vitro* (Levy *et al.*, 2000, Levy *et al.*, 2002). In this study, we evaluated and report the antagonism of natural origin yeasts against *P. expansum* by antibiosis.

Materials and methods

Selection and culturing of yeast strains

Forty-six yeast isolates with anti-*Penicillium* activity were screened. The isolates were from different sources (26 strains from apples, 7 from papaya, 5 from grape, 5 from pear, and 1 each from maize, maize silage and an anthill). A toxigenic (59 μg/ml patulin) single spored *P. expansum* strain #2 from apples, with an optimum growing temperature range of 7 to 21 °C), was selected as the test patulin producing organism. The selected strain #2 of *P. expansum* was subcultured each six months on Potato Dextrose Agar (PDA, 21 °C/120 hr) and stored at 4 °C.

Promising yeast strains were identified according Kurtzman and Fell (1998), and biochemically characterised using API 20 C AUX test (BioMerieux-Vitek, Hazelwood, MO). The yeast strains were subcultured each six months on sloped Potato Dextrose Agar (PDA, 21 °C/48 hr) and stored at 4 °C. The *killer* phenotype was characterised using *killer* sensitive strains of *Candida glabrata* NCYC 366, *Candida glabrata* NCYC 388, *Candida albicans* 12A, *Saccharomyces cerevisiae* Meyen ex Hansen NCYC 1006 and *Pichia kluyveri* CAY 270 (Polloneli *et al.*, 1983).

*Determination of anti-*P. expansum *activity of yeast isolates on solid media*

Yeast strains for antifungal assay were reactivated by shaking the culture in Yeast Medium broth (YM - 2% glucose, 0,5% yeast extract, 1% NaCl, 0,23% NaH_2PO_4, 0,5% $(NH_4)_2SO_4$) at 150 rpm and 25°C for 24 hr.

P. expansum strain #2 (10^5 spores) was pour-plated with 20 ml of Yeast Medium containing 1.7% agar and allowed to solidify. 100 µl of yeast suspension (3.0 x 10^6 cells) was introduced into an 8 mm central well in the YM-agar of each of three plates. The plates were incubated at 21 °C for 120 hr, and the fungal growth inhibition zone was expressed in mm diameter.

*Determination of anti-*P. expansum *activity of yeast isolates in liquid media*

Yeasts were reactivated by shaking YM broth cultures at 150 rpm and 25 °C for 24 hr. 100 µl yeast cell suspension (3.0 x 10^6 cells) was inoculated in each of five Erlenmeyer flasks containing 25 ml of YM broth, and incubated at 21 °C for various time periods under static conditions. After 24, 48, 72, 96 and 120 hrs incubation, one culture flask was centrifuged at 6,500 x g for 15 min and filter sterilised (Millipore 0.20 µm). Then 1.0 ml of sterile supernatant was added to each of three 120 x 13 mm culture tubes containing 1.0 ml of YM broth. The contents of each tube were inoculated with 10^5 spores of *P. expansum* strain #2. The tubes were incubated at 21°C for 120 hr. Percentage conidia germination (per 100 cells) was determined microscopically and fungal growth was determined by measuring hyphal length (average of 40 cells in µm).

Patulin production, degradation and analysis

Patulin was produced by growing *Penicillium expansum* strain #2 in ten Erlenmeyer flasks containing 25 ml of Malt Extract Broth (MEB) (Oxoid Brasilia Ltda, Sao Paulo, SP) for 15 days at 21 °C. Patulin was extracted by agitating a 1:1 culture broth to ethyl acetate mixture for 30 min. The extract was filtered through a bed of sodium sulfate, dried at 50 °C, and the pellet resuspended in chloroform. Identification and quantification of patulin was achieved by thin layer chromatography (TLC plates, MERCK Art. 5721, Darmstadt, Germany), using toluene+ethyl acetate+90% formic acid (50:40:10) as developing solvent (Scott *et al.*, 1970). The plates were treated with 0.5% 3-methyl-2-benzothialinone HCL in aqueous solution (Roth, Karlsruhe, Germany), heated at 130 °C for 15 min and analysed at 365 nm.

To assess patulin degradation, 588.4 µg of patulin was added to 25 ml of Malt Extract Broth (MEB) at pH 4.42. The broth was inoculated with a yeast strain (3.0 x 10^6 cells) and the flasks incubated at 25 °C for 15 days. Patulin residues were determined according to AOAC-Official method 974.18 (2000).

Statistical analysis

The data was analysed by the Tukey test using ANOVA/MANOVA program (STATISTICA version 5.0, 1995).

Results and discussion

*Anti-*P. expansum *activity of yeast isolates on solid media*

Conidia germination and inhibition of *P. expansum* strain #2 growth were expressed as percent inhibition of conidia germination or hyphal growth. Figure 1 shows anti-*P. expansum* activity of two selected yeast strains, *Candida guilliermondii* strain P3 and *Pichia ohmeri.* Over a 120 hour incubation period *Candida guilliermondii* strain P3 and *Pichia ohmeri*) showed higher inhibitory activity of *P. expansum*, measured in terms of hyphal growth and spore germination, when the supernatants were produced at 48 and 72 hrs. *C. guilliermondii* - P3 was more effective against conidia germination (~58% inhibition), while *P. ohmeri* showed better inhibition of micelial growth (~66% inhibition). The differences in the effects on different forms of *P. expansum* shown by the two yeasts strains suggested involvement of two different mechanism of action.

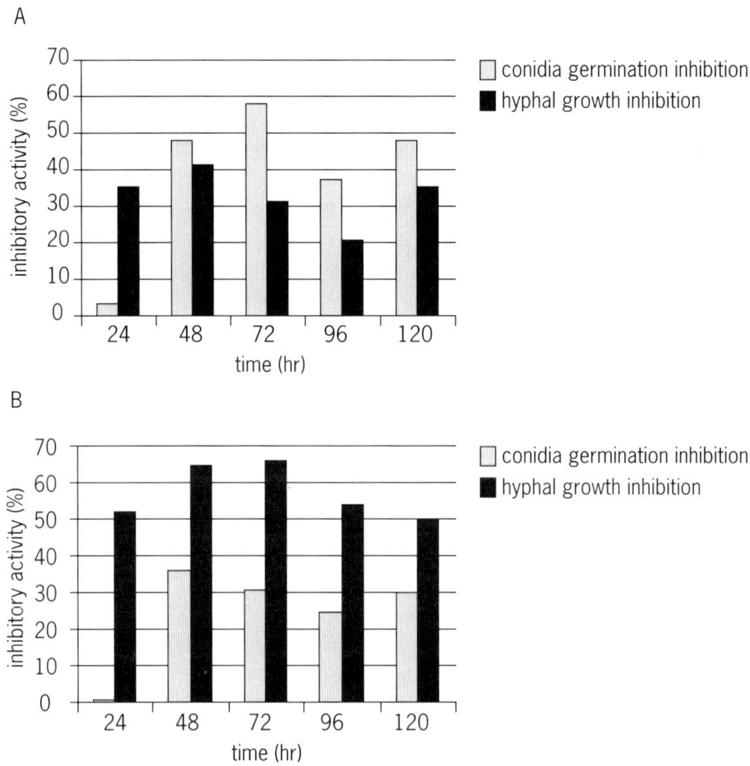

Figure 1. Percent inhibitory activity of culture supernatant of Candida guilliermondii *P3 (A) and* Pichia ohmeri *(B) against* P. expansum *(10^5 spores) germination and hyphal growth.*

*Anti-*P. expansum *activity of yeast isolates in liquid media*

The anti-*P. expansum* activity of the yeast isolates was presumed to be based on antibiosis. To verify this assumption, the activity of these yeasts was tested by examining the effect of the yeast culture supernatant in YM broth on *P. expansum* strain #2 (10^5 spores). Seventeen of the 46 strains were determined to be active against *P. expansum*. Seven of the active strains were isolated from maize silage (*Debaryomyces hansenii* var. *hansenii* strain C1 to C7), 1 from an anthill (*Pichia ohmeri*), 4 from papaya (*Candida guilliermondii* strains P1 to P4) and 5 from grape fruits (*Kloeckera* spp. strain G1; *Rhodotorula mucilaginosa* strain G2 to G5). Among the active isolates, *Candida guilliermondii* strain P3 and *Pichia ohmeri* showed higher antagonism by antibiosis, indicating probable production of effective extracellular compounds. Table 1 shows

Table 1. Inhibition of P. expansum *mycelial growth and conidial germination by culture supernatants of* Candida guilliermondii *strain P3 and* Pichia ohmeri.

Incubation (hours)	Mycelial growth (µm)		Conidial germination (%)	
	Control	Supernatant	Control	Supernatant
Candida guilliermondii				
24	93,10 ± 29,97^{bB}	60,32 ± 24,86^{bA}	69,50 ± 6,45^{aA}	67,38 ± 6,23^{bA}
48	91,40 ± 29,77^{bB}	53,51 ± 23,89^{abA}	81,17 ± 5,27^{abB}	42,00 ± 21,10^{aA}
72	76,42 ± 16,55^{aB}	52,51 ± 21,43^{aA}	84,83 ± 7,83^{bB}	35,50 ± 21,21^{aA}
96	95,97 ± 26,29^{bB}	76,17 ± 19,20^{cA}	85,17 ± 5,42^{bB}	53,25 ± 1,71^{abA}
120	94,19 ± 34,90^{bB}	60,63 ± 24,27^{bA}	86,50 ± 3,11^{bB}	44,75 ± 13,68^{abA}
Pichia ohmeri				
24	149,55 ± 58,57^{aB}	72,09 ± 33,80^{bA}	73,67 ± 9,16^{aA}	73,17 ± 8,79^{bA}
48	148,45 ± 58,00^{aB}	52,90 ± 24,21^{aA}	71,67 ± 5,68^{aB}	45,75 ± 6,82^{aA}
72	164,27 ± 59,64^{abB}	55,58 ± 30,47^{aA}	75,67 ± 7,81^{aB}	52,58 ± 10,41^{aA}
96	175,77 ± 57,04^{bB}	81,06 ± 45,40^{cA}	80,33 ± 11,50^{aB}	60,25 ± 17,44^{abA}
120	165,98 ± 62,41^{abB}	83,12 ± 43,68^{cA}	79,00 ± 6,90^{aB}	55,42 ± 17,13^{aA}

Each value is the mean of 120 data points for mycelial growth and 12 data points for conidial germination. Similar lower case letters in the same column indicate lack of statistically significant difference at p = 0.05 by Tukey´s test.; Similar upper case letters in the same row indicate lack of statistical significance at p = 0.05 by Tukey´s test.

the effects of *Candida guilliermondii* strain P3 and *Pichia ohmeri* on mycelial growth and conidial germination in liquid broth. The data indicates that optimal inhibitory activity is achieved within the first 24 hours.

Microscopic analysis showed that *D. hansenii* var. *hansenii* strain C7 consistently attached onto cell walls of *P. expansum* hyphae. Attachment of yeast cells on mycelium surface of the pathogen, followed by subsequent changes in hyphae viability suggested that the activity of the yeast was also based on hyperparasitism. This observation is in agreement with that reported in two previous studies. Arras *et al.* (1998) reported the attachment of *P. guilliermondii* onto *Penicillium italicum* mycelium, and Wisniewski *et al.* (1991), also reported attachment of *P. guilliermondii* onto *Botrytis cinerea* mycelium.

Patulin production and degradation

The 46 yeast isolates identified through the initial screening showed inhibitory activity against *P. expansum* strains based mainly on nutrient competition and hyperparasitism. Two strains, *Pichia membranifaciens* E.C. Hansen and *Sporobolomyces roseus* Kluyver & van Niel decreased the patulin level in broth by approximately 50% (Table 2). Moss and Long (2002) also showed a decrease in patulin level by *Saccharomyces cerevisiae* in a fermentative process of citron, transforming this toxin to E-ascladiol. The biodegradation effectiveness of patulin was lower in shaken flask cultures.

In the United States, *Candida oleophila* Montrocher (Aspire; Ecogen, Inc., Langhorne, PA) and *Pseudomonas syringae* (Bio-Save 110; EcoScience, Worcester, MA), whose mechanisms of inhibition are based on hyperparasitism (Sugar and Spotts, 1999), have already been commercialised as post-harvest bio-fungicides for apples.

The yeasts *P. ohmeri* and *C. guilliermondii* strain P3 have been shown to be positive for *killer* activity against *Saccharomyces cerevisiae* NCYC 1006 (National Collection of Yeast Cultures, Institute of Food Research, Colney, Norwich, United Kingdom). Among the 46 yeasts showing anti-*Penicillium* activity, 19 were identified as *killer* positive, suggesting the involvement of an exogenous compound. Walker *et al.* (1995) also reported susceptibility of yeasts and filamentous fungi in the presence of *killer* positive strains of *S. cerevisiae, P. anomala* and *Williopsis mrakii.* The *killer* phenotype is based on secretion of a low molecular mass protein or glycoprotein toxin capable of killing sensitive cells of the same or related yeast genera without direct cell-cell contact (Makower and Bevan, 1963). Previous studies indicated that the *S. cerevisiae* toxin was a protein which binds to a receptor on cell walls of sensitive yeasts, disrupting the electrochemical gradient across the cell membrane and hence the intracellular ionic balance (De La Pena *et al.*, 1981).

Table 2. Degradation of patulin by the yeasts Pichia membranifaciens *and* Sporobolomyces roseus *that also show anti-*Penicillium *activity (25 °C/15 days).*

Micro-organisms	Initial Patulin µg/25 ml	Initial pH	Final Patulin µg/25 ml	Final pH
Candida sake	588.4	4.42	294.2 ± 0.6	3.95
Pichia membranifaciens	588.4	4.42	290.0 ± 0.6	3.90
Sporobolomyces roseus	588.4	4.42	290.0 ± 0.9	3.62
Control	588.4	4.42	588.4	4.42

* Each value is the mean of triplicate determinations.

Indeed, the exogenous compound type of activity shown by *P. ohmeri* and *C. guilliermondii* strain P3 against *P. expansum* isolate #2 (antagonism by antibiosis) suggest a mechanism associated with the *killer* phenomenon. If validated through further study, use of *P. ohmeri* or *C. guilliermondii* holds promise as a bio-control tool against patulin producing *P. expansum*. The use of non-deleterious micro-organisms from the natural ecosystem, in particular non-mycotoxigenic yeasts, would be a safe choice of biological control to avoid the spoilage and/or mycotoxin production in fruits by post-harvest fungi.

Acknowledgments

CNPq, CAPES and Araucaria Foundation supported this research.

References

AOAC, 2000. Association of Official Analytical Chemists. Official method 974.18. Patulin in apple juice – Thin-layer chromatographic method. Official methods of analysis. Washington DC. *J. AOAC* 57: 61.

Arras, G., De Cicco, V., Arru, S., and Lima, G., 1998. Biocontrol by yeasts of blue mould of citrus fruits and the mode of action of an isolate of *Pichia gulliermondii. Journal of Horticultural Science & Biotechnology* 73: 413-418.

Castoria, R., De Curtis, F., Lima, G., Caputo, L., Pacifico, S., and De Cicco, V., 2001. *Aureobasidium pullulans* (LS-30) an antagonist of postharvest pathogens of fruits: study on its modes of action. *Postharvest Biology and Technology* 22: 7-17.

De la Peña, P., Barros, F., Gascón, S., Lazo, P.S., and Ramos, S., 1981. Effect of yeast killer toxin on sensitive cells of *Saccharomyces cerevisiae. The Journal of Biological Chemistry* 256: 10420-10425.

Fan, Q., and Tian, S., 2001. Postharvest biological control of grey mold and blue mold on apple by *Cryptococcus albidus* (Saito) Skinner. *Postharvest Biology and Technology* 21: 341-350.

FDA, 2002. Food and Drug Administration. In: http://www.cfsan.fda.gov/~dms/opa-micr.html (www.fda.gov/).

Hopkins, J., 1993. The toxicological hazards of patulin. *Food Chemistry Toxicology* 31: 455-459.

Hussein, H.S., and Brasel, J.M., 2001. Toxicity, metabolism, and impact of mycotoxins on humans and animals. *Toxicology* 167: 101-134.

Janisiewicz, W.J., 1996. Biological control of pathogens affecting temperate fruit crops. In: *V Sincobiol. Proceeding of V Sincobiol*, pp. 43-48.

Janisiewicz, W.J., Tworkoski, T.J., and Sharer, C., 2000. Characterizing the mechanism of biological control of postharvest diseases on fruits with a simple method to study competition for nutrients. *Phytopathology* 90: 1196-1200.

Kurtzman, C.P., and Fell, J.W., 1998. *The Yeasts, a Taxonomic Study*. 4 ed., 1055 pp.

Levy, R.M., Silva, R.S.F., Pagnocca, F.C., and Hirooka, E.Y., 2000. Factorial assay of inhibition activity of yeasts against *Penicillium* in apple fruits. *Brazilian Journal of Food Technology* 3: 145-150.

Levy, R.M., Hayashi, L., Carreiro, S.C., Pagnocca, F.C., and Hirooka, E.Y., 2002. Inhibition of mycotoxigenic *Penicillium* sp and patulin biodegradation by yeast strains. *Revista Brasileira de Armazenamento* 27: 41-47.

Lezi Sumbu, Z., Thonart, P., and Bechet, J., 1983. Action of patulin on a yeast. *Applied and Environmental Microbiology* 45: 110-115.

Lima, G., De Curtis, F., Castoria, R., and De Cicco, V., 1998. Activity of the yeasts *Cryptococcus laurentii* and *Rhodotorula glutinis* against postharvest rots on different fruits. *Biocontrol Science and Technology* 8: 257-267.

Makower, M., and Bevan, E.A., 1963. The physiological basis of the killer character in yeast. In: Genetics Today. *Proceedings of XI International Congress of Genetics*, pp. 202-203.

Moss, M.O., and Long, M.T., 2002. Fate of patulin in the presence of the yeast *Saccharomyces cerevisiae. Food Additives and Contaminants* 19: 387-399.

Polonelli, L.; Archibussi, C.; Sestito, M., and Morace, G., 1983. Killer system: a simple method for differentiating *Candida albicans* strains. *Journal of Clinical Microbiology*, 17: 774-780.

Scott, P.M.; Lawrence, J.W., and Van Walbeek, W., 1970. Detection of mycotoxin by thin-layer chromatography: application to screening of fungal extracts. *Journal of Apllied Microbiology* 20: 839-842.

Singh, J., and Faull, J., 1988. Antagonism and biological control. In: *Biocontrol of plant disease*, vol. 1, Nalini, N., and Mikerji, K. (Eds), pp. 167-169.

Sharma, R.P., 1993. Immunotoxicity of Mycotoxins. *Journal of Dairy Science* 76: 892-897.

Sugar, D., and Spotts, R.A., 1999. Control of postharvest decay in pear by four laboratory-grown yeasts and two registered biocontrol products. *Plant Disease* 83: 155-158.

Usall, J., Teixido, N., Torres, R., De Eribe, X.O., and Vinas, I., 2001. Pilot tests of *Candida sake* (CPA-1) applications to control postharvest blue mold on apple fruit. *Postharvest Biology and Technology* 21: 147-156.

Walker, G., Mcleod, A., and Hodgson, V., 1995. Interactions between killer yeast and pathogenic fungi. *FEMS Microbiology* 127: 213-222.

Wilson, C.L., and Wisniewski, M.E., 1994. *Biological control of postharvest plant diseases of fruits and vegetables: theory and practice.* Boca Raton: CRC Press, 465 pp.

Wisniewski, M., Biles, C., Droby, S., McLaughin, R., Wilson, C., and Chalutz, E., 1991. Mode of action of the postharvest biocontrol yeast, *Pichia guilliermondii*. I. Characterization of attachment to *Botrytis cinerea. Physiological and Molecular Plant Pathology* 39: 245-258.

WHO-IARC, 1996. World Health Organization-International Agency for Research on Cancer. Evaluation of certain food additives and contaminants. Forty-fourth report of the joint FAO/WHO Expert Committee on Food Additives. *WHO Technical Report Series* 859. Geneva: World Health Organization.

Wouters, M.F.A., and Speijers, G.J.A., 1996. Patulin. In: *Toxicological evaluation of certain food additives and contaminants.* World Health Organization, Geneva, pp. 337-402.

Methyl jasmonate effect on aflatoxin B_1 production by olives inoculated with *A. parasiticus* is concentration-dependent

P. Markaki, K. Velivassaki, D. Giannitsis and D. Galanopoulou

Laboratory of Food Chemistry, Department of Chemistry, University of Athens, Panepistimioupolis Zografou, 157 71 Athens, Greece; markaki@chem.uoa.gr; medgraph@otenet.gr

Abstract

Aflatoxin B_1 (AFB_1) is a carcinogenic and genotoxic metabolite produced by certain *Aspergillus* species on agricultural commodities. The biosynthesis of AFB_1 is affected by factors such as fatty acid hydroperoxides and the methylester of jasmonic acid methyl jasmonate (MeJA), a plant growth regulator. Fungal infection of a plant could trigger production of methyl jasmonate by the plant which may in turn influence the production of aflatoxin. The aim of this study was to investigate the effect of MeJA on AFB_1 production in Greek black olives inoculated with *A. parasiticus*. Addition of MeJA (1.51 mg/ g olives) stimulated AFB_1 production in both *A. parasiticus* inoculated and non-inoculated olives. When 0.015 mg MEJA/g olives was used, an inhibition of AFB_1 production in both types of olives was observed. These observations indicate the possibility of differing, perhaps concentration-dependent, mechanisms of the effect of MeJA on aflatoxin biosynthesis.

Keywords: aflatoxin B_1, *A. parasiticus*, methyl jasmonate, olives

Introduction

Aflatoxin B_1 (AFB_1) is a carcinogenic metabolite produced by certain *Aspergillus* species on agricultural commodities (Payne, 1992). AFB_1 biosynthesis is affected by lipid oxidation, and interactions with linolenic and linoleic acids and their metabolites (De Luca *et al.*, 1995). Linolenic and linoleic acids can be oxidized by plant-stress response lipoxygenase (LOX) (Gardner, 1995). Lipoxygenase isozymes catalyse the formation of hydroperoxy fatty acids which can be further metabolised by other plant enzymes to give a variety of jasmonic acid derivatives including methyl jasmonate (MeJA). Methyl jasmonate induces the expression of genes encoding for defensive proteins or certain secondary metabolites (Wasternack and Parthier, 1997). Therefore, a relationship between plant LOX-derived metabolites and aflatoxin production is possible, since fungal attack could initiate MeJA production by the plant.

Lipoxygenase activity has been detected in olives and found to contribute to desirable organoleptic properties (Williams *et al.*, 2000). In addition, it has been shown that LOX supports growth of molds and aflatoxin production (Eltem, 1996). This study investigated the effect of different MeJA concentrations on AFB_1 production in damaged black Greek olives inoculated with *A. parasiticus.*

Materials and methods

Preparation of spore inoculum

The aflatoxigenic strain *A. parasiticus* Speare (IMI 283883) was obtained from the International Mycological Institute (Surrey, United Kingdom). A stock culture was maintained on a slant of Czapek Dox Agar (CZA) (Merck, Damstadt, Germany) at pH 6.0-6.5, and 5 °C. A spore inoculum was prepared by growing *A. parasiticus* on CZA for 7 days at 30 °C. The spores were harvested aseptically using 10 ml of sterile 0.01% (v/v) Tween 80 solution. The spore suspension was centrifuged (1000 x g, 10 min) and resuspended in 10 ml sterile 0.05% Tween 80 solution twice to minimise AFB_1 carryover from the initial growth media. A serial dilution (1/10, $1/10^2$, $1/10^3$, $1/10^4$) of the spore preparation with 0.05% Tween 80 was prepared according to the method of Vergopoulou *et al.* (2001). Spore concentration was determined by the Surface Spread Plate technique using 0.1 ml of each dilution on four *Aspergillus flavus parasiticus* Agar (AFPA) plates and incubating at 30 °C for 2 days.

Olive treatment

Black "manaki" table olives from southern Greece were purchased from the local market. The olives were washed with water, cut into pieces and surface-sterilised by dipping and shaking in 70% ethanol for 1 min, washing with 1.25% NaOCl for 3 min and rinsing with sterile distilled water twice (Scott and Zummo, 1995). Fifteen gram portions of olives were aseptically placed in sterile flasks.

Inoculation and MeJA addition

For each day of observation, 24 flasks containing 15 g of olives were randomly assigned to eight groups (3 flasks/group) as shown in Table 1. All flasks were incubated under stationary conditions at 30 °C. On days 0, 3, 7, 9, 12 and 15, a set of 24 flasks was autoclaved (115 °C, 21 psi, 30 min) and the AFB_1 content of each flask measured.

Table 1. Experimental design.

MeJA (mg/flask)	0 (ethanol only)		0.0022		0.227		22.67	
MeJA (mg/g olives)	0		0.00015		0.015		1.51	
Group	1	2	3	4	5	6	7	8
Flasks	3	3	3	3	3	3	3	3
A. parasiticus (10^2 spores)	+	-	+	-	+	-	+	-

AFB_1 determination

The entire contents of a flask of olives were shaken for 10 min with 30 ml methanol-water (80:20). After filtration, 1 ml filtrate was mixed with 10 ml distilled water and the mixture transferred onto an Aflaprep® immunoaffinity column (Rhone Diagnostics, Glasgow). The column was washed twice with 10 ml distilled water (flow rate 6 ml/min) and allowed to dry by passing air through it. AFB_1 was eluted with 2 ml acetonitrile (flow rate 0.3 ml/min). The eluate was analysed immediately or stored at -20 °C but analysed within 24 hr.

Derivatisation and high performance liquid chromatography (HPLC) analysis

The eluate was evaporated to dryness on a water bath under a gentle stream of nitrogen. 200 µl of hexane and 200 µl of trifluoroacetic acid were added and the mixture heated at 40 °C in a water bath for 10 min. Residual derivatisation solution was evaporated to dryness under nitrogen. The dry derivatised residue was dissolved in 0.2 ml water+acetonitrile (9:1) and 40 µl analysed by HPLC using a Hewlett-Packard 1050 (Waldborn, Germany) equipped with a JASCO FP-920 fluorescence detector (CO Ltd, Japan) and an HP 3395 integrator. The HPLC column was a C_{18} Nova-Pak (Waters, 60 Å, 4 µm, 4.6x250 mm^2). The mobile phase was 0.45 µm filtered water+acetonitrile+methanol (360:72:54) and the flow rate was 1 ml/min. Detection of the hemiacetal derivative was carried out at λ_{ex}=364 and λ_{em}=425 nm.

Results and discussion

AFB_1 standard was purchased from Sigma (St. Louis, Missouri, USA). A Calibration curve of standard AFB_1, with satisfactory linearity (y=106974.2x-256663.0, R=0.9997), was established in the range 1.0 - 10.0 ng/ml in acetonitrile+water (9:1). A similarly satisfactory linearity (y=0.942x-0.117, R=0.998), was established with spiked olives.

The mean recovery was 94% (CV =% 6.1) and the limit of detection, based on a signal-to-noise ratio of 3:1 was calculated to be 0.15 ngAFB_1/g of olives.

All olive samples showed an initial (Day 0) residual contamination as shown in Table 2. Overall, at 0-0.015 mg MeJA /g olives, the AFB_1 levels did not substantially differ between Days 0 and 15 in fungi-inoculated and non-inoculated olives, except for a noted substantial transient increase on Days 9 and 12 in fungi-inoculated olives without MeJA. At 1.51 mg MeJA/g olives, AFB_1 content increased between Days 3 and 12 in both fungi-inoculated and non-inoculated olives; the increase was more pronounced in non-inoculated olives. The increase in aflatoxin levels in non-inoculated olives in the presence of 1.51 mg MeJA/g olives may be attributed to residual native aflatoxigenic fungal activity and perhaps, a concentration-dependent effect of methyl jasmonate. Presence of residual fungi could have resulted from the mild sterilisation employed that may not have destroyed all the native fungi. A more efficient sterilisation process could not be used because it would have dramatically affected the texture of the olives rendering them unsuitable for consumption. However, the foregoing may not explain why the aflatoxin levels were substantially higher in non-inoculated olives than in inoculated olives, or whether the postulated concentration-dependent effect of methyl jasmonate triggered the metabolic activity

Table 2. Aflatoxin B_1 (ng/flask) in damaged olives after addition of methyl jasmonate (MeJA) and tinoculation with Aspergillus parasiticus *and incubated over 15 days.*

MeJA(mg/g olives)	**Treatment**	**Day**					
		0	**3**	**7**	**9**	**12**	**15**
0	A	2.46± 0.13*	2.54 ±0.18	2.67±0.25	3.74± 1.51	2.81± 0.42	2.39 ±0.09
	B	2.52± 0.22	2.64± 0.30	2.91± 0.55	5.34± 0.42	4.32± 0.13	2.65± 0.29
0.0002	A	2.37 ±0.10	2.29 ±0.03	2.41± 0.14	3.12± 0.43	5.33 ±1.65	3.29 ±0.12
	B	2.55± 0.22	2.41 ±0.08	2.42± 0.16	3.51± 0.29	2.37 ±0.06	2.53 ±0.26
0.015	A	2.55± 0.16	2.65 ±0.417	2.68± 0.06	2.84 ±0.25	2.35 ±0.05	2.31± 0.08
	B	2.74± 0.08	2.85 ±0.25	3.01± 0.06	2.95 ±0.02	2.36± 0.04	2.26 ±0.30
1.51	A	2.34± 0.08	14.84 ±1.97	13.14± 0.11	8.21±0.11	8.42 ±0.04	5.73 ±0.04
	B	2.44± 0.14	5.03± 0.56	6.77± 2.36	6.22± 0.03	5.36± 0.04	2.57± 0.53

Treatment A = Olives not inoculated with *A. parasiticus*.
Treatment B = Olives inoculated with *A. parasiticus*.
*Standard deviation.

of native aflatoxigenic fungi more than that of the combination of native fungi and the introduced *A. parasiticus*. The AFB_1 content of olives in inoculated and non-inoculated olives declined to near baseline on Day 15 (Figure 1). The decline in aflatoxin content may be attributed to *in vivo* degradation of aflatoxin by the fungi (Smith and Harran, 1993).

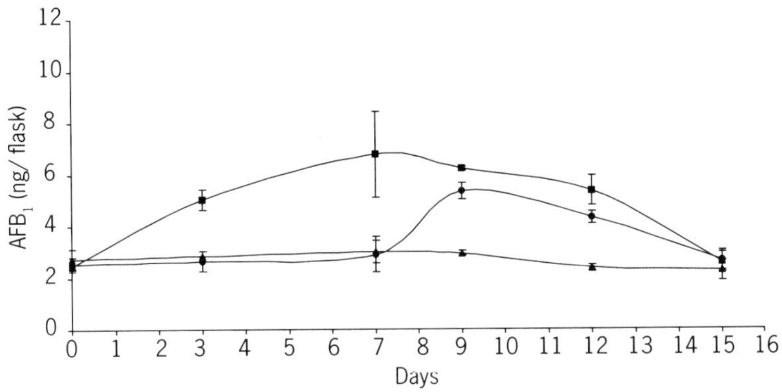

Figure 1. MeJA effect on AFB_1 production in olives inoculated with A. parasiticus *is concentration dependent. (a) (—▲—): 0.227 mg/flask MeJA added; there is a significant decrease of AFB_1 production after day 9 of incubation. (b) (—■—): 22.67 mg/flask MeJA added; there is a significant increase of AFB_1 between days 3 and 9 of incubation. (c) (—●—): Control olives. The effects were studied in three separate experiments with triplicate flasks. AFB_1 was determined after derivatisation to AFB_{2a}.*

Acknowledgements

This work was supported in part by a University of Athens, Special Account for Research grant (70/4/2508).

References

Burow, G.B., Nerbitt, T.C., Dunlap, J., and Keller, N.P., 1997. Seed lipoxygenase products modulate *Aspergillus* mycotoxin biosynthesis. *Molecular Plant-Microbe Interactions* 10: 380-387.

Daradimos, E., Markaki, P., and Koupparis, M., 2000. Evaluation and validation of two fluorometric HPLC methods for the determination of aflatoxin B_1 in olive oil. *Food Additives and Contaminants* 17: 65-73.

De Luca, C., Passi, S., Fabbri, A.A., and Fanelli, C., 1995. Ergosterol oxidation may be considered a signal for fungal growth and aflatoxin production in *A. parasiticus*. *Food Additives and Contaminants* 12: 445-450.

Eltem, R., 1996. Growth and aflatoxin B_1 production on olives and olive paste by molds isolated from Turkish-style natural black olives in brine. *International Journal of Food Microbiolology* 32: 217-223.

Gardner, H.W., 1995. Biological roles and biochemistry of the lipoxygenase pathway. *HortScience* 30:197-205.

Gonzalez-Aguilar, G.A., Buta, J.G., and Wang, C.Y., 2001. Methyl jasmonate reduces chilling injury symptoms and enhances colour development of "kent" mangoes. *Journal of the Science of Food and Agriculture* 81: 1244-1249.

Leontopoulos, D., Siafaka A., and Markaki, P., 2003. Black olives as substrate for *Aspergillus parasiticus* growth and AFB_1 production. *Food Microbiology* 20: 119-126.

Payne, G.A., 1992. Aflatoxin in maize. *Critical Revues in Plant Science* 10: 423-440

Samane, S., Tantaoui-Elaraki, A., and Essadaoui, M., 1991. Mycoflora of moroccan «Greek style» black olives II toxigenesis. *Microbiologie Alimentation Nutrition* 9: 335-352.

Scott, G.E., and Zummo, N., 1995. Size of maize sample needed to determine percent kernel infection by *Aspergillus flavus*. *Plant Disease* 79: 861-864.

Smith, J.E., and Harran, G., 1993. Microbial degradation of mycotoxins. *International Biodeterioration & Biodegradation* 32: 205-211.

Stubblefield, R.D., 1987. Optimum conditions for formation of aflatoxin-M_1-trifluoroacetic acid derivative. *Journal of the Association of Official Analytical Chemists* 70: 1047-1049.

Vergopoulou, S., Galanopoulou, D. & Markaki, P., 2001. Methyl jasmonate stimulates aflatoxin B_1 biosynthesis by *Aspergillus paraciticus*. *Journal of Agricultural and Food Chemistry,* 49(7): 3494-3498.

Wasternack, C. & Parthier, B., 1997. Jasmonate-signalled plant gene expression. *Trends in Plant Science,* 2: 302-307.

Williams, M., Salas, J. J., Sanchez, J. & Harwood, J. L., 2000. Lipoxygenase pathway in olive callus cultures. *Phytochemistry* 53: 13-19.

Yu, J,, Chang, P., Cary, J. W., Wright, M., Bhatnagar, D., Cleveland, T. E., Payne, G. A. and Linz, J. E., 1995. Comparative mapping of aflatoxin pathway gene clusters in *Aspergillus parasiticus* and *Aspergillus flavus*. *Applied Environmental Microbiology,* 61: 2365-2371.

Prevention of mould growth on black olives by heat treatment

D. Heperkan[1], F.K. Guler[1], G. Dalkılıc[1], H. Heperkan[2] and P.C. Vasavada[3]

[1]Istanbul Technical University, Chemical and Metallurgical Engineering Faculty, Department of Food Engineering, 34469 Maslak, Istanbul, Turkey; heperkan@itu.edu.tr
[2]Yıldız Technical University, Faculty of Mechanical Engineering, Besiktas-Istanbul, Turkey
[3]University of Wisconsin, Department of Animal and Food Science, River Falls, WI, USA

Abstract

The reduction of microbial load in olives is significant in terms of food safety and shelf life. The processes applied in olive production can vary from country to country and according to the type of olives. During the conventional production process of olives, a thin layer formed mainly by moulds such as *Penicillium* and *Aspergillus* is observed on the surface of the brine. Two *Penicillium* species, *P. citrinum* and *P. crustosum,* and two *Aspergillus* species, *A. niger* and *A. versicolor* were inoculated on black olives and after incubation, microwave-treated for different time periods (30, 60, 120, 180, 240 and 300 seconds). Comparison of mould counts before and after each treatment suggested that microwave treatment for 180, 240 and 300 seconds was effective in reducing the mould count and increasing shelf life. The highest inhibition for all mould strains was observed at 300 seconds.

Keywords: olives, mould, *Penicillium, Aspergillus,* microwave, prevention

Introduction

The reduction of microbial load in olives is significant in terms of food safety and shelf life. The processes applied in black olives production can vary according to the country of origin and type of olives. In Turkey, fermented olives are one of the most important products; table olive production makes approximately 30% of the total olive production. Black olives are sold on the market in different ways. Olives may be pasteurised in cans or jars at 70 – 80 °C as whole or sliced pieces. Alternatively, they may be sold directly after fermentation. It is customary among consumers to keep the olives in cold/hot water or even boil them, to reduce the bitter taste. They are then treated with different spices for flavour. Mould growth, which might cause spoilage and mycotoxin formation, can occur on the olives both during the fermentation and consumption stages.

Mould growth can cause softening of the olive tissue, and a mouldy taste and appearance, thus reducing the quality of the olives, shorten the shelf life as well as produce mycotoxins. The mycotoxins penicillic acid (Gourama and Bullerman, 1987), citrinin (Oral and Heperkan, 1999) and the aflatoxins (Yassa *et al.*, 1994) have been isolated from olives, and olive paste (Gourama and Bullerman, 1987) and olive oil (Toussaint *et al.*, 1997; Daradimos *et al.*, 2000).

Penicillium and *Aspergillus* are two commonly occurring genera of moulds on black olives (Fernandez *et al.*, 1997). During fermentation *Penicillium citrinum* and *P. crustosum* may grow on the surface of the olives (Oral and Heperkan, 1999). Mould growth may also be visually seen on one end of the black olives sold in transparent packages. In a recent study (Heperkan *et al.*, 2006), 46% of market-stage olives showed mould growth, mostly frequently *P. roqueforti*, *P. crustosum* and *P. viridicatum*.

Microwave energy is widely used in the food industry as well as in scientific research and medicine (Stolle and Schalch, 2000). Although microwave ovens as house appliances were first manufactured in 1955, their widespread usage only dates back to after 1984 (Decareau, 1985; Regier and Schubert, 2001). Several researchers have investigated the effect of microwaves on micro-organisms. Initially, it was proposed that the inactivation of the micro-organisms was due to non-thermal effects (Khalil and Villota, 1989; Kozempel *et al.*, 1998). Recent studies have shown that in the absence of heat, microwave energy did not inactivate micro-organisms; they suggest that microwave energy may complement or magnify thermal effects (Mertens and Knorr, 1992; Fellows, 2000; Kozempel *et al.*, 2000; Datta and Davidson 2001). The main objective of this study was to investigate the inactivation of common spoilage moulds in black olives by microwave treatment.

Materials and methods

Preparation of spore suspension

Four different strains of moulds: *Penicillium citrinum* CBS 252.55, *Penicillium crustosum* CBS 548.77, *Aspergillus niger* (11700) and *Aspergillus versicolor* (15825) were used in the study. The moulds were inoculated on Malt Extract Agar and incubated at 25 °C for 7 days. Spores were harvested using sterile 0.1% peptone water. Spores were counted using a thoma slide, and a spore suspension of 10^7/ml prepared.

Canned black olives (40±0.5 g) without seeds were placed in airtight containers and inoculated with 5 ml of a 10^7/ml spore suspension. An additional 5 ml of sterile distilled water was added and the contents allowed to incubate at 25 °C for 7 days.

Microwave treatment

Sterile distilled water (250 ml) at pH 6.5±0.2 was added aseptically to the incubation jars on day 7. The jars were placed at the centre of a rotating glass plate of a conventional microwave oven operating at 2450 MHz (144 Watt, Arçelik ARMD 560, Arçelik A.Ş., Bolu). The jars were microwaved for 30, 60, 90, 120, 180, 240 and 300 seconds. The microwaved jars were rapidly cooled in an ice-water bath. The microwave treatment was performed in triplicate.

Sample preparation, plating and enumeration

After cooling, the samples were homogenised for 2 min in a Stomacher (Stomacher 400, Seward Medical, London) and serially diluted using 0.1% peptone water with 0.5% Tween 80. Mould counts were enumerated using three different techniques: 3M Petrifilm (3M Microbiology Products, St. Paul), spread plating on Malt Extract Agar (MEA) (Pitt and Hocking, 1997) and Dichloran 18% Glycerol Agar (DG18) (Samson *et al.*, 2000). 3M petrifilms were incubated at 25 °C for 3 days and MEA and DG18 plates were incubated for 5 days.

Statistical analysis

Variances were analysed and compared by a general linear model procedure ($P<0.05$) with SPSS 10.0 for Windows statistical software (SPSS Inc., Illinois). Duncan's Multiple Range Test was used to determine significant differences ($P=0.05$) between treatment means.

Results and discussion

There were no differences in mould counts between 30 and 180 seconds. Table 1 shows the observed decrease in viable *P. citrinum, P. crustosum, A. versicolor* and *A. niger* spores as a result of microwave treatments beyond 180 seconds. Prolonged microwave treatment linearly reduced *P. citrinum* counts on all three types of growth media. A Duncan's Multiple Range test showed a significant difference ($p=0.05$) between treatment times 180, 240 and 300 seconds. An analysis of variance (ANOVA) of the effect of the type of culture media on *P. citrinum* showed a statistically significant ($p<0.05$) difference between the three media types. *P. citrinum* counts obtained with

Table 1. Effect of microwave treatment time on reduction of moulds.

Media and mould species	The number of moulds (log_{10}) Before treatment	180s After treatment	180s Difference[1]	240s After treatment	240s Difference	300s After treatment	300s Difference
P. citrinum							
MEA	6.93	2.72	4.21	1.43	5.50	0.73	6.20
DG18	6.93	2.47	4.46	1.53	5.40	0	6.93(0)
3M Petrifilm	6.93	1.05	5.88	0.53	6.40	0	6.93(0)
P. crustosum							
MEA	6.93	1.98	4.95	1.79	5.14	1.26	5.67
DG18	6.93	1.98	4.95	1.70	5.23	1.68	5.25
3M Petrifilm	6.93	0	6.93(0)	0	6.93(0)	0	6.93(0)
A. versicolor							
MEA	6.93	1.95	4.98	1.68	5.25	0	6.93(0)
DG18	6.93	2.25	4.68	2.17	4.76	0	6.93(0)
3M Petrifilm	6.93	0	6.93(0)	0	6.93(0)	0	6.93(0)
A. niger							
MEA	6.93	1.99	4.94	1.68	5.25	1.62	5.31
DG18	6.93	2.01	4.92	1.40	5.53	1.07	5.86
3M Petrifilm	6.93	0.72	6.21	0.63	6.30	0.81	6.12

[1]Difference (First inhibition): reduction in mould count after microwave treatment (dead + injured).

3M Petrifilm were lower compared to the other media. This was perhaps a result of the composition of 3M Petrifilm media.

Microwave treatment of *P. crustosum* and *A. versicolor* did not significantly decrease counts with time beyond 180 seconds. Statistically, there was a significant difference ($p<0.05$) in counts on MEA between 240 and 300 seconds for *P. crustosum* but not between 180 and 300 seconds on DG18 media and 3M Petrifilm. No viable mould counts were determined with 3M Petri film after 180 seconds. *A. versicolor* counts were comparable between 180 and 240 seconds but significantly decreased between 240 and 300 seconds on MEA and DG18. Similarly, there was no difference between viable counts obtained from all three media types beyond 180 seconds for *A. niger*.

Overall, viable counts obtained using the 3M Petrifilm were considerably lower than the ones obtained with MEA and DG18 media.

The results show that microwave treatment for 180 seconds yields a substantial reduction in mould counts of black olives. Prolonging microwave treatment to 300 seconds may afford more reductions in mould counts of *P. citrinum*, *P. crustosum* and *A. versicolor* but not *A. niger*. Reduction in the mould counts were between 4 and 7 logs depending on the treatment time. Few other studies have investigated the effects of microwaves on moulds. More *et al.*, (1992) found that 60 seconds microwave treatment of sorghum grain killed *Aspergillus* and *Penicillium* spp. In a study to determine the effect of microwave treatment on the growth of fungi on angel cakes, Zhang and Jianshao (2002), found no fungi for 60 days on cakes subjected to 2 minute microwave treatment and stored at a constant temp of 28 °C. On the contrary, several studies have investigated the effect of microwave heating on bacteria such as *Salmonella* and *Listeria* (Heddleson *et al.*, 1996), *Staphylococcus aureus* (Heddleson *et al.*, 1996; Yeo *et al.*, 1999), *E. coli* (Tajchakavit *et al.*, 1998) and *Clostridium sporogenes* (Welt *et al.*, 1994).

Conclusion

The utilisation of microwave ovens in the food and other industries is increasing because they reduce treatment time, energy consumption and labour. The rapid heating effect of microwaves is an indispensable advantage in today's technology. In addition, the temperature and treatment time can be adjusted easily. The results of this study are easily adaptable at the industrial and household level. Olives can be protected from most mould growth and the shelf life extended if they are treated in a microwave oven for about 5 minutes in a glass container filled with water just enough to cover the olives.

References

Daradimos, E., Markaki, P., and Koupparis, M., 2000. Evaluation and validation of two fluorometric HPLC methods for the determination of aflatoxin B1 in olive oil. *Food Additives and Contaminants* 17: 65-73.

Datta, A.K., and Davidson, P.M., 2001. Microwave and radiofrequency processing. *Journal Food Science Special Supplement*: Kinetics of microbial inactivation for alternative food processing technologies 1: 1-4, 32-41.

Decareau, R.V., 1985. Pasteurization and sterilization. In: *Microwaves in the Food Processing Industry*. Food Science and Technology Series, Schweigert, B.S. (Ed.), Academic Press, Florida, pp.182-202

Fellows, P., 2000. Dielectric, ohmic and infrared heating. In: *Food Processing Technology*. 2nd ed. Fellows, P. (Ed.) CRC Press, N.Y. pp. 365-384.

Fernandez, G.A., Diez, M.J.F., and Adams, M.R. (Eds.)., 1997. *Table Olives Production and Processing*. Chapman Hall, London.

Gourama, H., and Bullerman, L.B., 1987. Effect of oleuropein on growth and aflatoxin production by *Aspergillus parasiticus. Lebensmittel Wissenschaft und Technologie* 20: 226-228.

Heddleson, R.A., Doores, S., Anantheswaran, R.C., and Kuhn, G.D., 1996. Viability loss of *Salmonella* species, *Staphylococcus aureus*, and *Listeria monocytogenes* in complex foods heated by microwave energy. *Journal of Food Protection* 59: 813-818.

Heperkan, D., Meric, B.E., Sismanoglu, G., Dalkılıc, G., and Guler, F.K., 2006. Mycobiota, mycotoxigenic fungi and citrinin production in black olives. *Adv. Exp Med. Biol.* 571: 203-210.

Khalil, H., and Villota, R., 1989. The effect of microwave sublethal heating on the ribonucleic acids of *Staphylococcus aureus*. *Journal of Food Protection* 52: 544-548.

Kozempel, M. F., Annous, B.A., Cook, R.D., Scullen, O.J., and Whiting, R.C., 1998. Inactivation of microorganisms with microwaves at reduced temperatures. *Journal of Food Protection* 61: 562-585.

Kozempel, M. F., Cook, R.D., Scullen, O.J., and Annous, B.A., 2000. Development of a process for detecting nonthermel effects of a microwave energy on microorganisms at low temperature. *Journal of Food Processing and Preservation* 24: 287-301.

Mertens, B., and Knorr, D., 1992. Developments of nonthermal processes for food preservation. *Food Technolology* 46: 124-133.

More, H. G., Magan, N., and Stenning, B.C., 1992. Effect of microwave heating on quality and mycoflora of sorghum grain. *Journal of Stored Products Research* 28: 251-256.

Oral, J., and Heperkan, D., 1999. Penicillic acid and citrinin production in olives. In: *Food Microbiology and Food Safety into the Next Millenium.* Proceedings of the 17 th International ICFMH Conference. Tuijtelaars, A.C.J., Samson, R.A., Rombouts, F.M. & Notermans, S. (Eds.), Veldhoven, The Netherlands, 13-17 September. pp. 138-140.

Pitt, J.I., and Hocking, A.D. 1997. *Fungi and Food Spoilage.* 2nd ed. Blackie Academic & Professional, London.

Regier, M., and Schubert, H., 2001. Microwave processing. In: *Thermal Technologies in Food Processing.* Richardson, P. (Ed.), Woodhead Publishing Limited. CRC Press, New York. pp.178-207.

Samson, R.A., Hoekstra, E.S., Lund, F., Filtenborg, O., and Frisvad, J.C., 2002. Methods for the detection, isolation and characterisation of food-borne fungi. In: *Introduction to Food- and Airborne Fungi,* 6th ed. R.A. Samson, E.S. Hoekstra, J.C. Frisvad, and O. Filtenborg. (Eds.),CBS, Netherlands. pp.283-297.

Stolle, A., and Schalch, B., 2000. Action of microwaves. In: *Encyclopedia of Food Microbiology,* vol 2., Robinson, R., Batt, C.A., and Patel, P.D. (Eds.), Academic Press, U.K. pp.1036-1041.

Toussaint, G., Lafaverges, F., and Walker, E.A., 1997. The use of high pressure liquid chromatography for determination of aflatoxin in olive oil. *Archives Institute Pasteur, Tunis* 3-4:325-334.

Tajchakavit, S., Ramaswamy, H.S., and Fustier, P., 1998. Enhanced destruction of spoilage microorganisms in apple juice during continuous flow microwave heating. *Food Research International* 31: 713-722.

Welt, B. A., Tong, C.H., Rossen, J.L., and Lund, D.B., 1994. Effect of microwave radation on inactivation of *Clostridium sporogenes* (PA 3679) spores. *Applied and Environmental Microbiology* 60: 482-488.

Yassa, I. A., Abdalla, E.A.M., and Aziz, S.Y., 1994. Aflatoxin B1 production by moulds isolated from black table olives. *Annals of Agricultural Science* 39:525-537.

Yeo, C.B.A., Watson, I.A., Stewart-Tull, D.E.S., and Koh, V.H.H., 1999. Heat transfer analysis of *Staphylococcus aureus* on stainless steel with microwave radiation. *Journal of Applied Microbiology* 87:396-401.

Zhang, H., and Jianshao, H., 2002. Study on the microwave technology to inhibit the angel cake mould. *Food and Machinery* 2:18-19.

Influence of winemaking on ochratoxin A content in red wines

B. Grazioli, R. Galli, M.D. Fumi and A. Silva

Institute of Oenology and Food Engineering, Universita' Cattolica del Sacro Cuore, Via E. Parmense 84, 29100 Piacenza, Italy; angela.silva@unicatt.it

Abstract

Good viticulture, harvest and post harvest practices are sound management approaches for reducing ochratoxin A (OTA) contamination in wine. Available data do not conclusively point out the main factors that contribute to the occurrence of OTA in grapes during pre and post-harvest processing, and in wine. The aim of this work was to study the behaviour of OTA during the main stages of the winemaking process and identify critical control points (CCPs) in the process of making red wine. Trials were carried out at different wineries in the south of Italy, with two grape cultivars (Negroamaro and Primitivo) and different vintages (2001-2002-2003). The results showed that OTA was not produced during winemaking, but each stage in the process affected the content of OTA in the product to varying degrees. Maceration increased the concentration of OTA in the must by 32 to 44% while alcoholic and malo-lactic fermentation caused a 41 to 75% reduction in OTA in the wine. OTA monitoring during winemaking is necessary if OTA is present at harvesting because OTA can only be partially reduced but not completely eliminated. This information would be helpful in the development of an effective food safety program and so reduce the risk of OTA contamination to the consumer.

Keywords: ochratoxin A, red wine, winemaking, maceration, fermentation, clarification, critical control point

Introduction

Ochratoxin A (OTA) is a mycotoxin produced by some species of the fungi *Penicillium* and *Aspergillus*. These fungi grow in a wide range of commodities, most in cereals, but also in coffee, dried fruits and grapes (Battilani *et al.*, 2003). Ochratoxin A has also been found in grape juices, in must and in wines (Delage *et al.*, 2003). The concentration of OTA has been shown to be higher in red wines than in white and rose' wines (Pietri *et al.*, 2001). In Europe, there seems to exist an apparent correlation between level of OTA contamination of wines and the geographical area of production (Otteneder and Majerus, 2000). The data show that the further south the origin, the more frequent the occurrence and the greater the OTA concentration (Figure 1). Information on OTA persistence and transformation during processing, that would be helpful in the development of an effective food safety program, is not readily available. Such information would be important particularly that the European Commission is considering setting limits for wine (GUCE, 2002).

Materials and methods

Trial locality and sample collection

Trials were carried out at different Italian wineries in Puglia. The trials were conducted with two cultivars (Negroamaro and Primitivo) and on different vintages (2001-2002-2003). Samples were collected at crushing-destemming, during maceration and alcoholic fermentation, at drawing off, racking and during malo-lactic fermentation. Ninety (90) samples belonging to three vintages were collected at different red wine processing stages as shown in Figure 2. The number of duplicate samples collected at each stage is shown Table 1.

Geographical Area	OTA ug/L : min - max
A, B, C1	< 0.01 – 0.80
C2	< 0.01 – 2.60
C3	< 0.01 – 3.30

Figure 1. European vine-growing areas (A, B, C1, C2, C3), each of them with particular climatic conditions (modified from Otteneder et al.*, 2000).*

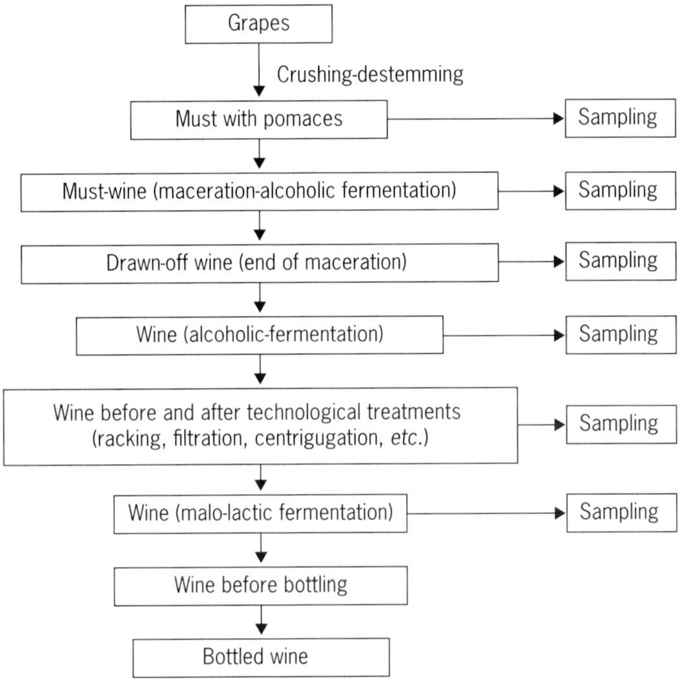

Figure 2. Red winemaking flow-chart, showing the sampling points.

Table 1. Red wine processing stages from which samples were collected.

Processing stage	Number of samples collected
Crushed grapes (must)	22
Drawn-off wine (after maceration)	22
After alcoholic fermentation	22
After first racking	16
After malo-lactic fermentation	8

Ochratoxin A analysis

Ochratoxin A analysis was done in duplicate after centrifugation and filtration of the sample according to the method of Visconti *et al.* (1999). Briefly, must and wine were diluted with a solution containing polyethylene glycol (PEG) and sodium hydrogen carbonate, filtered and cleaned up by Ochraprep® immunoaffinity columns (R-Biopharm Rhone LTD, Scotland). Ochratoxin A was eluted from the column with methanol and quantified by reversed-phase HPLC with fluorescence detection. The data were analysed by one-way ANOVA and Duncan's test using the statistical software SPSS version 11.0 (SSPS Inc., Chicago, IL).

Results and discussion

Results obtained from this study are shown in Figures 3, 4 and 5. Crushing (must) extracts OTA into the pomaces, and the maceration concentrates the toxin leading to a 32-44% increase in the level of OTA. Removal of pomaces and completion of the alcoholic fermentation resulted in small but varied decreases in OTA content. The decreases were 13.6% for Primitivo vintage 2001, and 1.2 and 9.5% for the 2002 and 2003 Negroamaro vintages, respectively. The temperature during the maceration and the alcoholic fermentation was between 20-32 °C. In order to start the alcoholic fermentation the must is inoculated with selected strains of *Saccharomyces cerevisiae.* Both operations seem to minimally affect OTA. The first racking operation caused a slight reduction of the OTA level in wine (Figures 4 and 5). Malo-lactic fermentation is the phase that causes a significant reduction ($P<0.05$) in OTA concentration. Figures 3 and 5 show reductions of 41 and 75% for the Primitivo 2001 and Negroamaro 2003 vintages, respectively. Malo-lactic fermentation in wines occurs by means of indigenous lactic acid bacteria.

Our results also indicate a substantial difference in the levels of OTA in the Negroamaro vintages of 2001 and 2003 compared to that of 2002 (Figure 6). In 2002 the concentration of the toxin in the vintage was much higher than in 2001 and 2003.

Overall, the results show that no OTA was produced during the making of the red wine vintages studied. Ochratoxin A present in the grapes passes into the juice during the crushing and the maceration. Therefore, if the grapes contained high levels of OTA, the must and the wine at drawing off would contain high levels of OTA too. The steps following maceration cause a gradual decrease of the toxin with a substantial reduction being effected by the malo-lactic fermentation stage. The reduction in concentration of OTA after the malo-lactic acid stage confirms the positive decontaminating effect of lactic acid bacteria that are responsible for the biological de-acidification of wines. These observations concur with patterns reported in previous trials (Silva *et al.*, 2003).

The percent reduction of OTA by means of lactic acid bacteria may however be very variable, depending on the inoculated strain and on the OTA concentration in

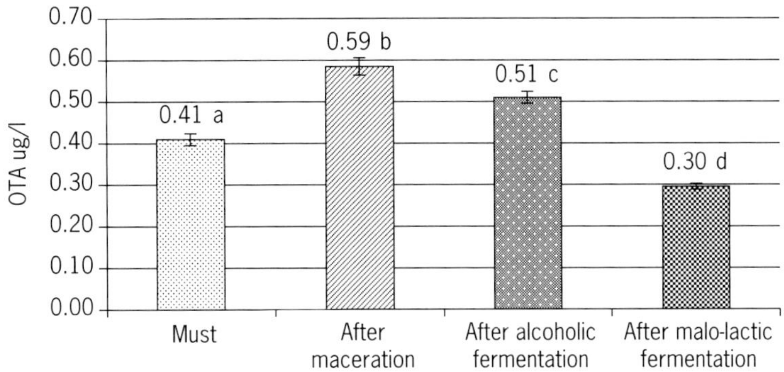

Figure 3. OTA average values ± SD during the winemaking of Primitivo (Puglia, Italy), vintage 2001. Means followed by the same letter are not significantly different (p≤0.05).

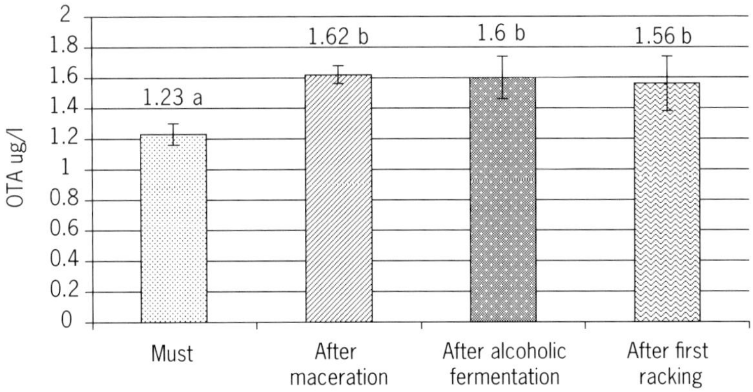

Figure 4. OTA average values ± SD during the winemaking of Negroamaro (Puglia, Italy), vintage 2002. Means followed by the same letter are not significantly different (p≤0.05).

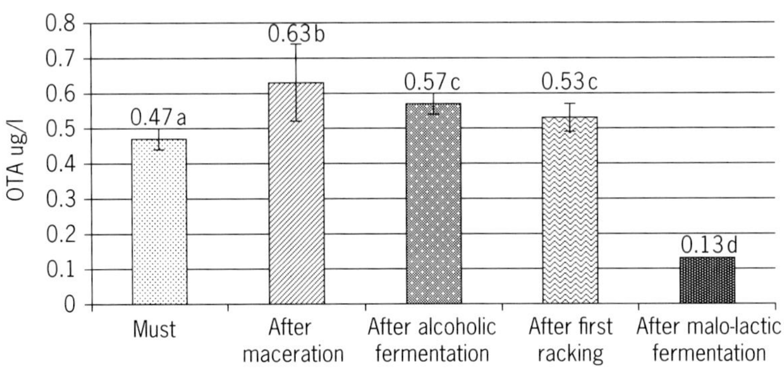

Figure 5. OTA average values ± SD during the winemaking of Negroamaro, vintage 2003. Means followed by the same letter are not significantly different (p≤0.05).

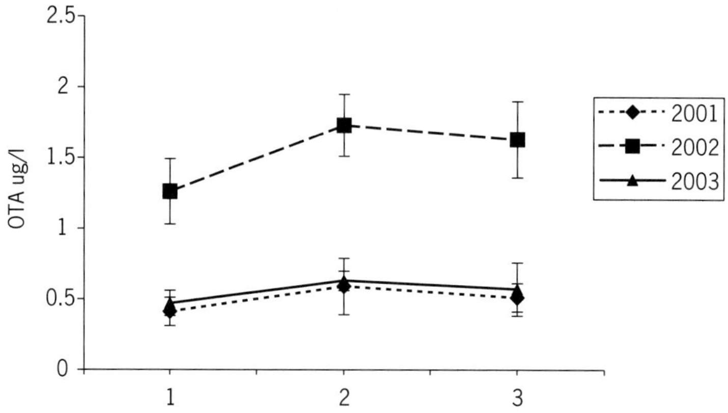

Figure 6. Trend of the OTA (mean ± SD) from crushed grapes to the wine after the alcoholic fermentation (Negroamaro). 1= must; 2= after maceration; 3= after alcoholic fermentation.

the wine. Since OTA can only be partially reduced during the winemaking process, but not completely eliminated, in vintages with low contamination, the malo-lactic fermentation can substantially reduce the risk of OTA. As for wines that are not subjected to malo-lactic fermentation, the OTA level could be decreased by corrective actions with chemical adjuvants which act as adsorbents. In particular, charcoal-based products (Atos, Micosorb, Toxical) have been shown to be very effective in reducing OTA in wine (Silva *et al.*, 2003). It is noteworthy to mention that the charcoal-based chemical adjuvants have a high affinity for OTA as well as the desirable polyphenolic compounds present in the wine. This negative effect is not very significant however, with amounts of charcoal lower than 10 g/hl. At present, the EU legislation allows the use of charcoal in must and wine, but not in red wines (GUCE, 1999).

Conclusions

Prevention through pre-harvest management is the best method to control mycotoxin contamination. However, if contamination of grapes by OTA occurred, the hazard associated with the toxin must be managed through post-harvest procedures. In the oenological practice, the steps of solid-liquid separations and malo-lactic fermentation are effective at reducing the content of OTA. The racking, centrifugation, filtration, *etc.* cause an adsorption of the toxin to the solids consequently reducing the content of OTA in the liquid fraction (wine). The mechanism by which malo-lactic fermentation decreases the OTA level in wine is presently not well understood. We can only postulate that it may be by either adsorption or degradation. If it were by adsorption the toxin would be eliminated with the lees, whereas if it occurred by degradation, it would be important to evaluate the toxicity of the metabolites.

As far as identifying winemaking OTA CCPs, the process can only minimise the hazard, but it cannot assure the complete control of the toxin. It is not possible to apply quick tests during winemaking to have immediate indications of the OTA level at each step. Analysis of the grapes just before harvesting can give an indication of the contamination and as such make it possible to foresee the reduction of the toxin during the winemaking process. If it is foreseen that the process unit operations may not sufficiently reduce the ultimate OTA level in the wine, use of adsorbent adjuvants should be considered.

Acknowledgments

This research activity was supported by EU, QLK1-CT-2001-01761-WINE-OCHRA RISK; Catholic University S. Cuore, Piacenza, Italy and part of a broader research project of the European Commission (Quality of Life Program, Key Action 1 on Food, Nutrition and Health) entitled "Risk assessment and integrated OTA management in grape and wine" (WINE-OCHRA RISK) that is coordinated by the Catholic University of Sacro Cuore, Piacenza, Italy.

References

Battilani, P., Pietri, A., Bertuzzi, T., Languisco, L., Giorni, P., and Kozakiewicz, Z., 2003. Occurrence of ochratoxin A producing fungi in grape grown in Italy, *Journal of Food Protection* 66: 633-636.

Delage, N., d'Harlingue A., Colonna Ceccaldi, B., and Bompeix, G., 2003. Occurrence of mycotoxins in fruit juices and wine. *Food Control* 14: 225-227.

GUCE, 1999. Gazzetta Ufficiale delle Comunità Europee, Regolamento CE N.1493 relativo a *Organizzazione comune del mercato vitivinicolo L179.* pp.23-59.

GUCE, 2002. Gazzetta Ufficiale delle Comunità Europee, Regolamento CE N.472/2002 recante modifica del Regolamento N. 466/2001 che definisce *I tenori massimi di taluni contaminanti presenti nelle derrate alimentari L75.* pp.18-20.

Otteneder, H., and Majerus, P., 2000. Occurrence of ochratoxin A (OTA) in wines: influence of the type of wine and its geographical origin. *Food Additives and Contaminants* 17: 793-798.

Pietri, A., Bertuzzi, T., Pallaroni, L., and Piva, G., 2001. Occurence of ochratoxin A in Italian wines. *Food Additives and Contaminants* 18: 647-654.

Silva, A., Galli, R., Grazioli, B., and Fumi, M.D., 2003. Metodi di riduzione di residui di Ocratossina A nei vini. *Industrie delle Bevande* 32: 467-472.

Visconti, A., Pascale, M., and Centone, G., 1999. Determination of ochratoxin A in wine by means of immunoaffinity column clean-up and high-performance liquid chromatography. *Journal Chromatography A* 864: 89-101.

A novel approach to deactivate ochratoxin A

G. Schatzmayr[1], *D. Schatzmayr*[1], *E. Pichler*[2], *M. Täubel*[1], *A.P. Loibner*[3] *and E.M. Binder*[4]

[1]*BIOMIN GmbH, Industriestrasse 21, 3130 Herzogenburg, Austria; gerd.schatzmayr@biomin.net*
[2]*Romber Labs Diagnostics GmbH, Technopark 1, 3430 Tulln, Austria*
[3]*IFA-Tulln, Department of Environmental Biotechnology, Konrad Lorenzstrasse 20, 3430 Tulln, Austria*
[4]*Erber AG, Industriestrasse 21, 3130 Herzogenburg, Austria*

Abstract

Ochratoxin A (OTA) is mainly produced by certain *Penicillium* and *Aspergillus* species in grains but there are also many reports about its occurrence in grapes, coffee beans and other commodities. OTA poses a high risk to the health of man and animals necessitating the need for measures that minimise its occurrence in foods and feeds. It has been shown that most adsorbents that are efficacious for the removal of aflatoxins, do not function as efficiently against ochratoxin A. A study was thus initiated to identify micro-organisms capable of biologically detoxifying ochratoxin A. Several OTA-detoxifying microbes were isolated and characterised. We report here results of further tests with the isolates to identify those strains with the most desirable characteristics for use as OTA detoxifying additives in animal feeds based on: rate of detoxification, influence of phenylalanine on OTA-deactivation activity, pathogenicity, possibility of fermentation and stabilisation, ability to function under gastrointestinal conditions as well as requirements for registration in the European Union. Identified "OTA-detoxifiers" belonged to the genera *Stenotrophomonas* and *Sphingomonas* (aerobic bacteria), *Eubacterium* (anaerobic bacteria) and *Trichosporon*, *Cryptococcus* and *Rhodotorula* (yeasts). A new yeast strain belonging to the genus *Trichosporon* (*T. mycotoxinivorans*) was selected as most appropriate. For the first time an additive capable of very efficiently detoxifying ochratoxin A has been identified.

Keywords: ochratoxin A, ochratoxin α, detoxification, deactivation, mycotoxins, *Trichosporon mycotoxinivorans*, feed additive

Introduction

Ochratoxin A is a secondary metabolite mainly produced by *Aspergillus ochraceus* and *Penicillium verrucosum* on several commodities prevalent in human and animal diets.

Ochratoxin A is the causative agent of kidney diseases in many types of domestic and wild animals that consume contaminated feed. High concentrations of dietary OTA can cause liver damage as well as intestinal necrosis and haemorrhage. In Denmark, a correlation was observed between prevalence rates of porcine nephropathy and frequency of OTA contamination in corresponding feed samples (Krogh, 1987). In a study with broiler chickens, OTA led to a reduced feed utilisation efficiency, increased liver, kidney, proventriculus, and gizzard weights, and altered serum chemistries (Kubena *et al.*, 1989). While the use of adsorbents such as aluminosilicates have been shown to be efficacious against aflatoxins (Kubena *et al.*, 1989, 1993; Huff *et al.*, 1992) the adsorption of other mycotoxins like ochratoxin A and diacetoxyscirpenol could not be achieved to a similar extent (Kubena *et al.*, 1993). An alternative approach is microbial or enzymatic degradation that could lead to less- or non-toxic metabolites. Pitout (1969) reported that carboxypeptidase A and α-chymotrypsin hydrolyse ochratoxin A to the non-toxic metabolite ochratoxin α (OTα) *in vitro* and Wegst and Lingens (1983) demonstrated the degradation of ochratoxin A by the aerobic bacterium *Phenylobacterium immobile*. Similarly, Hwang and Draughon (1994) reported that *Acinetobacter calcoaceticus* was able to degrade 10μg/ml of OTA to ochratoxin α at 25 and 30 °C. In a recent report, the degradation of OTA by *Aspergillus niger* in liquid and solid media was described by Varga *et al.* (2000). Chu *et al.* (1972) and Bruinick *et al.* (1998) showed that ochratoxin α is non-toxic or at least 500 times less toxic than OTA.

Recently, the development of an additive for the detoxification of trichothecene-containing feedstuffs has been reported (Binder *et al.*, 2001). The active biotransforming bacterium in this product is a new *Eubacterium*-species isolated from rumen fluid of cattle. Fed to animals, this microbe is able to detoxify trichothecenes in the small intestine prior to their absorption. In this study, micro-organisms derived from different habitats (Schatzmayr *et al.*, 2002, 2003) were investigated for their OTA-detoxification potential. The aim was to identify a strain most suitable for the development of a novel OTA-detoxifying feed additive.

Materials and methods

Mycotoxin production

Cracked wheat was inoculated with *Aspergillus ochraceus* (ATCC 22947) and *Petromyces albertensis* (ATCC 58745) and incubated for 4 weeks at 30 °C. The wheat was ground and ochratoxin A extracted from the wheat with chloroform. The solvent was evaporated and the residue dissolved in Tris-HCl. Ochratoxin α was produced by incubating OTA with carboxypepdidase A (Sigma, St. Lious, MO) in Tris/HCl solution and a photometric method was used to determine the concentration of OTα (Fuchs, 2002).

Degradation of OTA by active cells

Detailed procedures for preparation of culture media and the cultivation of bacteria and yeasts are described elsewhere (Schatzmayr *et al.*, 2002, 2003). Briefly, anaerobic bacteria (*Eubacterium ramulus* and *Eubacterium callanderi)* were grown in bottles containing complex media (MM10) for 3 to 5 days at 37 °C. The headspace of each bottle was flushed with nitrogen, carbon dioxide or hydrogen to obtain an anaerobic atmosphere. Aerobic bacteria (*Phenylobacterium immobile, Stenotrophomonas* spp., *Sphingomonas* spp. and *Stenotrophomonas nitritreducens)* were grown in shaking nutrient broth (Merck, Darmstadt, Germany) for 3 to 5 days at 25 °C, 30 °C or 35 °C depending on the most appropriate growth temperature for each strain. A yeast culture medium (YM) consisting of malt extract (10g/l), glucose (10g/l), yeast extract (5g/l) and peptone (5g/l) was used to grow the yeast strains (*Cryptococcus* spp., *Rhodotorula* spp. and *Trichosporon* spp.*)*. These strains were incubated at 25 °C in shaking flasks on a rotary shaker for 48 hours.

The mycotoxin-degrading activity of the different microbial groups (aerobic and anaerobic bacteria, and yeast) was investigated in the respective growth media and in minimal medium (MM) containing minerals, trace elements and vitamins but no carbon and energy sources. Twenty millilitres of growth culture media were concentrated by centrifugation at 3,210 x g, (Beckmann GS-6, Fullerton, CA, USA) and the pellet suspended in 20 ml minimal medium containing 200 μg/ml OTA. Samples were removed at 0.5, 24 and 48 hour time periods and assayed for OTA and OTα by high performance liquid chromatography (HPLC) as described below.

Effect of phenylalanine on strain activity

Since ochratoxin A consists of a phenylalanine moiety linked to an isocumarin derivate, the effect of the presence of phenylalanine (PA) in the microbial medium was studied. The test strains were grown in liquid medium for 48 hours, centrifuged and re-suspended in minimal medium containing 400 μg/l OTA and 2 g/l phenylalanine. Ochratoxin A and phenylalanine degradation were measured after 48 hours. To quantify phenylalanine, it was derivatised with o-phthaldialdehyde, separated with HPLC (Agilent Series 1100, Agilent Technologies, Palo Alto, CA, USA) and detected with fluorescence detection.

Degradation of OTA by lyophilisates

Pure cultures of mycotoxin deactivating microbes were grown in a fermenter with a working volume of 5 litres (Braun Biostat ED, Biotech International, Melsungen, Germany). The cells were harvested after a time period estimated to coincide with reaching the stationery phase and concentrated by centrifuging for 25 minutes at 16,000 x g (Beckman J2-MI, Fullerton, CA, USA). The pellets were frozen at –80 °C and freeze dried (Christ Beta 1-16, Christ Gefriertrocknungsanlagen GmbH., Osterode am Harz, Germany). Two grams of a lyophilisate were suspended in 18 ml of minimal media and serially diluted. The dilutions 10^{-2} and 10^{-3} were spiked with 400 μg/l OTA and incubated at 35 °C. Samples were drawn at 0.5, 24 and 48 hours and analysed for ochratoxin A and ochratoxin α.

Test model using pig ileum

Pig ileum was removed immediately after slaughter and transported to the laboratory under anaerobic (CO_2) conditions. Pieces, approximately 15 cm long, were cut from the ileum, tied at both ends and weighed. 2 gram of each lyophilised test strains was dissolved in 18 ml of MM. Then 1.5 ml of this suspension was injected into the gut piece. Thus the concentration of the test strains corresponded to a dilution step of 10^{-3}. After injecting 0.6 ml of a 10 mg/l solution of OTA into the ileum pieces they were placed in a 250 ml glass bottle and incubated at 37 °C under a CO_2 atmosphere. 1.5 ml samples for analytical determination of OTA and OTα were taken periodically using a syringe (3 to 4 times) within a 48 hr incubation period. Control batches were prepared without addition of the active strains, in order to assess the degradation activity of the original ileum microbes.

Analysis of mycotoxins

For quantification of ochratoxin A and its main metabolite ochratoxin α, 1.5 ml of the liquid sample was centrifuged. A 1 ml portion of the supernatant was transferred into a glass tube and acidified with 500 μl of a 1M phosphoric acid. Afterwards, OTA and OTα were extracted with 5 ml chloroform by shaking the closed tube vigorously for 1minute. After separation of the phases, 4 ml of the chloroform phase were transferred into another glass tube and evaporated to dryness under a stream of nitrogen. The residue was dissolved in acetonitrile+water+acetic acid (45:54:1) and analysed by HPLC with fluorescence detection (FLD). The HPLC unit was a HP 1050 (Hewlett Packard, Waldbronn, Germany), fitted with a HP ODS Hypersil 2.1x10 mm, 3μM column. The solvent system was: solvent A - acetonitrile+water+acetic acid

(45:54:1) and solvent B - water+acetic acid (99:1). The following gradient (solvent A mixed with solvent B) was employed:

minutes	0	2	10	25	28	35
% B	65	65	0	0	65	65

A flow rate of 0.2 ml/min was used.

Results and discussion

Degradation studies with active cells

In previous studies (Schatzmayr *et al.*, 2002, 2003) several OTA- degrading micro-organisms (anaerobic bacteria, aerobic bacteria, yeasts) were isolated and characterised (Table 1). A very important prerequisite for an OTA-detoxifying feed additive is that it acts very rapidly. For this reason the pre-selected "detoxifiers" were investigated in respective liquid media for their rate of degradation of 400 μg/l of ochratoxin A. Figure 1 shows that two yeast strains, *Trichosporon mycotoxinivorans* 115 and 178, and two aerobic bacteria strains, *Sphingomonas* 033-1 and 034-1, degraded OTA within 7.5 hours, whereas the anaerobic bacteria strain *Eubacterium* sp. (Due4_11) and the aerobic bacteria *Stenotrophomonas* spp. 041-9 needed 10 hrs to achieve total degradation of OTA. The aerobic bacteria strain *Phenylobacterium immobile* (Wegst and Lingens, 1983), degraded only 30% of the OTA by the end of the experiment (48 hours). The metabolite ochratoxin α was detected in all samples.

Effect of phenylalanine on strain activity

Since ochratoxin A consists of a phenylalanine moiety linked to an isocumarin derivate, it was important to determine whether phenylalanine, if present in the microbial medium, could inhibit OTA-degradation. Results in Table 1 show that two yeast strains (*Trichosporon mycotoxinivorans* 115 and *Trichosporon* spp. 178), and one aerobic bacteria strain (*Sphingomonas* spp. 034-1) could convert phenylalanine and cleave ochratoxin A into OTα. The yeasts *Cryptococcus* sp. (118) and *Rhodotorula* spp. (124) did not degrade OTA in the presence of phenylalanine, but did degrade phenylalanine to a certain extent. The aerobic bacteria strains *Sphingomonas* spp. (033-1) and *Stenotrophomonas* spp. (041-10) degraded the mycotoxin completely, but the amino acid just partially. The other two aerobic bacteria strains (*Stenotrophomonas* 041-9 and 043-1a) and the reference bacteria strain (*Phenylobacterium immobile*) completely degraded phenylalanine but did not metabolise all the OTA.

Table 1. OTA-degrading micro-organisms, respective numbers at the German Strain Culture Collection (DSMZ) for strains deposited, OTA-degradation rate in a phenylalanine-containing (2g/l) liquid medium, and remaining phenylalanine concentration after 48 hours of incubation.

Strain designation	**Related strain**	**Deposited at DSMZ**	**OTA-deg. %, 48hrs[1]**	**[PA] mg/l, 48hrs[2]**
F6	*Eubacterium ramulus*	no	n.a.	n.a.
Due4_11	*Eubacterium callanderi*	DSM 14197	n.a.	n.a.
115	*Trichosporon mycotoxinivorans*	DSM 14153	100	0
118	*Cryptococcus* spp.	DSM 14154	0	824
124	*Rhodotorula* spp.	DSM 14155	0	629
178	*Trichosporon* spp.	DSM 14156	100	0
033-1	*Sphingomonas* spp.	DSM 14170	100	212
034-1	*Sphingomonas* spp.	DSM 14167	100	0
041-9	*Stenotrophomonas nitritreducens*	DSM 14168	90	0
041-10	*Stenotrophomonas* spp.	DSM 14169	100	968
043-1a	*Stenotrophomonas*	no	40	0
DSM 1986	*Phenylobacterium immobile*	DSM 1986	80	0

[1]Ochratoxin A-degradation rate in % after 48 hours.

[2]Remaining phenylalanine concentration in liquid medium (mg/l) after 48 hours of incubation.

Degradation studies with lyophilisates

The yeast *Trichosporon mycotoxinivorans* 115 was tested at both the 10^{-2} and 10^{-3} dilutions whereas the other strains were only tested at the10^{-3} dilution. Lyophilisates of the aerobic bacterial strain *Stenotrophomonas nitritreducens* 041-9 and the yeast *T. mycotoxinivorans* 115 had the highest OTA-degrading activity (Figure 2). After 6 hours, degradation of OTA to OTα by *T. mycotoxinivorans* 115 was 100% in dilution 10^{-2} and 90% in dilution 10^{-3}. With one exception (*Trichosporon* sp., 178), all the other lyophilisates that were tested degraded OTA totally after 48 hours. The performance of the lyophilisates showed that two of the micro-organisms were able to regenerate very rapidly in liquid medium and to regain OTA-detoxification potential. It should further be noted that the lyophilisates (Figure 2) were tested at a much lower dilution (10^{-3}) than the active strains, which weren't diluted (Figure 1).

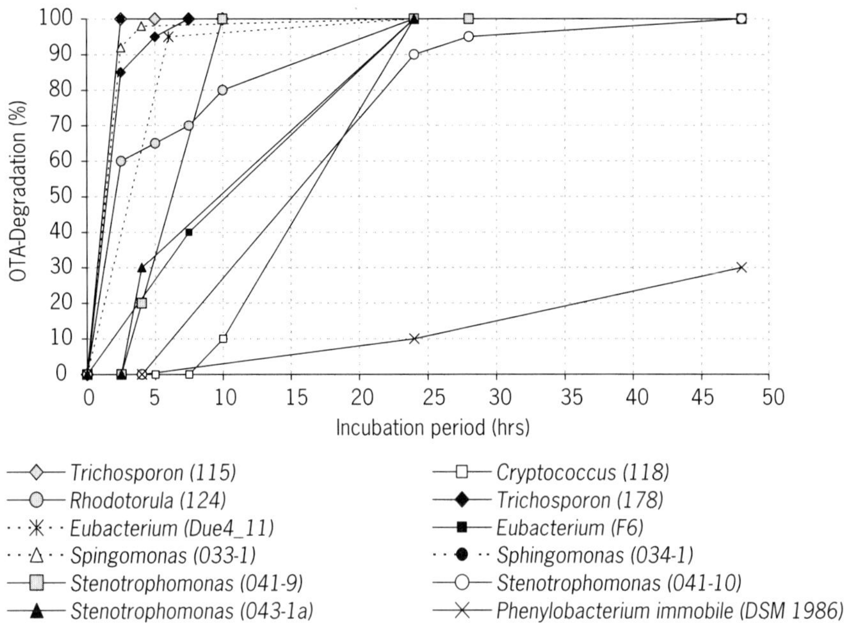

Figure 1. OTA-degradation rate of different OTA-detoxifiers within 48hrs of incubation in microbial culture media.

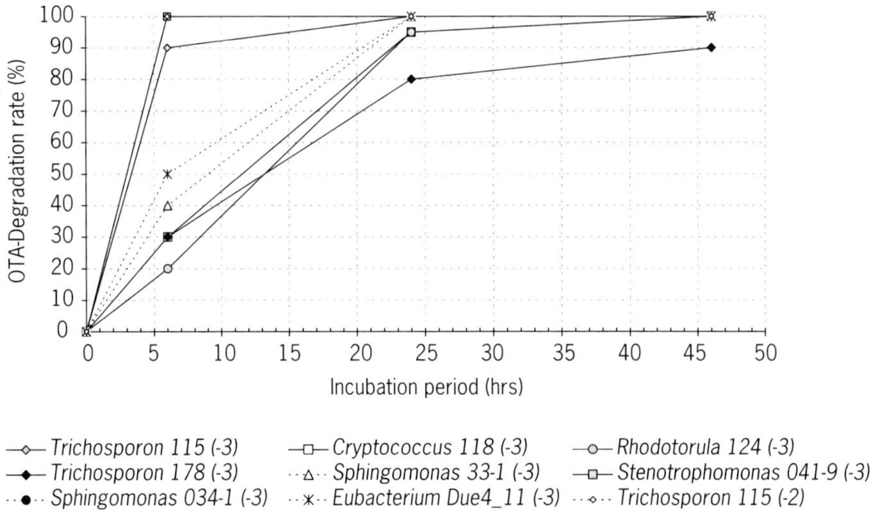

Figure 2. OTA degradation using lyophilised cells of the OTA-detoxifying micro-organisms in minimal medium during 48 hours of incubation.

In summary, the results show that *T. mycotoxinivorans* 115, *Stenotrophomonas* 041-9 and *Eubacterium* Du4_11 were the best degraders or detoxifiers of ochratoxin A. Therefore, these three strains were chosen for the *in vitro* test with pig ileum to evaluate their performance in an environment close to real conditions in a live animal. Figure 3 shows results of the *in vitro* pig ileum test with *Trichosporon mycotoxinivorans* 115, *Stenotrophomonas* 041-9 and *Eubacterium* Due4_11. The strains were able to degrade 90% (*T. mycotoxinivorans* 115, *Stenotrophomonas* 041-9) and 70% (*Eubacterium* Due4_11) of the OTA within 5 hr and achieved a 100 and 95% completion within 48 hr, respectively. The control pieces of intestine (without lyophilisates) showed a minor degradation capacity after 24 hours (10%) and 48 hours (20%). This could be due to the fact that intestinal contents of pigs harbour micro-organisms which are capable of metabolising OTA.

The highest deactivation rates were obtained with the yeast *T. mycotoxinivorans* 115 and the aerobic bacteria strain *Stenotrophomonas* 041-9. The decreased velocity of OTA-degradation beginning after six hours could be attributed to changes in the ileum microenvironment, possibly brought about by onset of spoilage (rotting) of

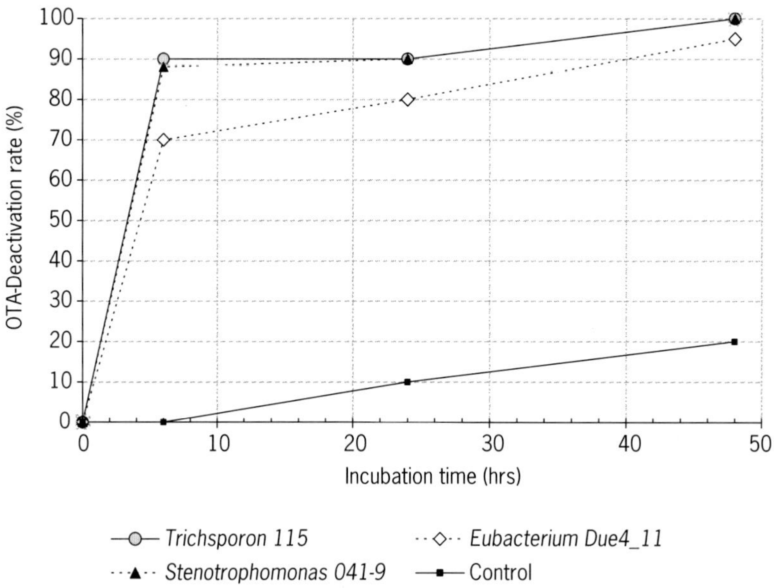

Figure 3. Comparison of lyophilisates (strains 115, Due4_11 and 041-9) and a control with respect to ochratoxin A deactivation using an in vitro model with pig intestine. Incubation period was 48 hours at 37 °C.

the ileum pieces. However, the most important time period is the first few hours because in animals OTA has to be deactivated in the ileum before it is absorbed. In tests to assess clearance of the OTA deactivating microbes, the bacteria strain *Stenotrophomonas* 041-9 showed resistance to many antibiotics of clinical relevance. The yeast *Trichosporon mycotoxinivorans* 115 was therefore, selected as the strain to be used as OTA-detoxifying feed additive. Preliminary feeding trials with pigs and poultry have shown that that *Trichosporon mycotoxinivorans* 115 degrades OTA in these animal species (Politis *et al.*, 2005).

In conclusion, this work has shown that addition of the yeast isolate *Trichosporon mycotoxinivorans* 115 to feeds is very promising as an efficacious new strategy to overcome ochratoxicosis in farm animals.

Acknowledgements

We thank the Austrian Industrial Research Promotion Fund for the financial support of this project.

References

Binder, E.M., Heidler, D., Schatzmayr, G., Thimm, N., Fuchs, E., Schuh, M., Krska, R., and Binder, J., 2001. Microbial detoxification of mycotoxins in animal feed. In: *Mycotoxins and Phycotoxins in perspective at the turn of the Millennium.* de Koe, W.J., Samson, R.A., van Egmond, H.P., Gilbert, J., and Sabino, M. (Eds). Proceedings of the 10th International IUPAC Symposium on Mycotoxins and Phycotoxins - 21-25 May, 2000 Guarujá, Brazil, pp. 271-277.

Bruinik, A., Rasonyi, T., and Sidler, C., 1998. Differences in Neurotoxic Effects of Ochratoxin A, Ochracin and Ochratoxin α *in vitro. Natural Toxins* 6: 173-177.

Chu, F.S, Noh, I., and Chang, Ch.C., 1972. Structural Requirements for Ochratoxin Intoxication. *Life Sciences* 11: 503-508.

Fuchs, E., 2002. *Study of the microbial degradation of the mycotoxin ochratoxin A by means of optimized chromatographic techniques as a contribution to the development of detoxifying agents for feeding stuffs.* Doctoral Thesis at University of Natural Resources and Applied Sciences, Vienna, pp. 59-61.

Huff, W.E., Kubena, L.F., Harvey, R.B., and Phillips, T.D., 1992. Efficacy of Hydrated Sodium Calcium Aluminosilicate to reduce the Individual and Combined Toxicity of Aflatoxin and Ochratoxin A. *Poultry Science* 71: 64-69.

Hwang C., and Draughon F.A., 1994. Degradation of Ochratoxin A by *Acinetobacter calcoaceticus. Journal of Food Protection* 57: 410-414.

Krogh, P., 1987. Ochratoxin in foods. In: *Mycotoxins in Food*, P. Krogh (Ed.). Academic Press, Harcourt Brace Jovanovisch, London, pp. 97.

Kubena, L.F., Harvey R.B., Huff W.E., Corrier D.E., Phillips T.D., and Rottinghaus, G.E. 1989. Influence of Ochratoxin A and T-2 Toxin Singly and in Combination on Broiler Chickens. *Poultry Science* 68: 867-872.

Kubena, L.F., Harvey, R.B., Huff, W.E., Elissalde, M.H., Yersin, A.G., Phillips, T.D., and Rottinghaus, G.E., 1993. Efficacy of a Hydrated Sodium Calcium Aluminosilicate to reduce the Toxicity of Aflatoxin and Diacetoxyscirpenol. *Poultry Science* 72: 51-59.

Politis, I., Fegeros, K., Nitsch, S., Schatzmayr, G., and D. Kantas, 2005. Use of Trichosporon mycotoxinivorans to suppress the effects of ochratoxicosis on the immune system of broiler chicks. *British Poultry Science* 46: 58-65.

Schatzmayr, G., Heidler, D. Fuchs, E., Loibner, A.P., Braun, R., and Binder E.M., 2002. Evidence of Ochratoxin A - Detoxification Activity of Rumen Fluid, Intestinal Fluid and Soil Samples as well as isolation of relevant microorganisms from these environments. *Mycotoxin Research* 18A: 183-187.

Schatzmayr, G., Heidler D., Fuchs, E., Mohnl, M., Täubel. M., Loibner, A.P., Braun, R., and Binder E.M., 2003. Investigation of different yeast strains for the detoxification of Ochratoxin A. *Mycotoxin Research* 19:124 -128.

Scott, P.M., 1998. Industrial and farm detoxification processes for mycotoxins. *Revue de Médecine vétérinaire* 149: 543-548.

Varga, J., Krisztina, R., and Teren, J., 2000. Degradation of ochratoxin A by *Aspergillus* species. *International Journal of Food Microbiology* 59: 1-7.

Wegst, W., and Lingens, F., 1983. Bacterial degradation of ochratoxin A. *FEMS Microbiology Letters* 17: 341-344.

Section VII:

Phycotoxins

Toxicology of yessotoxins and other DSP-related toxins

A. Tubaro and A. Giangaspero

DEMREP, University of Trieste, Via A. Valerio 6, 34127 Trieste, Italy; tubaro@units.it

Abstract

Yessotoxin (YTX) and its analogs are lipophilic polyether compounds initially presumed to belong to Diarrheic Shellfish (DS) toxins, since they were detected together with okadaic acid (OA) and its derivatives, the major toxins of this group. Unlike OA, YTXs are not diarrheic and their toxicological potential is unclear. Only few data on the acute and short term toxicity of YTX and its analogues are available, mostly after intra-peritoneal (i.p.) administration. However, the toxicity of YTXs after i.p. injection is quite different from that after oral administration, the common human exposure route. After i.p injection, different lethality data (LD_{50}), ranging from 80 to 750 μg/kg are reported for YTX. Despite these discrepancies, the observed symptoms are similar. Mice treated with both YTX and homo-yessotoxins (homoYTXs) were restless and jumped before death, but microscopy examination did not show significant changes in the main organs. Some ultrastructural alteration was observed in the heart tissue of mice i.p. treated with 1.0 mg/kg. After acute oral administration no lethality was detected for YTX (doses ranging from 1.0 to 54 mg/kg) and for homoYTXs (1-2 mg/kg). YTX and homoYTX (1-2 mg/kg) did not provoke any sign of toxicity. Only some ultrastructural changes in the myocardium, similar to those described after i.p. injection, were observed. After repeated oral administration (7 days) no mortality or other signs and haemato-clinical changes related to the treatment were observed in mice treated with YTX (2 mg/kg/day) or its analogues (1 mg/kg/day). Only by transmission electron microscopy were some changes (rounded mitochondria) in the myocardiocytes adjacent to capillaries observed. Although other studies are necessary to clearly define the toxicological profile of YTXs, their toxicological potential after oral exposure seems to be limited.

Keywords: yessotoxin, homoyessotoxin, 45-OH-homoyessotoxin, okadaic acid, acute toxicity studies, LD_{50}

Introduction

Yessotoxins (YTXs) are lipophilic polyether compounds initially presumed to belong to the Diarrheic Shellfish toxins group, since by the mouse bioassay they were often detected together with okadaic acid (OA) and its derivatives, the main diarrheic toxins. Contrary to OA and its derivatives, that provoke diarrhea and inhibit protein phosphatase 2A (PP2A), YTXs are not diarrheic and do not inhibit PP2A (Ogino *et al.*, 1997). Furthermore, whereas YTXs are produced by *Protoceratium reticulatum* (Satake *et al.*, 1997) and *Lingolodinium polyedrum* (Tubaro *et al.*, 1998), OA and derivatives are produced by *Dinophysis* ssp. and *Prorocentrum lima*. For these reasons, YTXs are no longer included in the Diarrheic Shellfish (DS) toxins family. Very often, positivity to mouse bioassay for DS toxins could be due to the presence of YTXs. Since positivity to the mouse bioassay triggers actions such as cessation of collection and selling of contaminated shellfish, ascertaining the actual toxicity of YTX is very important for both sanitary and economic reasons.

Acute toxicity studies

The human toxicological potential of YTXs has not yet been completely elucidated and no cases of human intoxication have been reported so far. Furthermore, only animal toxicological data from a small number of studies is available for yessotoxins, mostly after intra-peritoneal (i.p.) administration. The available studies, listed in Table 1, were focused on acute toxicity. LD_{50} after intra-peritoneal (i.p.) administration in mice ranged from 80 to more than 750 μg/kg body weight. Despite these discrepancies (probably due to strain, sex and weight differences in the used mice), the reported symptoms were similar. Mice treated with YTX were restless and jumped before death, but microscopy examination did not show changes in the main organs. Only slight tissue modifications of myocardiocytes was observed both by light (Aune *et al.*, 2002) and electron microscopy in mice i.p. treated with 1.0 – 10 mg/kg (Terao *et al.*, 1990; Aune *et al.*, 2002; Tubaro *et al.*, 2003).

Table 1. LD_{50} values for YTX after i.p. and per os *(p.o.) administration to mice.*

LD_{50} i.p. (μg/kg)	LD_{50} *p.o.* (mg/kg)	Investigators
286	0.5, (not lethal)	Terao *et al.*, 1990
80-100	1.0, (not lethal)	Ogino *et al.*, 1997
>750; <1,000	10, (not lethal)	Aune *et al.*, 2002
512	2, (not lethal)	Tubaro *et al.*, 2003

The toxicity of YTXs after i.p. injection is higher than that resulting from oral administration, the common human exposure route. After acute oral administration of 1.0 to 10 mg/kg, no lethality was detected for YTX; a dose as high as 54 mg/kg did not result in acute toxicity either (Towers, personal communication). Only some ultrastructural changes in the myocardium, similar to those described after i.p. injection, were observed. The difference between i.p. and *per os* (*p.o.*) effects can be explained by the reduced lipophilicity of YTX because of the sulfate group in the molecule. Usually the presence of sulfate groups reduces the absorption of the compounds from the digestive tract.

Two studies comparing the acute toxicity of two YTX analogs, homoyessotoxin (homoYTX) and 45-OH-homoyessotoxin (45-OH-homoYTX) and the parent compound (YTX) and okadaic acid, the main DS toxin have been conducted (Tubaro *et al.*, 2003). A summary of the results from the two studies is presented in Table 2. After i.p. administration, homoYTX and YTX showed similar toxicity (LD_{50} = 444 μg/kg and 512 μg/kg, respectively), while 750 μg/kg of 45-OH-homoYTX did not cause death. The toxicity of YTXs was lower than that of OA (LD_{50} = 225 μg/kg).

YTXs did not induce any histological alteration in the main organs of the i.p. treated mice. On the contrary, OA provoked previously reported main alterations in the gastro-intestinal system as well as at the hepatic level (Tubaro *et al.*, 2003).

No lethal effects, differences in behaviour or histological modifications were observed after *p.o.* acute administration of YTX and its analogues, but ultrastructural alterations of myocardiocytes were observed, with damage limited only to some areas near capillaries (Figure 1). *P.o.* acute OA treatment (1 mg/kg) induced some lesions such as vacuolar degeneration of the epithelium associated with acute inflammation of the submucosa, in the forestomachs of 3 of 5 treated mice. No apoptotic changes

Table 2. LD_{50} values after i.p. and p.o. *administration of some YTX analogues, in comparison to okadaic acid (Tubaro et al., 2003).*

Toxin	LD_{50} i.p. (μg/kg)	LD_{50} p.o. (mg/kg)
YTX	512	2.0, (not lethal)
HomoYTX	444	1.0, (not lethal)
45-OH- homoYTX	>750	1.0, (not lethal)
Okadaic acid	225	>1.0; <2.0

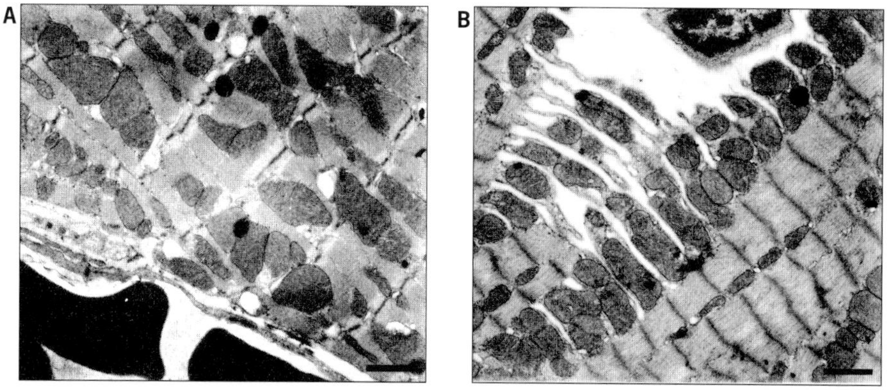

Figure 1. Electron micrographs of myocardium (scale mark: 2 μm).(A) Control; (B) YTX (2 mg/kg).

at cardiac level were observed in the myocardium of i.p or *p.o.* treated mice, using *in situ* TUNEL staining (Tubaro *et al.*, 2003).

Toxicity after repeated oral administration

After repeated oral administration for 7 days, no mortality, weight loss, signs of toxicity or cumulative effects related to the treatment were observed in mice treated with YTX (2 mg/kg/day) or its analogues (1 mg/kg/day), doses that are approximately 100 times the highest foreseeable human exposure. Only by transmission electron microscopy were some changes in the myocardiocytes of limited areas adjacent to capillaries observed. In particular, packages of rounded mitochondria and alterations of the cell boundaries were noted but without any increase in lactate dehydrogenase, an index of cardiac damage. Neither were there apoptotic changes at cardiac level in the myocardium of treated mice.

On the contrary, repeated oral administration of 1 mg/kg OA over the same 7-day period resulted in diarrhoea, body weight loss, reduced food consumption and the death of 2/5 mice after 5 days (Tubaro *et al.*, 2004).

Mechanism of action

Although several *in vitro* studies are available, the mechanism of action of YTX has not yet been delineated. *In vitro,* YTX is able to exert different biological effects at concentrations ranging form 10^{-5} to 10^{-10} M. Effects on cell adhesion and E-cadherin fragmentation are observed at 10^{-9} M (Pierotti *et al.*, 2003, Ronzitti *et al.*, 2004), while calcium movements and cAMP levels are affected by YTX at concentrations ranging from 10^{-5} to 10^{-7} M (de la Rosa *et al.*, 2001; Alfonso *et al.*, 2003; Malagoli and Ottaviani, 2004). Cytotoxic and apoptogenic effects are reported at concentration ranging from 10^{-5} to 10^{-10} (Ogino *et al.*, 1997; Malaguti *et al.*, 2002; Leira *et al.*, 2002). Although caspase 3 activation leading to apoptotic changes is reported at very low concentration *in vitro* (Malaguti *et al.*, 2002, Leira *et al.*, 2002), no apoptotic changes at cardiac level were observed *in vivo* in the myocardium of treated mice, the only organ affected by YTX (Terao *et al.*, 1990; Aune *et al.*, 2002; Tubaro *et al.*, 2003 and 2004). This apparent discrepancy can be ascribed to the reduced YTX absorption after *per os* administration. It should also be noted that *in vitro* interactions may not always be reproduced *in vivo*.

Conclusions

From the limited data available, the toxicological potential of YTX, homoYTX and 45-OH-homoYTX seems to be lower than that of okadaic acid, both after acute intraperitoneal injection and, particularly, after oral administration. Although more studies are necessary to clearly define the potential risk of YTXs, their toxicological effects after oral exposure seem to be limited. Pharmacokinetic data will be useful to ascertain the toxicological potential of YTXs.

Acknowledgments

This work was partially supported by a grant of the Italian Ministry of Instruction, University and Research.

References

Alfonso, A., De la Rosa, L., Vieytes, M.R., Yasumoto, T., and Botana, L.M., 2003. Yessotoxin, a novel phycotoxin, activated phosphodiesterase activity. Effect of yessotoxin on cAMP levels in human lymphocytes. *Biochemical Pharmacology* 65: 193-208.

Aune, T., Sorby, R., Yasumoto, T., Ramstad, H., and Landsverk, T., 2002. Comparison of oral and intraperitoneal toxicity of yessotoxin towards mice. *Toxicon* 40: 77-82.

De la Rosa, L., Alfonso, A., Filarino, N., Vieytes, M.R., and Botana, L.M., 2001. Modulation of cytosolic calcium levels of human lymphocytes by yessotoxin, a novel marine phycotoxin. *Biochemical Pharmacology* 61: 827-833.

Leira, F., Alvarez, C., Vieites, J.M., Vieytes, M.R., and Botana, L.M., 2002. Characterization of distinct apoptotic changes induced by okadaic acid and yessotoxin in the BE(2)-M17 neuroblastoma cell line. *Toxicology in Vitro* 16: 23-31.

Malagoli, D., and Ottaviani, E., 2004. Yessotoxin affects fMLP-induced cell shape changes in *Mytilus galloprovincialis* immunocytes. *Cell Biology International* 28: 57-61.

Malaguti, C., Ciminiello, P., Fattorusso, E., and Rossini, G.P., 2002. Caspase activation and death induced by yessotoxin in HeLa cells. *Toxicology in Vitro* 16: 357-363.

Ogino, H., Kumagai, M., and Yasumoto, T., 1997. Toxicologic evaluation of yessotoxin. *Natural Toxins* 5: 225-259.

Pierotti, S., Malaguti, C., Milandri, A., Poletti, R., and Rossini, G.P., 2003. Functional assay to measure yessotoxins in contaminated mussel samples. *Analytical Biochemistry* 312: 208-216.

Ronzitti, G., Callegari, F., Malaguti, C., and Rossini, G.P., 2004. Selective disruption of the E-cadherin-catenin system by an algal toxin. *British Journal of Cancer* 90: 1100-1107.

Satake, M., McKenzie, L., and Yasumoto, T., 1997. Identification of *Protoceratium reticulatum* as the biogenic origin of Yessotoxin. *Natural Toxins* 5: 164-167.

Terao, K., Ito, E., Oarada, M., Murata, M., and Yasumoto, T., 1990. Histopathological studies on experimental marine toxin poisoning-5. The effects in mice of yessotoxin isolated from *Patinopecten yessoensis* and of a desulfated derivative. *Toxicon* 28: 1095-1104.

Tubaro, A., Sidari, L., Della Loggia, R., and Yasumoto, T., 1998. Occurrence of homoyessotoxin in phytoplankton and mussels fron Northern Adriatic Sea. In: *Harmful algae,* Reguera, B., Blanci, J., Fernandez, M.L. and Wyatt, T., (Eds.), Xunta de Galicia and Intergovernmental Oceanic Commission of UNESCO, Grafisant, Santiago de Compostela, pp. 470-472.

Tubaro, A., Sosa, S., Carbonatto, M., Altinier, G., Vita, F., Melato, M., Satake, M., and Yasumoto, T., 2003. Oral and intraperitoneal acute toxicity studies of yessotoxin and homoyessotoxins in mice. *Toxicon* 41: 783-792.

Tubaro, A., Sosa, S., Altinier, G., Soranzo, M.R., Satake, M., Della Loggia, R., and Yasumoto, T., 2004. Short-term oral toxicity of homoyessotoxins, yessotoxin and okadaic acid in mice. *Toxicon* 43: 439-445.

Brevetoxin metabolism in shellfish associated with neurotoxic shellfish poisoning

H. Ishida[1], *A. Nozawa*[2], *H. Nukaya*[1], *L. Rhodes*[3], *P. McNabb*[3], *P.T. Holland*[3] *and K. Tsuji*[1]

[1]*School of Pharmaceutical Science, University of Shizuoka, 52-1 Yada, Shizuoka 422-8526, Japan; ishida@u-shizuoka-ken.ac.jp*
[2]*Pharmaceutical R&D Department, POLA Chemical Industries, Inc., 560 Kashio-cho, Totsuka-ku, Yokohama, Kanagawa 244-0812, Japan*
[3]*Cawthron Institute, 98 Halifax St. East, Private Bag 2, Nelson, New Zealand*

Abstract

Previously, we isolated brevetoxin and/or metabolites PbTx-3, BTX-B5 and BTX-B1 from toxic cockle, *Austrovenus stutchburyi*, PbTx-2 and PbTx-3 from Pacific oyster, *Crassostrea gigas*, following an outbreak of neurotoxic shellfish poisoning (NSP) in New Zealand. In this study, procedures for quantitative determination of PbTx-2, PbTx-3, BTX-B1 and BTX-B5 were developed by liquid chromatography coupled with tandem mass spectrometry (LC-MS/MS). In mass spectrometry with an electrospray ionisation interface operating in the positive or negative ion mode, the protonated ions $[M+H]^+$ of PbTx-2 (m/z 895), $[M+H]^+$ of PbTx-3 (m/z 897), $[M-H]^-$ of BTX-B5 (m/z 909), and $[M-Na]^-$ of BTX-B1 (m/z 1016) were generated abundantly, when 0.1% formic acid+acetonitrile was used as the mobile phase for column chromatography. The product ions of m/z 877, 725, 111 and 80 from PbTx-2, PbTx-3, BTX-B5 and BTX-B1 were identified, respectively, allowing unambiguous confirmation of these toxins by selective reaction monitoring LC-MS/MS analysis. This LC-MS/MS method was successfully applied to determine these toxins in the NSP-associated toxic shellfish. High levels of PbTx-3 and BTX-B5 were found in *C. gigas*, of PbTx-3, BTX-B1 and BTX–B5 in *A. stutchburyi*, and of PbTx-2, PbTx-3 and BTX-B5 in greenshell mussel (*Perna canaliculus*), contaminated in the field. In conclusion, these bivalves could commonly metabolise PbTx-2 to PbTx-3 and BTX-B5 and accumulate these metabolites temporally on exposure to brevetoxin producing *Karenia* spp.

Keywords: neurotoxic shellfish poisoning, brevetoxins, PbTx-2, PbTx-3, BTX-B1, BTX-B5, liquid chromatography-tandem mass spectrometry, *Austrovenus stutchburyi*, *Perna canaliculus*, *Crassostrea gigas*, *Karenia brevis*

Introduction

A serious shellfish poisoning occurred in New Zealand during the summer of 1992/93. As profiles of symptoms, mostly affecting the gastrointestinal and/or nervous systems, were similar to those observed in a Neurotoxic Shellfish Poisoning (NSP) episode in North Carolina in 1991, it was suggested that the New Zealand event could be NSP as well (Bates *et al.*, 1993), and a mouse bioassays on contaminated shellfish according to the APHA-ether extraction protocol gave positive results.

It is well known that some *Karenia (*formerly *Gymnodinium)* species of marine phytoplankton produce brevetoxins, such as PbTx-1, PbTx-2 and PbTx-3 (Baden and Tomas, 1988; Chou and Shimizu, 1982). Neurotoxic Shellfish Poisoning was for sometime suspected to be caused by brevetoxins accumulated directly in shellfish. This was first confirmed by an isolation of brevetoxins and their metabolites, PbTx-3, BTX-B1 and BTX-B5, from toxic cockles harvested during the shellfish poisoning outbreak in New Zealand (Ishida *et al.*, 1995). PbTx-2 and PbTx-3 were isolated from toxic oysters and BTX-B2, B3 and B4 from toxic greenshell mussels (Ishida *et al.*, 1996; Morohashi *et al.*, 1995, 1999; Murata *et al.*, 1998).

From these studies, it was speculated that New Zealand shellfish metabolise brevetoxin in a species-specific manner (Figure 1). For example, PbTx-2 may be mainly converted by cockles to PbTx-3, BTX-B5 and BTX-B1 via BTX-B5, by greenshell mussel to BTX-B2, B3 and B4, and by oysters to PbTx-3. However, both reduction and oxidation of the terminal aldehyde terminus of PbTx-2 to afford PbTx-3 and BTX-B5 seem to be common in all the shellfish. It is very important to clarify the metabolism of brevetoxins in shellfish for not only foreseeing NSP, but also finding effective treatment of NSP. Therefore, we have carried out further studies on the metabolism of brevetoxins in shellfish associated with NSP.

Materials and methods

Samples

Commercial shellfish, *Austrovenus stutchburyi* (Cockle), *Perna canaliculus* (Green Mussel) and *Crassostrea gigas* (Pacific Oyster), collected from New Zealand in 2001 were used as toxin-free materials. Toxic *A. stutchburyi, P. canaliculus* and *C. gigas,* harvested in January (from Whangarei and Coromandel) and February (from Tiki Road) in Northland, and stored at –30 °C, were received from Cawthron Institute, Nelson, New Zealand.

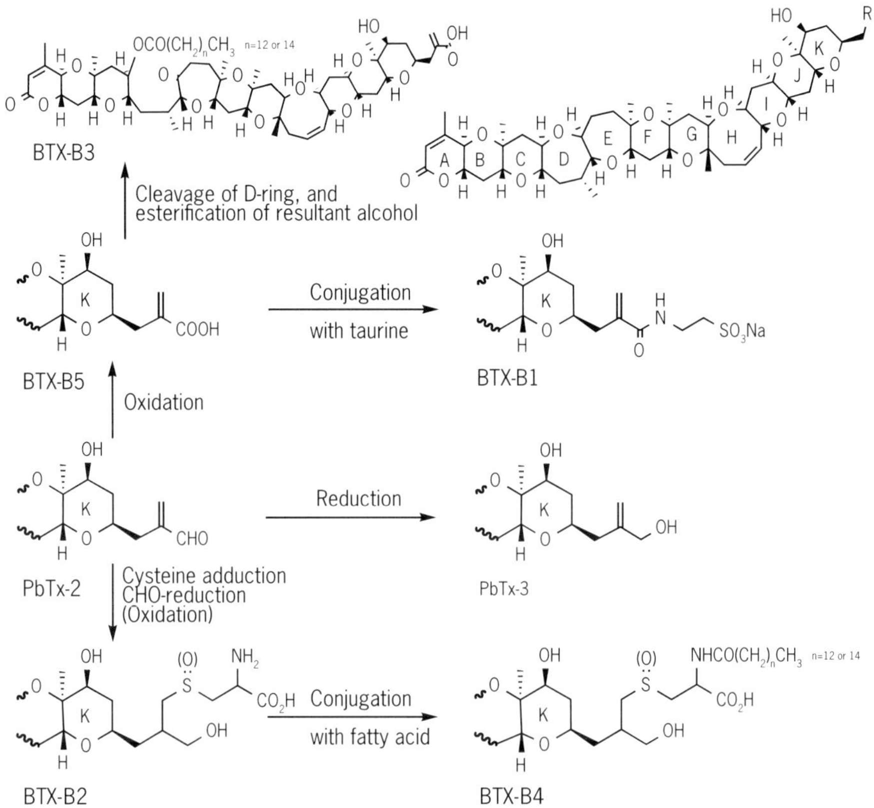

Figure 1. Plausible metabolic pathway of PbTx-2 in shellfish.

Reagents

All HPLC or analytical-grade solvents were purchased from Wako Pure Chemical Industries Ltd. (Osaka, Japan). BTX-B1 and BTX-B5 standards were isolated from toxic *A. stutchburyi* according to the method of Ishida *et al.* (1995, 2004a). PbTx-2 and PbTx-3 standards were purchased from LATOXAN (Valence, France).

Extraction for LC-MS/MS and mouse bioassay

Shellfish homogenates (0.5 g) were extracted under reflux, with stirring, for 20 minutes with either (1) 2 ml 80% MeOH, (2) 2 ml acetone, (3) 2 ml diethyl ether (Nozawa *et al.*, 2003), or (4) 2 ml dichloromethane (CH_2Cl_2). Extracts (1), (2) and (3) were partitioned between 2 ml CH_2Cl_2 and 2 ml water. The dichloromethane phase was further partitioned between 0.5 ml *n*-hexane and 0.5 ml 80% MeOH. The 2 ml CH_2Cl_2 extract (4) was similarly partitioned between 0.5 ml *n*-hexane and 0.5 ml 80% MeOH, and the methanolic phase thus obtained used for LC-MS/MS analysis of PbTx-2, PbTx-3 and BTX-B5. The methanolic phase obtained from fractionation of extract (1) was used for LC-MS/MS analysis of PbTx-3, BTX-B1 and BTX-B5, and a mouse bioassay.

Mouse bioassay

The 80% methanolic phases obtained from the fractionation of extracts (2) and (3), and a portion of the final methanolic fraction of extract (1) from the paragraph above, were used in the mouse bioassay. The methanolic fraction of extract (2) was only used directly whereas the methanolic fractions of extracts (1) and (3) were used directly and also fractionated further on a silica gel column, and the subsequent eluates tested separately. The silica gel columns were first eluted with CH_2Cl_2+MeOH (90:10) (CM) followed by CH_2Cl_2 +MeOH+H_2O (65:45:10) (CMW).

The mouse bioassays were conducted according to the National Research Council Guidlines on best practices for animal toxicity testing (NRC, 1992, 1996). Each test sample was suspended in 1% Tween 60 in physiological saline and administered by intraperitoneal injection to ddY mice weighing 18–22 g. The toxicity in mouse units (MU) was extrapolated from the dose-response table provided in the official method protocol (Delaney, 1985). Mouse units of PbTx-2, PbTx-3, BTX-B1 and BTX-B5 were converted by using the conversion factors of 3.2, 3.6, 4.0 and 4.0 μg/MU, respectively, based on the mouse bioassay data.

LC-ESI-MS/MS equipment and conditions

ESI-MS/MS

Mass spectrometric measurements were performed with an API 3000 LC-MS/MS System (Applied Biosystems, Forster City, CA, U.S.A.) and a Quattro Ultima triple-stage quadrupole MS system (Micromass, Cheshire, UK) equipped with an electrospray ionisation (ESI) interface. Masslynx NT (version 3.5) and Analyst (version 1.3) softwares were used for data acquisition. Full-scan mass spectra in both positive and negative modes were acquired in single MS over the mass range of m/z 300–1200 by direct infusion of toxins in the eluent with a flow rate of 0.2 ml/min. 0.1% formic acid–acetonitrile was used as eluents.

Full-scan collision-induced dissociation (CID) spectra were acquired by colliding the quadrupole 1 (Q1)-selected precursor ion at a nitrogen pressure of approximately

3 x 10^{-5} torr in the collision cell (Q2) and scanning the second quadrupole mass spectrometer Q3 in the range of m/z 50–1100. A precursor ion for CID and a diagnostic product ion for the toxin were identified to carry out selective reaction monitoring (SRM) LC-MS/MS analysis. The source and dissolving temperatures were maintained at 100 and 500 °C, respectively. The electrospray capillary voltage and cone voltage were optimised for each toxin.

Liquid chromatography

Liquid chromatography was performed on an Agilent 110 LC system (Agilent Technologies, Waldbronn, Germany). Chromatographic separation of PbTx-2, PbTx-3, BTX-B5 and BTX-B1 was carried out on a Cadenza CD-C18 (3 x 150 mm, 3 μm, Imtact) column. The column temperature was kept at 45 °C. A gradient, 20 to 75% acetonitrile containing 0.1% formic acid, was run between 0 and 20 min and held at 70% until 30 min. The gradient was then increased linearly to 100% acetonitrile containing 0.1% formic acid at 30 min and held for 5 min to wash the column. The column was returned to initial conditions at 35 min and the column conditioned for 10 min before the next injection. The flow rate from 0 to 30 min was 0.3 ml/min and that from 30 to 35 min was 0.5 ml/min.

Under the above LC SRM conditions, the linearity of dose-response was examined at four known dose levels by injecting 50% MeOH solution of PbTx-2, PbTx-3, BTX-B1 or BTX-B5. Peak areas were used to express peak intensity. The minimum detection levels for matrix-free and for matrix pure PbTx-2, PbTx-3, BTX-B1 or BTX-B5 were estimated from the data. These brevetoxin standards were dissolved in 50% MeOH at the concentrations of 2, 8, 80 and 800 ng/ml, respectively.

Quantitative determination of PbTx-2, PbTx-3 BTX-B5 and BTX-B1 by LC-MS/MS

Ten microlitres of the final 80% methanolic fraction from either 80% MeOH or CH_2Cl_2 extract partitioning of each shellfish homogenate was used for LC-MS/MS analysis of PbTx-2, PbTx-3, BTX-B1 and/or BTX-B5. Reference toxin solution (10 μl) was injected before and after each sample analysis. The peak area was used for calibrating toxin amounts in tissues.

Recovery experiment

Recoveries of PbTx-2, PbTx-3, BTX-B1 or BTX-B5 were determined by either using whole shellfish meat (n=3) tissue homogenates of non-toxic *A. stutchburyi*, *P. canaliculus* and *C. gigas* or the corresponding 80% MeOH and CH_2Cl_2 extracts. The tissue homogenates or extracts were spiked with the toxins at concentrations of

2, 8, 80 and 800 ng/g, respectively. The homogenates or extracts were fractionated as described in the extraction for LC-MS/MS and mouse bioassay above. The determination limit was calculated from the signal-to-noise ratio.

Results and discussion

Optimisation of LC-ESI-MS/MS conditions

In mass spectrometry with an electrospray ionisation interface operating in the positive or negative ion mode, the protonated ions $[M+H]^+$ of PbTx-2 (m/z 895), $[M+H]^+$ of PbTx-3 (m/z 897), $[M-H]^-$ of BTX-B5 (m/z 909), and $[M-Na]^-$ of BTX-B1 (m/z 1016) were generated abundantly, when 0.1% formic acid – acetonitrile was used as the mobile phase for column chromatography. The product ions of m/z 877, 725, 111 and 80 from PbTx-2, PbTx-3, BTX-B5 and BTX-B1 were identified, respectively.

In order to achieve maximum selectivity as well as quantitative analysis, SRM was then implemented. Based on the MS/MS fragmentations of PbTx-2, PbTx-3, BTX-B1 and BTX-B5, combination precursor-product ions of m/z 895 – 877 for PbTx-2, of m/z 897 – 725 for PbTx-3, of m/z 1016 – 80 for BTX-B1 and of m/z 909 – 111 for BTX-B5 were considered to be suitable for confirmatory analysis (Ishida *et al.*, 2004a, b). Typical chromatograms from the SRM LC-MS/MS analysis of standard PbTx-2, PbTx-3 BTX-B1 and BTX-B5 (80 ng/ml) are shown in Figure 2. Good linearity (R^2>0.99) of the peak area with increasing amounts of both toxins was confirmed over the range 0.02 to 8 ng per injection (2–800 ng/g of whole shellfish meat). The determination limits of PbTx-2 and BTX-B5 in the spiked shellfish, based on a signal-to-noise ratio of 5, were 2, 0.2, 0.4 and 0.4 ng/g of whole shellfish meat, respectively.

Recoveries of PbTx-2, PbTx-3, BTX-B1 and BTX-B5

The toxins PbTx-2, PbTx-3, BTX-B5 and BTX-B1 were spiked into tissue homogenates at four dose levels and extracted with either 80% MeOH or CH_2Cl_2. The extracts were subjected to fractionation(s) with CH_2Cl_2/H_2O and/or n-hexane/80% MeOH as indicated in the extraction for LC-MS/MS and mouse bioassay. Methanolic extract recoveries of PbTx-3, BTX-B5 and BTX-B1 from spiked tissue homogenate, estimated from the LC-MS/MS data, were determined for *A. stutchburyi* (90–104, 92–112 and 96–113%), *P. canaliculus* (83–98, 82–98 and 84–93%) and *C. gigas* (81–99, 69–75 and 99–105%). Dichloromethane recoveries of PbTx-2 from the various tissue homogenates were lower: *A. stutchburyi* (42–54%), *P. canaliculus* (40–44%) and *C. gigas* (46–50%).

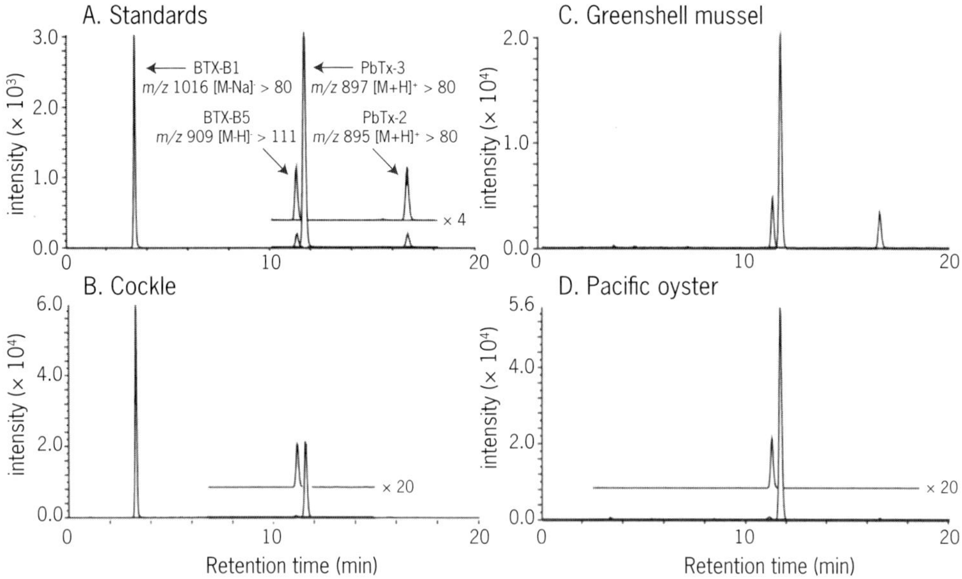

Figure 2. SRM LC-MS/MS chromatograms of 80 ng/ml standards (A), fractions from 80% MeOH and CH_2Cl_2 extracts of toxic cockles (B), greenshell mussels (C) and Pacific oysters (D) from field with positive and negative ESI, respectively.

Extraction and fractionation method and toxicity of shellfish extracts in the mouse bioassay

Whole meat homogenates of cockles, greenshell mussels and Pacific oysters were partitioned as delineated in the section on the extraction for LC-MS/MS. All the final methanolic fractions from the three initial extraction solvents tested [(1) - 80% MeOH), (2) - acetone), and (3) – diethyl ether)] showed apparent NSP toxicity in the mouse bioassay (Table 1). The toxicities of fractions from the 80% MeOH and acetone extracts of each shellfish were similar, but were 2-fold greater than those of the 80% MeOH fractions from the diethyl ether extracts.

Only fractions from 80% MeOH extracts and diethyl ether extracts were subject to further silica gel column fractionation with CH_2Cl_2+MeOH (10:1) (CM) and CH_2Cl_2+MeOH+water (65:40:5) (CMW). The toxicities of the CM eluates were very similar for all shellfish species (Table 1). In contrast, the toxicity of the CMW eluate from the 80% MeOH extract was higher than that from the diethyl ether extract.

Previously Hannah *et al.* (1995) and Naar *et al.* (2002) reported that acetone was more effective in extracting lipid-soluble toxins from NSP-associated toxic shellfish

Table 1. Comparative toxicity of the final methanolic fractions of extracts (1), (2) and (3) and the CM and CMW column eluates of extracts (1) and (3).

Shellfish (sample date)	Extraction solvent	Toxicity (mouse units/g)		
		80% MeOH	CM[1] eluate	CMW[2] eluate
Greenshell mussels (January-93)	80% MeOH	2.9	0.4	2.4
	Acetone	2.7	-	-
	Ether	1.2	0.5	0.6
Cockle (January-93)	80% MeOH	1.4	-	0.8
	Acetone	1.2	0.2	-
	Ether	0.5	-	0.2
Pacific oyster (February-93)	80% MeOH	1.2	0.2	0.5
	Acetone	1.4	0.5	-
	Ether	0.7	-0.5	0.2

[1]CH_2Cl_2+MeOH (10:1) eluate.
[2]CH_2Cl_2+MeOH-Water (65:40:5) eluate.
- Not done.

than diethyl ether. Our findings support this observation, as 80% MeOH and acetone were superior to diethyl ether in extracting lipid-soluble toxins.

The mouse toxicities of highly polar CH_2Cl_2+MeOH+water (65:40:5) eluates from the 80% MeOH and acetone extracts were higher than those from diethyl ether extracts, while those of the less polar CH_2Cl_2+MeOH (10:1) eluates were nearly identical for all extracts and shellfish species. These results indicate that the acetone and 80% MeOH extraction methods are effective in extracting highly polar lipophilic toxins from tissues of NSP-associated toxic shellfish.

Analysis of toxic shellfish

The profiles of toxins in the shellfish samples were obtained by application of the SRM LC-MS/MS method (Figure 2). The identifications were unambiguously confirmed by the comparisons of sample chromatographic peak retention times and responses with authentic standards. The contents of PbTx-2, PbTx-3, BTX-B5 and BTX-B1 (ng/g of whole shellfish meat) in the tested samples from shellfish are shown in Table 2 (Ishida *et al.*, 2004b; Nozawa *et al.*, 2003). PbTx-2 and PbTx-3 were found

Table 2. Concentrations of PbTx-3 and BTX-B1 in 80% MeOH fraction from 80% MeOH extract and of PbTx-2 and BTX-B5 in 80% MeOH fraction from CH_2Cl_2 extract of whole meat of toxic New Zealand shellfish.

Shellfish (sample date)	Concentration (ng/g)			
	PbTx-2	PbTx-3	BTX-B5	BTX-B1
Greenshell mussel (January-93)	320 (±20)	375 (±36)	655 (±38)	14 (±2)
Cockle (January-93)	N.D.	291 (±18)	102 (±6)	1192 (±33)
Pacific oyster (February-93)	12 (±3)	1014 (±57)	123 (±15	3 (±1)

Results are mean (±SD) (n=3); N.D.= not detected.

in the toxic oyster and PbTx-3, BTX-B5 and BTX-B1 in the toxic cockle, as reported previously (Ishida *et al.*, 1995, 1996). It was further found that the toxic cockle was also contaminated with BTX-B5, and the toxic mussel with PbTx-2, PbTx-3, BTX-B1 and BTX-B5. The amounts of PbTx-2, PbTx-3, BTX-B1 and/or BTX-B5 found in the New Zealand toxic shellfish demonstrated that the positive results in mouse bioassay were to some degree due to high levels of these parent toxins and metabolites. About 29% of the total mouse toxicity of *A. stutchburyi* could be accounted for by PbTx-3, BTX-B1 and BTX-B5, 26% of that of *C. gigas* by PbTx-2, PbTx-3 and BTX-B5, and 13% of that of the *P. canaliculus* by PbTx-2, PbTx-3 and BTX-B5.

The lipophilic toxins PbTx-3 and BTX-B5 and the polar lipophilic toxin BTX-B1 are certainly involved in the NSP-associated toxicity of *A. stutchburyi,* as are PbTx-3 and BTX-B5 in that of *C. gigas,* while PbTx-2, PbTx-3 and BTX-B1 seem to play minor roles in the case of *P. canaliculus.* Murata *et al.* (1998) and Morohashi *et al.* (1995, 1999) showed that polar lipophilic toxins, BTX-B2, BTX-B3 and BTX-B4 were responsible for much of the toxicity of NSP-contaminated *P. canaliculus.* PbTx-2, PbTx-3, BTX-B2, BTX-B3, BTX-B4 and BTX-B5 could account for most of the toxicity of *P. canaliculus.*

The mouse bioassay is the only method approved for NSP regulatory application. However, an alternative method would be highly desirable, because the bioassay is less sensitive and specific for brevetoxins, especially metabolites. The LC-MS/MS method unambiguously confirmed the presence of PbTx-2, PbTx-3, BTX-B1 and/or BTX-B5 in all of the tested toxic shellfish samples, at concentrations ranging from 3 to 1,192 ng per gram of whole shellfish meat, thereby accounting in part for the lethality of the 80% MeOH fractions from all of the extracts in the mouse bioassay. This suggests that our LC-MS/MS method provides analytically useful data for estimating the toxicity of shellfish by quantitative determination of PbTx-2, PbTx-3, BTX-B1 and/or BTX-B5 in these shellfish. PbTx-3 and BTX-B5 were detected at significant levels in all three shellfish, indicating that these toxins are useful as markers for monitoring shellfish toxicity after algal blooms involving *Karenia* species.

These findings from chemical studies on shellfish involved in the 1992–1993 outbreak of NSP in New Zealand indicate that bivalves metabolise brevetoxins in species-specific ways (Ishida *et al.*, 1995, 1996, 2004a, b; Morohashi *et al.*, 1995, 1999; Murata *et al.*, 1998; Nozawa *et al.*, 2003). Structural modification of dinoflagellate toxins by vector animals seems to be common among shellfish and fish, as exemplified by oxidation of ciguatoxins by fish (Yasumoto and Murata, 1993; Satake *et al.*, 1997) for detoxification. Thus, it is strongly suggested that all three New Zealand bivalves have a common ability, more or less, to metabolise PbTx-2 to the primary metabolites, the reduction product PbTx-3 and the oxidation product BTX-B5, and to transform BTX-B5 to BTX-B1, and then to excrete these metabolites into seawater. More work will be needed to fully evaluate the metabolites of brevetoxins in shellfish including the *S*-oxidised and terminal aldehyde-reduced *S*-cysteine adduct, BTX-B2 and its *N*-myristoyl and *N*-palmitoyl conjugates, BTX-B4.

Conclusion

Procedures for quantitative determination of PbTx-2, PbTx-3, BTX-B1 and BTX-B5 by liquid chromatography-tandem mass spectrometry were developed and successfully applied to toxic New Zealand shellfish harvested in the summer of 1992/93. High levels of PbTx-2, PbTx-3 and BTX–B5 were detected in *C. gigas*, PbTx-3, BTX-B1 and BTX-B5 in *A. stutchburyi*, and PbTx-2, PbTx-3 and BTX-B5 in *P. canaliculus*. Trace levels of BTX-B1 were found in *P. canaliculus* and *C. gigas*. PbTx-3 and BTX-B5 could be useful as markers for monitoring shellfish toxicity after a *K. brevis* red tide event. All three bivalves seem to have the ability to metabolise brevetoxins (PbTxs) in both species-specific and common ways. The bivalves all produce the PbTx-2 reduction product PbTx-3 and the oxidation product BTX-B5 to various extents.

References

Baden, D.G., and Tomas, C.R. 1988. Variations in major toxin composition for six clones of *Ptychodiscus brevis*. *Toxicon* 26: 961-963.

Bates, M., Baker, M., Wilson, N., Lane L., and Handford S., 1993. Epidemiological overview of the New Zealand shellfish toxicity outbreak. *The Royal Society of New Zealand Miscellaneous Series* 24: 35–40.

Chou, H.-N., and Shimizu, Y., 1982. A new polyether toxin from *Gymnodinium breve* Davis. *Tet. Lett.* 23: 5521–5524.

Delaney, J.E., 1985. Chapter 4: Bioassay procedures for shellfish toxins. In: *Laboratory procedures for the examination of seawater and shellfish*. 5th Ed. Greenburg, & Hunt, (Eds.). American Public Health Association Inc., Washington, DC., pp. 66–78.

Hannah, D.J., Till, D.G., Deverall, T., Jones, P.D., and Fry, J.M., 1995. Extraction of lipid-soluble marine biotoxins. *J. AOAC. Int.* 78: 480-483.

Ishida, H., Muramatsu, N., Nukaya, H., Kosuge, T., and Tsuji, K., 1996. Study on neurotoxic shellfish poisoning involving the oyster, *Crassostrea gigas*, in New Zealand. *Toxicon* 34: 1050–1053.

Ishida, H., Nozawa, A., Hamano, H., Naoki, H., Fujita, T., Kaspar, H. F., and Tsuji, K., 2004. Brevetoxin B5, a new brevetoxin analog isolated from cockle, *Austrovenus stutchburyi*, the marker for monitoring shellfish neurotoxicity in New Zealand. *Tet. Lett.* 45: 29–33.

Ishida, H., Nozawa, A., Nukaya, H. and Tsuji, K., 2004, Comparative concentrations of brevetoxins PbTx-2, PbTx-3, BTX-B1 and BTX-B5 in cockle, *Austrovenus stutchburyi*, greenshell mussel, *Perna canaliculus*, and Pacific oyster, *Crassostrea gigas*, involved neurotoxic shellfish poisoning in New Zealand. *Toxicon*, 43: 779-789.

Ishida, H., Nozawa, A., Totoribe, K., Muramatsu, N., Nukaya, H., Tsuji, K., Yamaguchi, K., Yasumoto, T., Kaspar, H., Berkett, N., and Kosuge, T., 1995. Brevetoxin B1, a polyether marine toxin from the New Zealand shellfish, *Austrovenus stutchburyi*. *Tet. Lett.* 36: 725–728.

Morohashi, A., Satake, M., Murata, K., Naoki, H., Kaspar, H.F., and Yasumoto, T., 1995. Brevetoxin B3, a new brevetoxin analog isolate from the greenshell mussel *Perna canaliculus* involved in neurotoxic shellfish poisoning in New Zealand. *Tet. Lett.* 36: 8995–8998.

Morohashi, A., Satake, M., Naoki, H., Kaspar, H.F., Oshima, Y., and Yasumoto, T., 1999. Brevetoxin B4 isolated from *Perna canaliculus*, the major toxin involved in neurotoxic shellfish poisoning in New Zealand. *Nat. Toxins* 7: 45–48.

Murata, K., Satake, M., Naoki, H., Kaspar, H.F. and Yasumoto, T., 1998. Isolation and structure of a new brevetoxin analog, brevetoxin B2, from greenshell mussels from New Zealand. *Tet. Lett.* 54: 735–742.

Naar, J., Bourdelais, A., Tomas, C., Kubanek, J., Whitney, P.L., Flewelling, L., Steidinger, K., Lancaster, J., and Baden, D.G., 2002. A competitive ELISA to detect brevetoxins from *Karenia brevis* (formerly *Gymnodinium breve*) in seawater, shellfish, and mammalian body fluid. *Environ. Health Perspect.* 110: 179-185.

Nozawa, A., Tsuji, K., and Ishida, H., 2003. Implication of brevetoxin B1 and PbTx-3 in neurotoxic shellfish poisoning in New Zealand by isolation and quantitative determination with liquid chromatography-tandem mass spectrometry. *Toxicon* 42: 91–103.

NRC (National Research Council), 1992. *Recognition and alleviation of pain and distress in laboratory animals*. Washington DC, National Academy Press.

NRC (National Research Council), 1996. *Guide for the care and use of laboratory animals*. 7th edition, Washington DC. National Academy Press.

Satake, M., Ishibashi, Y., Legrand, A-M., and Yasumoto, T., 1997, Isolation and structure of ciguatoxin-4A, a new ciguatoxin precursor, from cultures of dinoflagellate *Gambierdiscus toxicus* and the parrotfish *Scarus gibbus*. *J. Am. Chem. Soc.* 108: 514–515.

Yasumoto, T., and Murata, M., 1993, Marine toxins. *Chem. Rev.* 93: 1897–1909.

Electro-separation methods for the analysis of phycotoxins

A. Gago-Martínez, P. de la Iglesia, E. Vaquero, J.M. Leão and J.A. Rodríguez-Vázquez

Departamento de Química Analítica y Alimentaria, Facultad de Química, Edificio de Ciencias Experimentales, Campus Universitario, Universidad de Vigo, 36310 Vigo, Spain; anagago@uvigo.es

Abstract

The search for simple and fast methods to be used routinely for the analysis of phycotoxins is one of the main interests of researchers working in the field of marine toxins. Electro-separation methods are usually fast and simple contributing to a reduction in the consumption of solvents as well as increasing efficiency when compared to other separation methods such as liquid chromatography. This work shows examples of the application of capillary electrophoresis for the analysis of phycotoxins such as yessotoxins and amnesic shellfish poisoning toxins. Capillary electrophoresis methods previously developed in our laboratory showed the potential of this technique for such applications. The present work was mainly focused on improving sensitivity, one of the main drawbacks associated with this technique, by optimising the electrophoretic conditions. Different modes of stacking with pre-concentration have been applied leading to significant improvements in sensitivity. Capillary electrophoresis under these improved conditions showed potential for efficient determination of these toxins at concentrations below regulatory limits. This technique can be considered as the basis for alternative methods for the analysis of such phycotoxins.

Keywords: capillary electrophoresis, sample stacking, phycotoxins, yessotoxins, Amnesic Shellfish Poisoning, domoic acid

Introduction

The search for simple and rapid methods to be applied in the routine analysis of phycotoxins is one of the main interests of researchers working in the field. Capillary electrophoresis (CE) has provided simple and fast separations for a wide range of analytes in the pharmaceutical, environmental and food analysis fields (Altria *et al.*, 2001; Frazier *et al.*, 2003; Dabek-Zlotorzynska *et al.*, 2003). The major advantages of CE over other liquid-phase separation techniques such as liquid chromatography are its higher efficiency, shorter analysis times, lower operational costs, minimal sample and reagents consumption, and easy operation. Capillary electrophoresis has already been applied as an alternative for the analysis of algal toxins in previous work (Piñeiro *et al.*, 1999; Gago-Martínez *et al.*, 2003a, b). Nevertheless, no methods have been developed for the routine application of CE to regulatory analysis. This may be explained by the generally low sensitivity provided by CE as a consequence of the small-scale separations which represent a major drawback of this technique to the analysis of algal toxins. Therefore, improvements to sensitivity are one of the main requirements in order to carry out regulatory analysis of phycotoxins in food or environmental samples.

Several analytical approaches including conventional "*off-line*" pre-concentration strategies can be used in order to overcome the lack of sensitivity in CE. However, the sensitivity of electroseparation techniques can still be enhanced while using "*on-line*" pre-concentration when UV detectors are used (Osbourn *et al.*, 2000). The present work outlines the development of stacking procedures and application of CE technique to the analysis of yessotoxins (YTXs) and amnesic shellfish poisoning (ASP) toxins.

Materials and methods

Toxins and standard samples

Yessotoxin standards, phytoplankton samples from an algae culture of *Protoceratium reticulatum* and scallop samples naturally contaminated with ASP toxins from Ireland, were kindly provided by Dr. Kevin James, from the Cork Institute of Technology, Ireland. A research grade analytical standard of yessotoxin was also provided by Paul McNabb from Cawthron Institute, Nelson, New Zealand. Samples of hepatopancreas from the Greenshell mussel (*Perna canaliculus)* naturally contaminated with YTXs were provided by Dr. David Stirling, from the Institute of Environmental Science and Research Limited, Wellington, New Zealand. A certified calibration solution (DACS-1D) and mussel tissue reference material (MUS-1D) containing 87.7 μg/ml and 36.6 μg/g of DA, respectively, were purchased from National Research Council (NRC), Halifax, Canada. Domoic acid (DA) calibration solutions were prepared from DACS-1D using acetonitrile + water (1:9).

Chemicals and reagents

Acetonitrile and methanol (HPLC grade) were obtained from Merck (Darmstadt, Germany). Ultra-pure water was produced using a Milli-Q (Millipore, Bedford, MA, USA) water purification system. Sodium hydroxide pellets, di-sodium hydrogen phosphate (anhydrous salt) for analysis ACS-ISO and ortho-phosphoric acid 85% were purchased from Panreac (Barcelona, Spain). Phosphate 50mM buffer pH 7.0 was purchased from Agilent Technologies (Waldbronn, Germany). Background electrolyte (BGE) for yessotoxins analysis was 50mM methanol + sodium phosphate (40:60) at pH 8.5. BGE for ASP toxins analysis was 25 mM phosphate buffer at pH 7.0. All buffer solutions were filtered through Millex®-HV 0.45 μm polyvinil dilen-fluoride (PVDF) membranes (Millipore, Bedford, MA, USA) and degassed before CE analysis.

Apparatus

Separations were performed on an HP^{3D}CE system (Agilent, Waldbronn, Germany) equipped with a UV diode-array detection (DAD) system and a HP ^{3D}Chemstation for instrument control, data acquisition and data analysis.

Sample preparation and analysis

Extraction and Solid phase extraction (SPE) clean-up for YTX toxins
Extraction and clean-up conditions previously described by Yasumoto and Takizawa (1997) were slightly modified. Yessotoxins were extracted by homogenising (Ultra-Turrax T25, Janke & Kunkel, IKA-Labortechnik, Staufen, Germany) 1g of shellfish hepatopancreas or 1 ml algae extract with 9 ml methanol + water (8:2 v/v) for 3 minutes. The slurry was centrifuged at 3,000 rpm for 10 minutes.

Solid phase extraction (SPE) cleanup was carried out using LC-18 cartridges (3 ml, 500 mg, Supelco, Bellefonte, PA, USA). 5 ml of the centrifugation supernatant was mixed with 15 ml 0.02 M phosphate buffer at pH 5.8 was loaded onto an LC-18 cartridge previously conditioned with 20 ml methanol. The cartridge was washed with 6 ml methanol + water (20:80 v/v) and the toxins eluted with 6 ml methanol + water (70:30 v/v). After evaporation to dryness and redissolving in 500 μl methanol, the extract was ultra-filtered through a Ultrafree®-MC 0.45μm mini-filter (Millipore, Bedford, MA, USA) at 14,000 rpm. The purified extract was stored at –18 °C until analysis.

CE analysis of YTXs

The initial experimental conditions for this study were based on information kindly provided by professor Yasumoto (Japan Food Research Laboratories, Tokyo, Japan). Slight modifications were made to the protocol advised by professor Yasumoto. Electrophoretic analysis was performed using 48.5 cm x 50 μm i.d (40 cm effective length) coated fused-silica capillaries obtained from Agilent Technologies (Waldbronn, Germany). The UV detector was set at 230 nm wavelength. Before injections, the capillary was rinsed with 0.1 N NaOH for 3 min followed by ultrapure water (2 min) and background electrolyte (BGE) 50 mM methanol+sodium phosphate (40:60) at pH 8.5 (5 min). Further equilibration by application of 30 kV for 5 min was used to improve the repeatability. Between injections the capillary was rinsed with BGE for 5 min and then equilibrated by application of 30 kV for 5 min.

Extraction and Solid phase extraction (SPE) clean-up for ASP toxins

Sample extraction and clean-up was carried out under the conditions proposed by Quilliam *et al.* (1989) with minor modifications. Sixteen (16 ml) milliliters methanol+water (1:1 v/v) were added to 4g of pulverised tissue. The sample was homogenised for 3 min with Ultra-Turrax T25 at 10,000 rpm and then centrifuged at 4,500 rpm for 10 min. The supernatant was filtered through a Millex®-HV 0.45 μm filter (Millipore, Bedford, MA, USA). Clean-up was carried out using SPE-C18 cartridges, previously conditioned with 6 ml acetonitrile and 6 ml water. After loading 2 ml sample extract on the cartridge, the ASP toxins were eluted with 3 ml acetonitrile + water (1:9 v/v). The cartridges were not allowed to go dry during the procedure. Purified extracts were stored at 4 °C until analysis.

CE analysis of ASP toxins

CE analysis was performed using uncoated fused-silica capillaries, 62.5 cm (54 cm effective length) x 50 μm i.d. (Composite Metal Services, The Chase, Hallow, UK). UV detection was set at wavelength of 242 nm. Before analysis, the capillary was initially conditioned with a three-step protocol consisting of flushing 0.1 N NaOH solution (10 min) followed by ultrapure water (5 min) and background electrolyte 0.025 M phosphate buffer at pH 7.0 (20 min), and equilibrating by application of 30 kV (2 min). Between injections the capillary was rinsed with running buffer for 5 min and then equilibrated by application of 30 kV for 5 min. The capillary temperature was fixed at 25 °C. A stacking procedure based on large-volume sample stacking (LVSS) was also developed for ASP toxins in order to enhance the sensitivity for this application. The sample, in MeCN+H_2O (1:9), was loaded onto the capillary by pressure of 50 mbar over a 75 seconds time period and then stacked by application of -5 kV during 130 seconds. The separation was then performed at 30 kV.

Results and discussion

CE analysis of YTXs

Sample buffer: Field-amplified Sample Stacking (FASS) mode

The simplest way to carry out pre-concentration based on a stacking mode is to dissolve the sample in a low conductivity buffer (Chien and Burgi, 1991). The electric field strength will be higher over the low conductivity buffer zone than the BGE zone. Under field amplified conditions, the charged analytes will migrate rapidly and then slow down when reaching the buffer interface leading to a narrow band of stacked analytes. In order to optimise this pre-concentration mode, sample buffer was prepared by diluting the BGE in methanol with the aim of reducing the sample conductivity. A range of methanol concentrations (20 to 90%) was studied and the results show increased sensitivity when the sample buffer conductivity was decreased (Figure 1). Efficiency losses were observed above methanol:BGE ratios of 70:30 for the sample solution. The losses could be due to the hydrodynamic backflow with a laminar flow profile generated at the buffer boundaries because the local electroosmotic velocity in the sample plug is greater than the bulk electroosmotic flow (EOF) velocity in the separation compartment.

Sample pH: pH-mediated stacking

The stacking process is greatly influenced by sample pH (Zhao and Lunte, 1999) since the stacking efficiency is limited by the hydrodynamic flow created at the

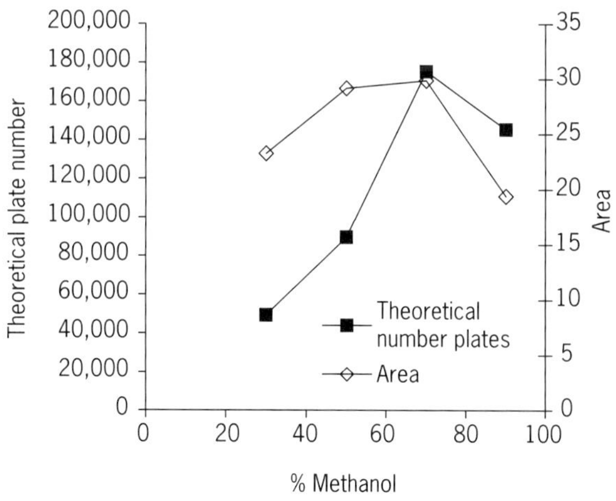

Figure 1. Influence of sample buffer composition on peak efficiency and area for the analysis of 5 μg/ml standard solution of YTX. Sample buffer was prepared by diluting BGE with methanol.

buffer boundaries. A substantial increase in the efficiency was achieved when the sample was prepared in acidic media (Figure 2). The increase in efficiency could be explained by the isotachophoretic behaviour of H_3O^+ and OH^- in the sample buffer and background buffer respectively, as well as the decrease in EOF in the sample compartment under amplified field strength at pH 3.0. The decrease in EOF contributes to reduced hydrodynamic flow at the buffers boundaries.

Reverse polarity: Large-volume Sample Stacking (LVSS) mode

Large-volume sample stacking (LVSS) is an improved stacking mode (Chien and Burgi, 1992), which allows sample buffer removal from the capillary by applying reverse polarity. This procedure allows the separation to be carried out under higher field strength and therefore contributes to achieving shorter and more efficient analysis. This mode has been optimised for YTX enrichment and effective sample buffer removal was obtained by applying -25 kV reverse polarity for 40 s (Figure 3). Analyte losses from the capillary were observed with reverse polarity times of greater than 40 seconds.

Stacking voltage was also studied because the field-amplified conditions can make it different from run voltage initially applied (Vinther and Soeberg, 1991). The best results for this particular application were obtained with relatively high voltage of -25 kV (data not shown), due to the band broadening observed under low pH conditions in the sample compartment. After the optimisation of the stacking conditions, a nearly 20-fold increase in sensitivity was achieved for YTX analysis (Figure 4). These improved conditions provided the sensitivity required in the EC (Regulation EC No 853/2004, L 139/55).

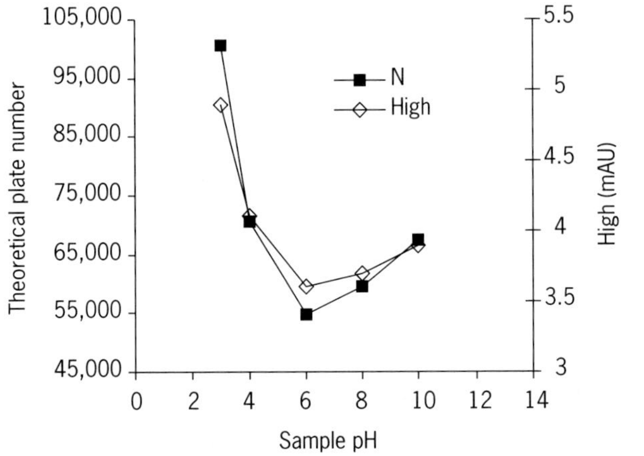

Figure 2. Effect of sample buffer pH on peak efficiency (N) and peak height (High) for 5 μg/ml standard solution of YTX. Sample buffer was held at MeOH+BGE (70:30). BGE.

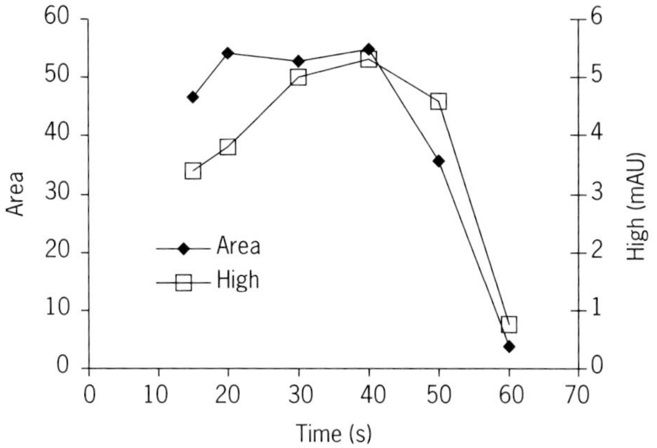

Figure 3. Effect of time for sample buffer removal from capillary on YTX peak height and area. Sample buffer was MeOH+BGE (70:30) adjusted to pH 3.0.

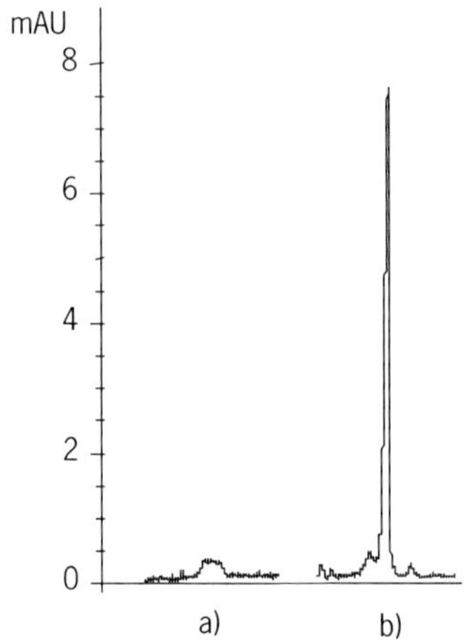

Figure 4. Effect of the stacking procedure on the sensitivity for YTX. 12 μg/ml standard solution loaded hydrodynamically into capillary by application of 50 mbar pressure during 20 s: a) dissolved in BGE. b) Dissolved in MeOH+BGE (70:30) adjusted at pH 3.0, and then removed from the capillary by applying reverse polarity of -25 kV for 40 s.

The optimised conditions were applied to the analysis of naturally contaminated samples. Figure 5 shows the efficiency of CE for the analysis of YTX in different matrices. Under the optimised conditions, only parent YTX was detected in the phytoplankton sample (Figure 5b). However, in contaminated mussel samples, several peaks with very similar electrophoretic behaviour were detected (Figure 5c) in addition to YTX.

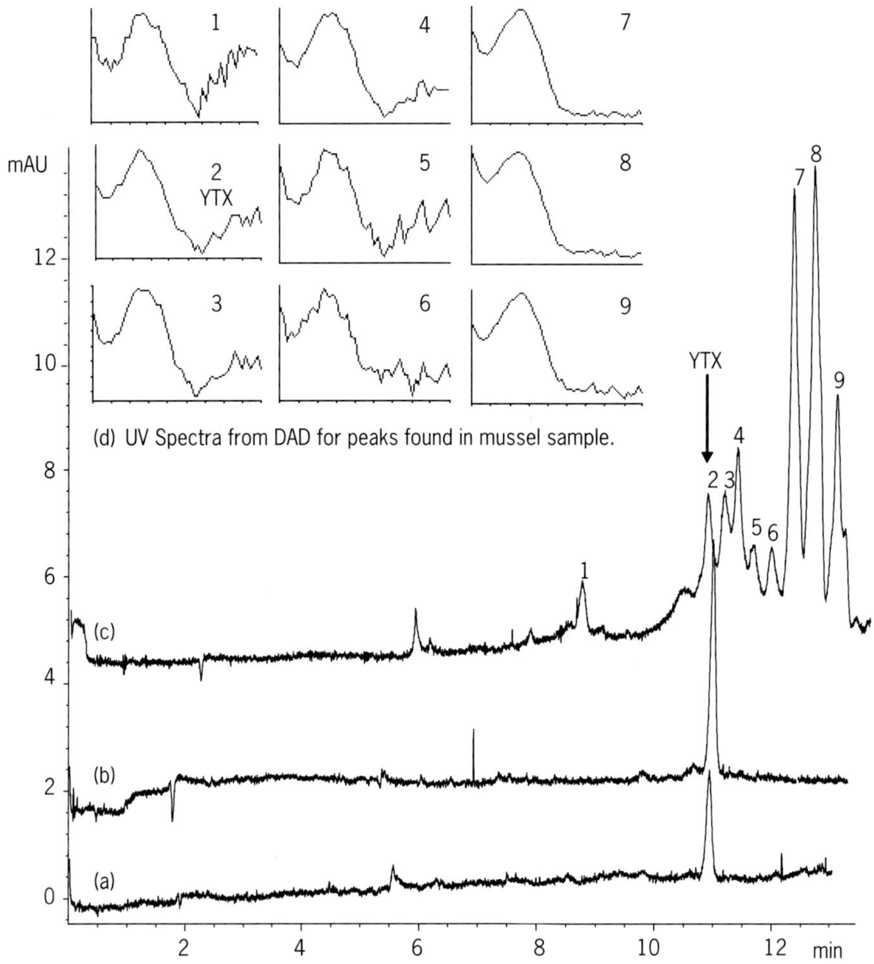

Figure 5. CE-UV/DAD analysis of (a) Yessotoxin standard 2.0 μg/ml, (b) phytoplankton algal extract from P. reticulatum *naturally contaminated with 4.1 mg/l YTX (c) mussel hepatopancreas sample from* P. canaliculus *naturally contaminated with 2.1 mg/kg YTX and (d) UV Spectra obtained from peaks found in hepatopancreas mussel sample by using UV/DAD detector. BGE: sodium monohydrogen phosphate:methanol (60:40) 50 mM adjusted at pH 8.5. Sample solution: methanol+BGE (70:30) adjusted to pH 3.0. Injection: 50 mb for 20 s. Sample Stacking: -25 kV for 40 s. Run: 30 kV, 25 °C. Detection: UV/DAD at 230 nm.*

UV/DAD spectra (Figure 5d) provided a preliminary confirmation of the presence of some YTX analogues which have been previously reported for mussel samples (Ciminiello *et al.*, 2002; Fernandez-Amandi *et al.*, 2002). All yessotoxin analogues which have diene groups (including the most common 45-hydroxyYTX, homoYTX and 45-hydroxyhomoYTX) can be detected by UV detection. Nevertheless, further confirmation is still required and future work will focus on coupling CE to mass spectrometry detection.

CE analysis of ASP toxins

The procedure allowed the concentration of the analyte into a narrow band, enhancing the sensitivity 5 times over that by the conventional CE method for this particular application (Figure 6). Figure 7 shows an example of the application of the procedure to naturally contaminated samples, in this case for contaminated scallops. The efficiency and sensitivity of CE is clearly enhanced under the optimised stacking conditions, allowing the determination of domoic acid at levels below the regulatory limit (20 mg/kg). Table 1 presents the main evaluation parameters obtained for the CE-UV/DAD methods developed for both YTX and DA determination.

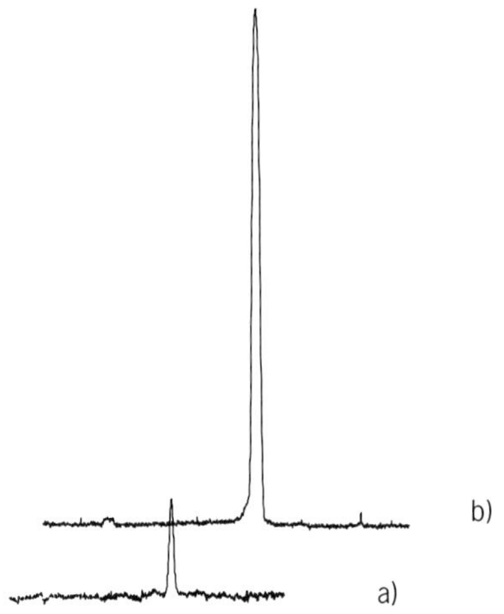

Figure 6. Effect of stacking on the sensitivity for domoic acid. 2.5 μg/ml standard solution of DA in MeCN+H_2O (1:9). Injection conditions: a) 50 mbar pressure for 9 s. b) 50 mbar pressure for 75 s and then applying reverse polarity of -5 kV for 130 s.

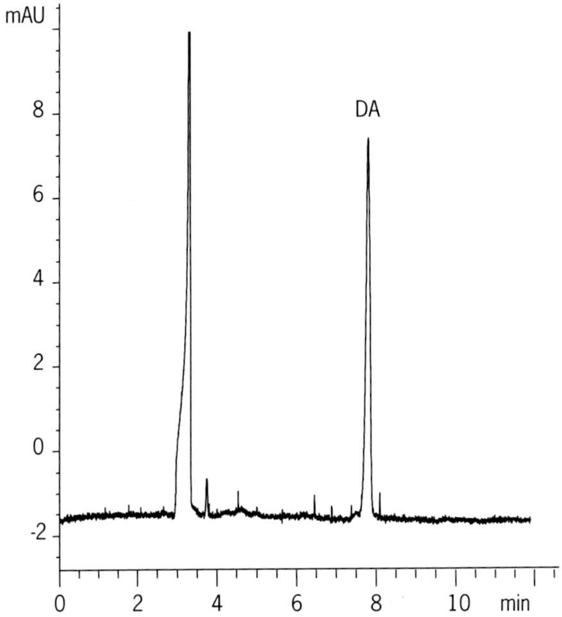

Figure 7. CE-UV/DAD analysis of hepatopancreas scallop sample naturally contaminated with 3.12 mg/kg DA. BGE 25 mM phosphate buffer at pH 7.0. Sample solution: MeCN+H_2O (1:9). Injection: 50 mbar during 75 s. Sample Stacking: -5 kV during 130 s. Run: 30 kV, 25 °C. Detection: UV/DAD at 242 nm.

Table 1. Main evaluation parameters obtained for CE-UV/DAD methods for both YTX and DA determination.

		YTX	**DA**
LOD (μg/ml)		0.3	0.05
LOQ (μg/ml)		1.00	0.18
Repeteability	% RSD (area)	3.2	5.7
	% RSD (migration times)	2.6	3.6
Linear range (μg/ml)		1.0-6.0	0.25-2.0
Linearity (r^2)		0.9975	0.9912

Conclusions

The results obtained in this work show the potential of CE under these improved stacking conditions as an alternative to conventional HPLC methods for a simple and efficient determination of phycotoxins present in biological matrices. The sensitivity of this technique has been clearly enhanced by using an on-line pre-concentration mode based on Large-Volume Sample Stacking.

Acknowledgements

The authors of this work want to acknowledge the financial support from the Spanish Ministry of Education, EU FEDER Ref.: 1FD 97-0095, and the Spanish Fishing Company Rampesca S.A. We would also like to thank Dr. Kevin James, Cork Institute of Technology, Cork, Ireland, for providing YTX standard and phytoplankton cells; Paul McNabb, Cawthron Institute, Nelson, New Zealand, for providing YTX standard and Dr. David Stirling, Institute of Environmental Science and Research Limited, Wellington, New Zealand for providing naturally contaminated mussel digestive gland samples. We are indebted to Professor T. Yasumoto, Japan Food Research Laboratories, Tokyo, Japan for providing the initial CE conditions applied in this study.

References

Altria, K.D., Chen, A.B., and Clohs, L., 2001. Capillary electrophoresis as a routine analytical tool in pharmaceutical analysis. *LC-GC Europe* 14: 736, 738, 740 -744.

Chien, R.L., and Burgi, D.S., 1991. Optimization in Sample Stacking for High-Performance Capillary Electrophoresis. *Analytical Chemistry* 63: 2042-2047.

Chien, R.L., and Burgi, D.S., 1992. Sample Stacking of an Extremely Large Injection Volume in High-Performance Capillary Electrophoresis. *Analytical Chemistry* 64: 1046-1050.

Ciminiello, P., Dell´Aversano, C., Fattorusso, E., Forino, M., Magno, S., and Poletti, R., 2002. Direct detection of yessotoxin and its analogues by liquid chromatography coupled with electrospray ion trap mass spectrometry. *Journal of Chromatography A* 968: 61-69.

Dabek-Zlotorzynska, E., Chen, H., and Ding, L., 2003. Recent advances in capillary electrophoresis and capillary electrochromatography of pollutants. *Electrophoresis* 24: 4128-4149.

Fernandez-Amandi, M., Furey, A., Lehane, M., Ramstad, H., and James. K.J., 2002. Liquid chromatography with electrospray ion-trap mass spectrometry for the determination of yessotoxins in shellfish. *Journal of Chromatography A* 976: 329–334.

Frazier R.A, and Papadopoulou A., 2003. Recent advances in the application of capillary electrophoresis for food analysis. *Electrophoresis* 24: 4095-4105.

Gago-Martínez, A., Leao, J.M., Piñeiro, N., Carballal, E., Vaquero, E., Nogueiras, M., and Rodríguez-Vázquez, J.A., 2003. An application of capillary electrophoresis for the analysis of algal toxins from the aquatic environment. *International Journal of Environmental Analytical Chemistry* 83: 443-456.

Gago-Martínez, A., Piñeiro, N., Aguete, E.C., Vaquero, E., Nogueiras, M., Leao, J.M., Rodríguez-Vázquez, J.A., and Dabek-Zlotorzynska, E., 2003. Further improvements in the application of high performance liquid chromatography, capillary electrophoresis and capillary electrochromatography to the analysis of algal toxins in the aquatic environment. *Journal of Chromatography A* 992: 159-168.

Osbourn, D.M., Weiss, D.J., and Lunte, C.E., 2000. On-line preconcentration methods for Capillary Electrophoresis. *Electrophoresis* 21: 2768-2779.

Piñeiro, N., Leao, J.M., Gago-Martínez, A., and Rodríguez-Vázquez, J.A., 1999. Capillary Electrophoresis with diode array detection as an alternative analytical method for paralytic and amnesic shellfish toxins. *Journal of Chromatography A* 847: 223-232.

Quilliam, M.A., Sim, P.G. McCulloch, A. W., and McInnes, A.G., 1989. High-performance liquid chromatography of domoic acid, a marine neurotoxin, with application to shellfish and plankton. *International Journal of Environmental Analytical Chemistry* 36: 139-154.

Vinther, A., and Soeberg, H., 1991. Temperature Elevations of the Sample Zone in Free Solution Capillary Electrophoresis. *Journal of Chromatography* 559: 27-42.

Yasumoto, T., and Takizawa, A., 1997. Fluorometric Measurements of Yessotoxins in Shellfish by High-pressure Liquid Chromatrography. *Bioscience, Biotechnology, Biochemistry* 61: 1775-1777.

Zhao, Y., and Lunte, C.E., 1999. pH-mediated Field Amplification On-Column Preconcentration of Anions in Physiological Samples for Capillary Electrophoresis. *Analytical Chemistry* 71: 3985-3991.

Monitoring microcystin and physicochemical parameters: Role of agricultural–aquaculture activity in the water quality

M. Kamogae[1], H. Hashimoto[2], C.G. Pádua[2], F.S. Soares[2], A.P.F.L. Bracarense[2], M.J.S. Yabe[2], E.Y.S. Ono[2], M.C. Bittencourt-Oliveira[3], C.L. Sant'ana[4], T. Tsutsumi[5], S. Nagata[6], Y. Ueno[6], K.-I. Harada[7] and E.Y. Hirooka[2]

[1]UNOPAR, Northern Parana University, Parana, Brazil
[2]State University of Londrina, Center of Agricultural Sciences,Department of Food & Drug Technology,P. O. Box 6001, 86051-990 Londrina/PR, Brazil; hirooka@uel.br; mjyabe@uel.br
[3]ESALQ-USP, University of São Paulo, Brazil
[4]Botanical Institute, São Paulo, Brazil
[5]National Inst. Health Science, Tokyo, Japan
[6]Science University of Tokyo, Japan
[7]Laboratory of Environmental Sciences, Faculty of Pharmacy, Meijo University, Nagoya, Japan

Abstract

The intensity of agricultural activity in the state of Parana, Brazil, has been a growing concern due to eutrophication of the aquatic environment. Occurrence of the microcystins (MCs), physicochemical (eutrophication) parameters and *Cyanophyceae* (cyanobacteria) in the Itaipu lake and Capivara dam in the southwest and north of the state of Parana, respectively, were monitored. 56 fresh water (June 1999 to October 2000) and 14 bloom (February 2000 to August 2001) samples were collected from Lakes Itaipu and Barra do Irai, and Capivara dam, while 39 aquaculture samples were collected (October 2002 to October 2003) from three distinct aquaculture systems in northern Parana. The samples were analysed for microcystins using a monoclonal antibody (anti-microcystin M8H5 mAb) based IC-ELISA, HPLC and MALDI-TOFMS, or a commercial ELISA kit. The 56 fresh water samples and 39 aquaculture samples were also assayed for *Cyanophyceae* (cyanobacteria) and physicochemical parameters. *Mycrocystis* spp., *Anabaena* spp., *Oscillatoria* spp. were detected in almost every freshwater sample. 17 (53.12%) samples from Capivara and Tres Bocas dams had MCs from non-detectable to 1.7 μg/l range (mean, 0.17 μg/l), while 22 (91.26%) samples from Itaipu contained MCs from non-detectable to 240.0 μg/l. Microcystin-LR (MC-LR) was detected in 7 bloom samples (max. 48.13 μg/100 g dry cell, Itaipu, April 2000), besides 3 MC-YR positive samples and 3 MC-

RR positive samples. The monitoring of aquacultures indicated significant changes in physicochemical parameters (max. total N_2, 6.09 mg/l and max. PO_4, 11.14 mg/l), when the three distinct aquaculture systems were compared, *i.e.* provision water, discharge water and reservoir with higher fish population. Fresh water for human consumption that is exposed to aquaculture discharge may be hazardous hence preventative and control measures would prevent both ecosystem and human health hazard.

Keywords: microcystins, eutrophication, water quality

Introduction

The microcystins (MCs) are cyclic heptapeptide toxins produced in eutrophic fresh water during high proliferation of phytoplanktonic communities of *Cyanophyceae*-'blooms' (Reynolds, 1981). *Cyanophyceae* will proliferate in low turbulence water when the temperature is 18 to 30 °C and the pH is 6 to 8 (Carmichael, 1996). The intensity of agricultural activity in the state of Parana, Brazil, has been a growing concern due to eutrophication of the aquatic environment. Eutrophication affects the physical and chemical characteristics of the water mainly affecting the content of phosphorus (as PO_4) and nitrogen (N), decreasing dissolved oxygen (DO) and water quality, leading to death of some species of fish.

The Parana river traverses the Brazilian fluvial basin, the most important industrial and agricultural region in the country. Progressive urbanisation, associated with agribusiness and industrial development in the north of Parana, has stimulated favourable conditions for *Cyanophyceae* growth, as evidenced by eutrophication observed in Capivara and Tres Bocas dams. Aquaculture has also been increased, generating a chemical balance favourable for *Cyanophyceae* proliferation (Hashimoto *et al.*, 2003).

The occurrence of MCs in Parana State has been investigated since 1995 when the focus of the investigations was on potential risks from microcystins detected in Itaipu lake (Hirooka *et al.*, 1999, Kamogae, 2002). In this study, MC levels, *Cyanophyceae* and physicochemical characteristics were monitored in several water bodies of Parana State. The water bodies monitored were Capivara, Tres Bocas and Barra do Irai dams and Itaipu lake. Itaipu lake and Capivara dam are located in the Parana river basin in the State of Parana, Brazil. Microcystins were monitored using indirect competitive enzyme linked immunosorbent assay (IC-ELISA), QuantiPlate® ELISA kit, and high performance liquid chromatography (HPLC) and matrix assisted laser desorption/ionisation-time of fight mass spectrometry (MALDI-TOFMS).

Material and methods

Sampling sites

The samples were collected from Capivara and Tres Bocas dams in the north, Itaipu lake in the southwest and Barra do Irai lake in the south of Parana State. The sampling of Capivara dam was carried out at two sites (Capivara and Paranatur) in Primeiro de Maio county and a single site (Ponte Caida) in Sertanopolis county. Tres Bocas dam was sampled from Londrina county. Sampling of Itaipu lake was carried out within the municipal districts of Santa Helena, São Miguel of Iguaçu and Santa Terezinha of Itaipu. 56 freshwater samples were collected from Lake Itaipu, Capivara and Tres Bocas dams from June, 1999 to October, 2000. In addition, 14 bloom samples were collected from Lake Itaipu, and Barra do Irai and Capivara dams from February 2000 to August 2001. Furthermore, 39 samples were collected from 3 aquacultures, Esperança (aquaculture A), Cafezal stream (aquaculture B) and Tres Bocas Dam (aquaculture C) in Londrina county, northern Parana State, Brazil from October, 2002 to October, 2003. The aquaculture samples were collected from three distinct points of each aquaculture system, *i.e.* provision water (point 1), reservoir with higher fish population (point 2) and discharge water (point 3).

Sample preparation

Samples for total MCs determination were immediately mixed with sodium-azide (0.1% w/v) and refrigerated at 4 °C; samples for *Cyanophyceae* microscopy were treated with lugol (1%) and maintained at room temperature, while those for physicochemical analysis were maintained at 4 °C without addition of any preservatives.

ELISAs for detection of total microcystins

The 56 freshwater samples collected June 1999 to October 2000 were analysed for total microcystins using a M8H5 monoclonal antibody based IC-ELISA according to the method of Nagata *et al.* (1996). Briefly, 95% purity microcystin-LR (MC-LR) conjugated to bovine serum albumin (MCLR-BSA) was added to the microtiter plate wells followed by the addition of sample mixed with the M8H5 monoclonal antibody. After overnight incubation, anti-mouse IgG–goat-peroxidase (1:5000 in PBST with 0.5% gelatin) was added, and after incubation, the reaction was terminated by the addition of 3,3'-5,5'tetrametil-benzidine (TMBZ). Total microcystin content was determined spectrophotometrically at 450 nm (Bio-Tek Instruments, Inc., High Land Park, USA). The limit of quantification was 0.05μg/l.

The 39 aquaculture samples collected October 2002 to October 2003 were analysed for total microcystins using the commercial test kit QuantiPlate® ELISA from EnviroLogix Inc., (Portland, ME, USA). The QuantiPlate® is a competitive format ELISA test kit with a limit of quantification of 0.147μg/l.

Physicochemical parameters of water samples

The 56 freshwater samples collected June 1999 to October 2000 and 39 aquaculture samples (October, 2002 – October, 2003) were analysed for physicochemical parameters: temperature, alkalinity, pH, total nitrogen, nitrate, nitrite, ammonia, total phosphate, turbidity, dissolved oxygen, organic matter, conductivity, sulfate and chlorophyll *a* (American Public Health Association, 1989). Total nitrogen was determined by the Kjeldahl method and chlorophyll *a* was determined according to the method of Lorenzen (1967).

Cyanophyceae *identification*

The 56 freshwater samples (June, 1999 to October, 2000) and 39 aquaculture samples (October, 2002 to October, 2003) were analysed for *Cyanophyceae* presence (identity). Samples preserved in lugol 1% were counted in a sedimentation chamber by the Utermohl method, and density expressed as cells/ml (Wilovert microscope, 400x, WESCO - Western Scientific Co., California, USA). The *Cyanophyceae* were identified based on Komarek and Anagnostidis (1986).

Analysis of MCs in bloom samples by HPLC and Matrix Assisted Laser Desorption/Ionisation-Time Of Flight Mass Spectrometry (MALDI-TOFMS)

The 14 'bloom' samples collected February, 2000 to August, 2001 were analysed for microcystins only using HPLC and MALDI-TOFMS according to the method of Harada *et al.* (1996). Lyophilised bloom samples were extracted with 5% acetic acid, centrifuged and the supernatant applied onto ODS - octadecyl silanised cartridge (Baker, Phillipsburgh, NJ, USA) for clean-up. The eluate was dried and MCs determined by HPLC with a photodiode array (PDA) detector (238 nm) under isocratic condition using a 150 X 4.6 mm Cosmosil 5C18-AR column (Nacalai Tesque, Kyoto, Japan). MALDI-TOFMS was performed as described by Imanishi and Harada (2004). Sample eluates were loaded onto the crystallised α-cyano-4hidroxy-cinnamid acid (CHCA) matrix solution on the MALDI sample plate. After drying, mass spectra were acquired on an Applied Biosystems Voyager DE-PRO mass spectrometer (Framingham, MA, USA) operated in the reflectron delayed-extraction mode.

Statistical analysis

The overall relationships among physicochemical parameters data of the 39 aquaculture samples were analysed by multivariate analysis applying standard recognition method through main components analysis. Orthogonal Varimax rotation, with maximised variance weight, was used to improve the component data analysis (Scarmínio, 1989).

Results and discussion

The 56 freshwater samples collected June 1999 to October 2000 from Itaipu lake, and Capivara and Tres Bocas dams, were assayed for *Cyanophiceae* and for total MC contamination. Potentially toxigenic species of *Cyanophiceae*, including *Anabaena* spp., *Microcystis* spp. and *Oscillatoria* spp., were detected at almost every sampling site (Table 1). In terms of *Cyanophiceae* population density, the highest (6,887 cells/ml) was recorded at the Primeiro de Maio sampling site of Capivara dam. Earlier studies have shown that *Cyanophiceae* cell proliferation was mainly dependent on the nitrogen and phosphorus level in the water body (Carmichael, 1996). Our evaluation of eutrophication profiles was therefore, focused on these two parameters. Table 2 shows that the Primeiro de Maio site had the highest total nitrogen (89 mg/l) and phosphorus (0.32 mg/l) recorded. Together with the high temperature (32 °C) and chlorophyll-*a* content (49.53 mg/l) of this sampling site, the environment was the most favourable for *Cyanophyceae* proliferation and microcystin production.

Inspite of the high cyanobacteria counts at Primeiro de Maio, overall microcystins levels in freshwater collected in the northern Parana (Capivara and Tres Bocas dams) only ranged from non-detectable to 1.7 μg/l with mean contamination levels of 0.099 to 0.304 μg/l and a 50% positive sample rate. These MC levels were far much less than those observed from sampling sites of Itaipu lake in the southwest where levels were up to 240 μg/l. At one Itaipu lake sampling site, Santa Terezinha de Itaipu, all samples were positive for MCs and the mean contamination level was the highest recorded (30.236 μg/l). At other Lake Itaipu sampling sites, more than 80% of the samples contained MCs, with mean contamination of 3-17 μg/l (Table 1). The differences in microcystin content observed in this study could be attributed to two major factors. First, although the physicochemical parameters recorded for Capivara dam were generally similar to those from Itaipu lake, there were occasions of much cooler environmental temperature at the time of sampling in Itaipu lake (3-6 °C) than Capivara dam (16-18 °C) (Table 2). Secondly, the data for Itaipu lake includes bloom samples which would normally have higher microcystin levels. And indeed,

Table 1. MCs by IC-ELISA in freshwater from Itaipu lake, Capivara and Tres Bocas dams, State of Parana, Brazil (June 1999 to October 2000).

Sampling sites		Number of samples	Count (cell/ml) (range)	*Cyanophiceae*	MC (µg/l)	MC average (µg/l)	MC Positive (%)
Itaipu lake	Santa Helena	7	6.0 – 4,583.3	*Microcystis* spp., *Anabaena* spp.	ND – 21*	3.113	87.5
	São Miguel do Iguaçu	7	0.4 – 832.2	*Microcystis* spp., *Anabaena* spp.	60 – 86*	17.112	87.5
	Santa Terezinha de Itaipu	7	33.2 – 1,100.9	*Microcystis* spp., *Anabaena* spp.	<50 – 240*	30.236	100.0
Capivara dam	Primeiro de Maio	7	98.9 – 6,887.2	*Microcystis* spp., *Anabaena* spp., *Oscillatoria* spp.	ND – 1.7	0.304	62.5
	Capivara	7	14.1 – 1,786.4	*Microcystis* spp., *Anabaena* spp., *Oscillatoria* spp.	ND –1.2	0.249	50.0
	Sertanopolis	7	43.2 – 1,327.3	*Anabaena* spp., *Oscillatoria* spp.	ND -0.38	0.099	50.0
Tres Bocas adam	Londrina	7	4.0 – 5,722.1	*Oscillatoria* spp.	ND -0.134	0.040	50.0

* - bloom samples are included.

IC-ELISA - detection limit of 0.050 µg/l, using monoclonal antibody produced by M8H5 hybridoma.

ND - not detected.

Table 2. Physicochemical analysis of freshwater in Itaipu lake, Capivara and Tres Bocas dam, State of Parana, Brazil (June 1999 to October 2000).

Site		Number of samples	T (°C)	Alkalinity (mg/l)	pH	Total Nitrogen (mg/l)	DO (mg/l)	COD (mg/l)	PO_4 (mg/l)	Chlorophyll a (mg/l)
Itaipu lake	Santa Helena	7	5 - 31	9.6 - 16	6.8 - 7.5	3.59 - 47.94	1.2 - 7.1	4.3 - 53.9	0.0 - 0.14	0.85 - 23.76
	São Miguel de Iguaçu	7	3 - 28	11.6 - 19	6.6 - 7.4	2.19 - 59.93	2.5 - 6.9	1.5 - 27.8	0.0 - 0.25	1.15 - 9.72
	Santa Teresinha de Itaipu	7	6 - 29	8.0 - 11.3	6.7 - 7.4	2.99 - 65.92	2.6 - 7.0	3.0 - 16.4	0.0 - 0.11	0.0 - 10.04
Capivara dam	Primeiro de Maio	7	18 - 32	9.0 - 14.6	7.1 - 7.6	2.59 - 89.0	2.7 - 6.0	0.8 - 20.0	0.04 - 0.032	0.5 - 49.53
	Capivara	7	18 - 30	0.0 - 17.6	7.1 - 7.5	2.39 - 77.9	2.1 - 6.8	4.1 - 19.5	0.0 - 0.17	0.0 - 25.36
	Sertanopolis	7	16 - 29	6.0 - 15.0	6.8 - 8.1	2.39 - 83.9	2.7 - 6.7	4.87 - 5.5	0.0 - 0.30	0.6 - 41.3
Tres Bocas dam	Lordrina	7	17 - 31	13.3 - 8.5	6.6 - 7.4	2.29 - 61.92	2.4 - 8.2	4.8 - 4.67	0.0 - 0.18	0.0 - 7.95

DO – Dissolved oxygen.

COD – Chemical Oxygen Demand [COD is used to estimate the amount of organic pollutants in water, which indicates the mass of oxygen consumed per liter. In this case, COD was used to measure the eutrophication level of fresh water].

all samples positive for microcystins in the HPLC/MALDI-TOFMS assay (Table 3) were from the month of April 2000 when a *Cyanophiceae* bloom had occurred.

Seven of the 14 'bloom' samples analysed by HPLC/MALDI-TOFMS were found to contain microcystin-LR (MC-LR), one of the most common and most toxic among the microcystins (range, 8.47 to 48.13 μg/100 mg dry cell). In addition, 3 contained microcystin-RR (MC-RR) and 3 contained microcystin-YR (MC-YR) (Table 3), indicating predominance of MC-LR producing strains. When the HPLC-MS data (Table 3) is compared to that from the IC-ELISA (Table 1), some discrepancies in the occurrence of microcystins in the samples tested are evident. The discrepancies may be a result of false positive reactions in the IC-ELISA system, a high sensitivity feature common with most immunoassays. Although high sensitivity is desirable, the disadvantage of immunoassay has been the risk of undesired cross-reactions, resulting in higher than actual values and false-positive data.

Table 3. Microcystin analysis of lyophilised bloom samples by HPLC and MALDI-TOF (February 2000 to July 2001). (The number of samples per site depended on frequency of blooming).

Location	Site	Sampling date	MC-RR (μg/100g dry cell)	MC-YR (μg/100g dry cell)	MC-LR (μg/100g dry cell)
Itaipu lake	Santa Helena	March 2000	(-)	7.93	11.56
		April 2000	(-)	(-)	(-)
	Santa Teresinha de Itaipu	May 2000	(-)	(-)	48.13
		July 2000	(-)	(-)	20.28
	São Miguel de Iguaçu	May 2000	(-)	(-)	(-)
		May 2000	(-)	(-)	(-)
Capivara dam	Primeiro de Maio	February 2000	(-)	(-)	(-)
		May 2000	(-)	(-)	(-)
		February 2000	(-)	(-)	(-)
	Capivara	April 2000	4.08	(-)	8.75
		October 2000	(-)	(-)	16.67
Barra do Iraí	Curitiba	July 2001	32.14	38.71	21.15
		July 2001	45.14	52.47	(-)
Aquaculture	Londrina	February 2000	(-)	(-)	8.47

The occurrence of potentially toxigenic *Cyanophyceae* in a water body emphasises the importance of monitoring for eutrophication of agri-industrial areas, where changes in the ecosystem by activities further inland or farther away could trigger bloom proliferation. For example, in Parana state, Brazil, aquaculture has been encouraged for the production of low-cost healthy food. However, such activity could favour eutrophication and increase micro algal proliferation. When fish in a water body come in contact with cyanotoxins, undesired consequences to public health may arise due the probable contamination of edible portions of the fish.

Monitoring of fish ponds (aquaculture, Londrina, Parana state) from October 2002 to October 2003 showed eutrophication along with *Cyanophyceae* blooms. Although at low concentration (average, 0.275 μg/l), MCs were detected in 95% of the samples tested (data not shown). Samples from aquaculture B (Figure 1) showed levels >1μg/l, *i.e.* above the guideline for drinking water recommended by WHO (1998) and the Brazilian Sanitary Vigilance Agency - ANVISA (2004). Figure 1 also illustrates that high MC concentration in aquaculture B (December 2002 and January 2003) occurred transiently at peak times, followed by a rapid decrease to non-quantifiable levels. Therefore, the MC level detected in Itaipu lake (Table 1) may be a result of bloom related peak cyanobacteria proliferation periods that would normally have been followed by a decrease in the level. Where local freshwater for human consumption is exposed to aquaculture discharge, preventative measures based on heightened awareness and regular monitoring and control would prevent human health hazards.

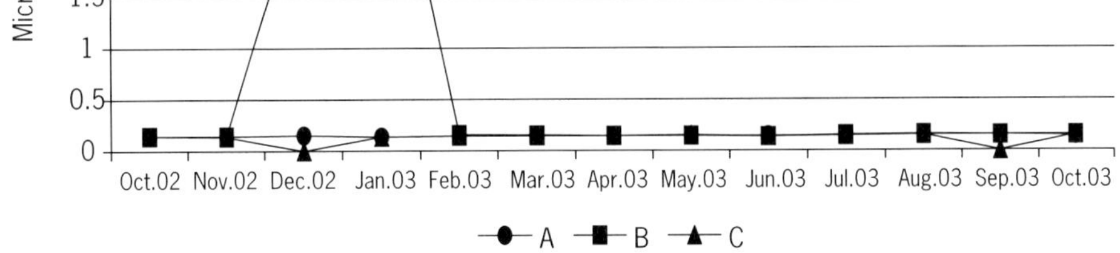

Figure 1. Microcystin level in aquaculture A, B and C monitored by IC-ELISA in Londrina county at northern Parana state, October 2002 to October 2003 (QuantiPlate® ELISA test, EnviroLogix Inc., detection limit of 0.147 μg/l).

Acknowledgments

CNPq, CAPES, JICA and Araucaria Foundation supported this research.

References

Agostinho, A.A.; Lulio Jr., H.F., and Petrere Jr., M., 1994. Itaipu reservoir (Brazil): impacts of the impoundment of the fish fauna and fisheries. In: COWX. I.G. (ed.). *Rehabilitation of Freshwater Fisheries*. Osney Mead. Oxford: Fishing New Books, pp.171-184.

American Public Health Association, 1989. *Standard Methods for the Examination of Water and Wastewater*. American Health Association, American Water Works Association, and Water Pollution Federation. 18th ed., Washington, DC, 491 pp.

ANVISA, 2004. Agência Nacional de Vigilância Sanitária. Ministério da Saúde. *Portaria 518*, 2004.

Beasley, V.R., Cook, W.O., Dahlem, A M., Hooser, S.B., Lovell, R.A., and Valentine, W.M., 1998. Algae intoxication in livestock and waterfowl. *Vet. Clin. North AM. [Food Anim Pract]* 5: 345-361

Carmichael, W.W., 1996. Toxic Microcystis and environment. In:. Watanabe, M., Harada, K-I., Carmichael, W.W., and Fujiki, H. (Eds.). *Toxic Microcystis*. CRC Press, Boca Raton. Fl., pp. 1-11.

Dawson, R.M., 1989. The toxicology of microcystins. *Toxicon* 36: 953-962.

Dittmann, E., Neilan, A., Erhard, M., Döhren, H., and von Börner, T., 1997. Insertional mutagenesis of a peptide synthetase gene that is responsible for hepatotoxin production in the cyanobacterium *Microcystis aeruginosa* PCC 7806. *Molecular Microbiology* 26: 779-787.

Hashimoto, E.H., Ono, E.Y.S., Santos, M.A., Hayashi, C., Bracarense, A.P.F.R.L., and Hirooka, E.Y., 2003. Bromatology and fumonisin/aflatoxin contamination in feed of aquaculture, Londrina, Parana-Brazil. *Semina* 24: 117-125.

Harada, K., Murata, H., Qiang, Z., Suzuki, M., and Kondo, F., 1996. Mass spectrometric screening method for microcystin in cianobacteria. *Toxicon* 34: 701-710.

Hirooka, E.Y., Pinotti, M.H., Tsutsumi, T., Yoshida, F., and Ueno, Y., 1999. Survey of microcystins in water between 1995 and 1996 in Parana, Brazil using ELISA. *Nat. Toxins* 7: 103-109.

Imanish, S., and Harada, K.-I., 2004. Proteomics approach on microcystin binding proteins in mouse liver for investigation of microcystin toxicity. *Toxicon* 43: 651-59.

Kamogae, M., 2002. *Monitoramento de microcistinas e estudo de parâmetros que favorecem o seu desenvolvimento nas represas de Itaipu, Capivara and Tres Bocas dam.* Doctorate Thesis, State Universtity of Londrina-UEL, Londrina-PR, 111 pp.

Komárek, J., and Anagnostidis, K., 1986. Modern approach to the classification system of cyanophytes 2 - Chrococculus. *Arch. Hidrobiol.* 43(suppl.): 157-226.

Nagata, S., Tsutsumi, T., Hasegawa, A., Yoshida, F., and Ueno, Y., 1996. Enzyme immunoassay for direct determination of microcystins in environmental water. *J. AOAC International*.80: 408-417.

Reynolds, C.S., Jaworski, G.H.M., Cmiech, H.A., and Leedale, G.F., 1981. On the annual cycle of the blue-green algae *Microcystis aeruginosa* Kútz emend. Elekin, *Proceedings and Philosophical Transactions of the Royal Society of London, B. Biological Sciences* 293: 419-477.

Scarmínio, I., 1989. *Desenvolvimento de um Sistema Quimiométrico para Microcomputadores e algumas Aplicações*. Doctorate Thesis. State University of Campinas-Unicamp. Campinas, SP.

Tencalla, F., and Dietrich, D., 1997. Biochemical characterization of microcystin toxicity in rainbow trout (*Oncorhynchus mykiss*). *Toxicon* 35: 583-595.

WHO, 1998. *Guidelines for drinking-water quality*. Second edition, Addendum to volume 2, Health criteria and other supporting information. World Health Organization, Geneva.

Multiresidue LS-MS analysis of ASP and DSP toxins in shellfish: Validation and laboratory QA/QC data

P.T. Holland, P.S. McNabb, A.I. Selwood and R. Van Ginkel

Cawthron Institute, 98 Halifax Street East, Nelson 7001, New Zealand; patrick.holland@cawthron.org.nz

Abstract

Progress is reviewed with validation and routine application of an LC-MS/MS method for highly sensitive and specific determination of ASP, DSP and other lipophilic algal toxins and metabolites in shellfish. Tissue homogenates are blended with 90% aqueous methanol and the centrifuged extracts cleaned-up with a hexane wash. LC-MS uses reversed phase gradient elution with an acidic buffer, electrospray (+ & - ion modes) and triple quadrupole MS-MS detection. The method is quantitative for six key toxins where reference standards are available: azaspiracid-1 (AZA1), domoic acid (DA), gymnodimine (GYM), okadaic acid (OA and OA-esters, following hydrolysis), pectenotoxin-2 (PTX2) and yessotoxin (YTX). Relative response factors are used to estimate concentrations for a range of related toxins and metabolites. The extensive within-lab validation of the method was augmented by comparative studies on toxin extractability and stability, and a small inter-laboratory study. Successive extractions showed that a single extraction with 90% methanol was more effective than a single extraction with 80% methanol for AZA1, YTX and esters of OA and DTX2. There was little effect of time between homogenisation and extraction on levels of most toxins (6 min to 22 hours at 20 °C). However, low levels of PTX2 and 7-epiPTX2 seco acid significantly increased with time. Steaming shellfish prior to tissue homogenisation caused a 2-fold decrease in measured YTX. Eight laboratories generally obtained consistent inter-lab data for toxins in shellfish extracts. Between-lab reproducibility was satisfactory for seven key toxins (HORRAT 0.8 – 1.9). The new method is fit for the purpose of regulating ASP and DSP toxins in shellfish and has been in routine use for over two years. QA/QC data is summarised for 259 batches of shellfish samples from New Zealand public health and industry monitoring programmes. The precision of the method was sustained over extended periods and was consistent with estimates of the uncertainty of measurement from the validations.

Keywords: marine biotoxins, shellfish, LC-MS, validation, inter-laboratory study, QA/QC

Introduction

New Zealand has a large export aquaculture industry based mainly on Greenshell™ mussel (*Perna canaliculus*). Risks from hazardous algal blooms are managed by a comprehensive program that includes routine phytoplankton monitoring and shellfish flesh testing. Over the past 3 years, Cawthron has worked with industry and the New Zealand Food Safety Authority to implement new test methods for shellfish toxins that have superseded traditional mouse-based assays and provided advantages in speed, specificity and sensitivity. These needs were particularly urgent for the broad group of lipophilic 'DSP' toxins where, in addition to okadaic acid and dinophysis toxins, there have been rapid advances in knowledge on yessotoxins, pectenotoxins and several groups of nitrogen-containing polyether toxins (Draisci *et al.*, 2000; Mackenzie *et al.*, 2002). The legislation of the European Union (EU, 2002) has highlighted the problems with mouse bioassays for these toxins in shellfish.

The advantages of LC-MS have been recognised for DSP toxins and methods have been published (Draisci *et al.*, 1999; Goto *et al.*, 2001; Quilliam *et al.*, 2001; Quilliam, 2003). However, the coherent combination of a range of fully validated new test methods to provide comprehensive and high-throughput results has not been documented for any routine marine biotoxin screening program. This paper reviews progress with implementation of an LC-MS method for ASP and DSP toxins within a new program for management of marine biotoxins in New Zealand (Holland *et al.*, 2004), including the full validation and inter-laboratory study (McNabb *et al.*, 2005) and performance over two years of routine operation.

Materials and methods

ASP and DSP toxins by LC-MS/MS

High purity reference standards were available, for six 'Quan' toxins: domoic acid (DA; NRC, Halifax, Nova Scotia, Canada), okadaic acid (OA; NRC, Halifax), pectenotoxin-2 (PTX2; NRC, Halifax), yessotoxin (YTX; AgResearch, Hamilton, New Zealand) gymnodimine (GYM; AgResearch) and azaspiracid-1 (AZA1; Marine Institute, Ireland). The following 'Qual' toxins were partially validated using small amounts of standard materials or crude extracts of contaminated shellfish: desmethyl-azaspiracid (AZA2), methyl-azaspiracid (AZA3), isodomoic acid-C, dinophysistoxin-1 & -2 (DTX1, DTX2), PTX6, PTX11, PTX2-secoacid (PTX2sa, 2 epimers), 45-OHYTX, carboxyYTX, 1a-homoYTX.

A Waters 2790 LC coupled to Quattro Ultima triple quadrupole MS (Waters-Micromass, Manchester) was operated with a Luna 5 µm C18(2) 150 x 2 mm column (Phenomenex, CA) and 0.2 ml/min water+acetonitrile+buffer gradient (13% to 77% acetonitrile at constant buffer composition of 4 mM ammonium hydroxide + 50 mM formic acid). Injection volume was 10 µl and the run time was 25 minutes. Electrospray ionisation (positive/negative ion switching) with multiple reaction monitoring (MRM) was used for detection of each toxin using a specific daughter ion from the molecular anion or cation. See McNabb *et al.* (2005) for LC-MS parameters. The system was calibrated daily using a set of five Quan standard mixes (5 – 200 ng/ml in 80% methanol). Qual toxins were calibrated using relative response factors (RRFs) to the related Quan toxins.

Whole shellfish tissues were homogenised and a subsample (2g) was blended with 18 ml methanol+water (9:1 v/v) and centrifuged. An aliquot was washed with hexane and the lower layer transferred to an autosampler vial for LC-MS. Ester forms of OA/DTX1/DTX2 were determined after conversion to parent toxins by alkaline hydrolysis of the crude methanolic extract (Mountfort *et al.,* 2001).

Extractability

Extractability of incurred toxins was established using homogenised whole mussel tissues naturally contaminated by a range of toxins. Sub-samples were extracted using either 2.0 g with 18.0 ml 90% methanol or 4.0 g with 16.0 ml 80% methanol (three replicates each). Following centrifugation, all liquid was decanted off the pellet. The tube plus pellet was weighed and the pellet re-extracted with the same solvent mixture. This was repeated once more, and toxin levels in the three successive extracts were determined by the LC-MS method. The quantities of toxins in the extracts were corrected for carryover of liquid in the pellet.

Effects of sample pre-treatment

A bulk sample of Greenshell™ mussel contaminated by several toxins was divided in half and shucked fresh and following steaming. The whole flesh was taken from 15 specimens and homogenised. Sub-samples of steamed shell fish homogenate were extracted immediately. Sub-samples of fresh homogenates were stored for varying periods at 20, and -20 °C prior to extraction. The toxin levels were determined by LC-MS (duplicate sub-samples).

Inter-laboratory study

Ten participating laboratories received samples as aliquots of five calibration standard mixtures and four mussel extracts (blank and three different samples containing a range of toxins). Data was received from eight laboratories consisting of calibrations and solution concentrations for toxins found in the mussel extracts (ng/ml, triplicate injections). The concentration data was analysed for precision according to ISO 5725 (1994).

QA/QC data

Each batch of 10-20 routine samples included the following QC samples run through the full method: a blank shellfish homogenate, a fortified blank (five Quan toxins at 0.1 mg/kg each and a duplicate of a selected sample in the batch. The QC data from 259 batches of routine shellfish samples tested by the ASP/DSP LC-MS method were examined for precision.

Uncertainty of measurement

The equation for the concentration of a toxin in shellfish was formulated as:

$$L = F_{rep} \times F_{hom} \times \frac{A \times RF \times V}{F_{rec} \times M \times 1000} \text{ mg/kg}$$

Where:

F_{rep} = Repeatability factor for toxin by LC-MS (1.0 with associated uncertainty).

F_{hom} = Sample homogeneity factor (1.0 with associated uncertainty).

A = Area counts for LC-MS peak of toxin.

RF = Response factor for toxin (area counts/ml/ng) – slope of linear calibration function for Quan toxins; assigned for Qual toxins, with associated uncertainty.

V = Volume of extraction solvent (18 ml) + sub-sample moisture.

F_{rec} = Recovery factor for toxin with associated uncertainty.

M = Sub-sample weight (2.0 g).

Estimates of the uncertainty of measurement were made following procedures outlined by Eurachem (2000) and used with a previous LC-MS method (Holland *et al.,* 2003). In summary, the uncertainty components of this equation were assigned using the precision and other data from the within-laboratory validation of the method (Holland and McNabb, 2002; McNabb *et al.,* 2005). An overall estimate of the method uncertainty for each toxin was made by combining the components using the square root of the summed variances.

Results and discussion

The within-lab validation of the method established excellent short-term performance for the determination of the lipophilic toxins and domoic acid in shellfish (Holland and McNabb, 2002; McNabb *et al.*, 2005). The small inter-laboratory study gave acceptable precision for most toxins (HORRAT 0.8-1.9) with excellent data for the key toxins OA and DTX2 (Holland and McNabb, 2003; McNabb *et al.*, 2005). This was encouraging for routine use of the method considering that most laboratories had little experience with the method and that a range of instruments were used including some with single quadrupole mass analysers.

Additional studies were carried out on the stability and extractability of naturally incurred toxins. Figure 1 summarises the recovery of a number of toxins from two samples using three sequential extractions with either 80% or 90% methanol. A single extraction with either solvent was efficient for PTX2 seco acid, OA, DTX2 and polar metabolites of YTX. However, AZA1 and ester forms of OA and DTX2 gave lower recoveries with 80% methanol. YTX was not efficiently extracted by either solvent (mean recovery 75% by a single extraction).

In the sample pre-treatment experiments, low levels of GYM, OA, OA-esters, PTX2 and YTX (15-100 µg/kg) and higher levels of PTX2-SA in a mussel homogenate showed no significant changes after 72 hours at -20 °C. Similarly levels of most toxins did not change over 22 hours at 20 °C. However, the level of PTX2 approximately doubled and 7-epiPTX2-Sa increased from 16% to 25% of PTX2-SA, although PTX2-SA did not increase significantly (Figure 2). These changes indicate that digestive enzymes released into the homogenate are capable of not only facilitating lactone hydrolysis and epimerisation but also the back reaction to PTX2. Alternatively other unidentified conjugates of PTX2 are being transformed to PTX2. The changes in toxin levels are minor in normal operation of the method where the time-frame between homogenisation and extraction is 10-30 minutes or if homogenates are stored at -20 °C prior to analysis. Steaming of mussels has been proposed as an alternative to fresh shucking for analysis. However, tests on this sample showed that levels of extractable YTX decreased from 84 to 48 µg/kg (57%) after steaming, although levels of the other toxins were similar to those in fresh homogenate (taking into account moisture loss with steaming).

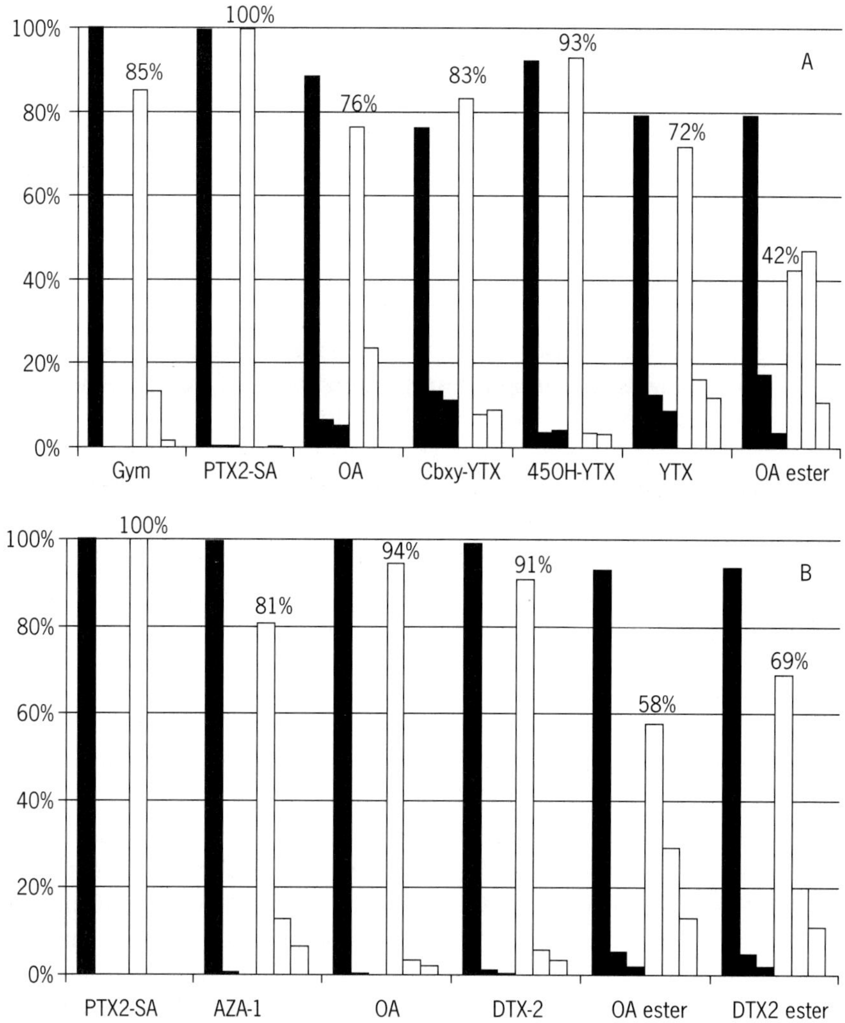

Figure 1. Extraction of toxins from contaminated mussel. Percentage of total amount of toxin extracted from whole tissue homogenates by three successive extractions with aq. methanol: ■ 2g + 18 ml 90% methanol; □ 4g + 16 ml 80% methanol. A. New Zealand Greenshell™ mussel; B. Irish blue mussel.

The precision characteristics of the method are dominated by the repeatability component, similar to other LC-MS methods for biotoxins validated in our laboratory (Holland *et al.*, 2003). This is attributed to short-term variations in levels

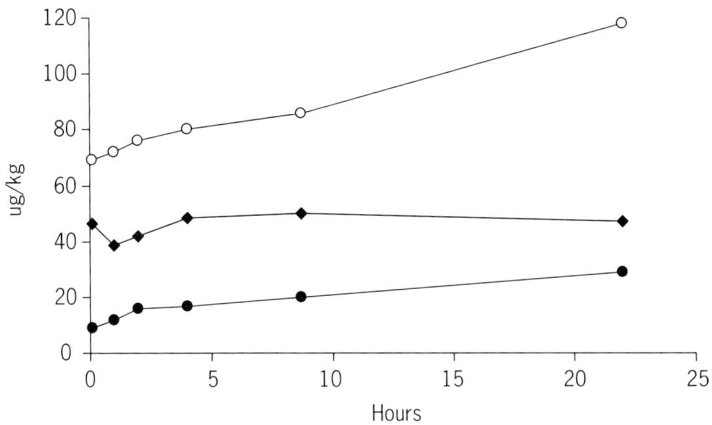

Figure 2. Effect of time of standing (20 °C) of contaminated Greenshell™ mussel homogenate on levels of pectenotoxins extracted by 90% methanol and determined by LC-MS. ● PTX2; ◆ PTX2-SA (÷10); ○ 7-epiPTX2-SA.

of coeluting components influencing the ES ionisation efficiency of the toxins. To better understand the longer term performance of the method, an analysis was made of the uncertainty of measurement (Table 1). The overall uncertainty was estimated by adding to the repeatability further uncertainties for the recovery, calibration, homogeneity and variation in ES enhancement/suppression. This provided a good to fair correspondence to the reproducibility data gathered from the within-lab studies (Table 1). However, the measured reproducibilities (RSDs 8% to 24% depending on toxin and level) do not reflect some of the true uncertainties arising from calibration such as purity of Quan standards or use of assigned response factors for Qual toxins. There was also a good correspondence for OA/DTX2 and YTX between the estimated method uncertainty and the RSDs found in the interlaboratory study (Table 1) but not for some of the other toxins.

The QA/QC data gathered during routine operation of the method provide a good guide to long term performance. For 259 batches of shellfish analysis tested over two years, the mean recoveries were very consistent for the five Quan toxins in the spike QC sample. The grand mean recoveries were 84% to 103% with RSDs of 17% to 20% (Table 2). This is very similar to the reproducibility data gathered in the short term within-laborartory studies and demonstrate that this validation performance can be maintained.

Table 1. Uncertainty of measurement for LC-MS analysis of toxins in shellfish and measured reproducibility from within- and inter-laboratory studies. Estimated uncertainty 'U' as coefficient of variation.

	Repeatability [a]	Recovery		Calibration		ESI e/s [b]	Total [c]	Reproducibility W [a]	Reproducibility I [d]
	U	F	U	F	U	U	U	CV	CV
0.05 – 0.25 mg/kg									
DA	0.170	1.0	0.056	1.0	0.032	0.10	0.21	0.24	
GYM	0.096	0.91	0.035	0.95	0.042	0.05	0.12	0.10	0.42
PTX2	0.159	1.0	0.049	0.95	0.042	0.10	0.23	0.16	0.48
AZA1	0.112	0.98	0.032	0.95	0.042	0.10	0.16		
OA	0.145	0.92	0.051	1.0	0.032	0.05	0.17	0.17	0.20
YTX	0.142	0.76	0.046	0.95	0.042	0.10	0.19	0.15	
45OHYTX	0.067	0.79	0.022	1.0	0.20	0.10	0.24	0.08	
0.5 – 2 mg/kg									
DA	0.046	1.0	0.044	1.0	0.032	0.10	0.13	0.05	0.23
GYM	0.076	0.93	0.022	0.95	0.042	0.05	0.11	0.08	
PTX2	0.089	1.0	0.023	0.95	0.042	0.10	0.18	0.10	
AZA1	0.096	0.98	0.032	0.95	0.042	0.10	0.15		0.34
OA/DTX2	0.096	0.91	0.039	1.0	0.032	0.05	0.12	0.10	0.12
YTX	0.126	0.77	0.044	0.95	0.042	0.10	0.17	0.13	0.18
PTXsa [e]	0.074	1.51	0.034	1.0	0.33	0.15	0.37	0.12	

[a]F = 1.0. Repeatability and reproducibility data from within-laboratory validation (Holland & McNabb 2002; McNabb *et al.*, 2005).
[b]ESI enhancement/suppression variations during running of batch (assumed F=1.0).
[c]Including U = 0.03 for homogeneity of sample as prepared for sub-sampling (F = 1.0).
[d]Reproducibility data from inter-laboratory study (Holland and McNabb 2003; McNabb *et al.*, 2005).
[e]Sum of epimers.

Table 2. Recoveries of toxins for QC samples (blank shellfish fortified at 0.10 mg/kg) run with every batch of routine samples over two years. Mean recovery for each 3 month interval (% ±SD). N is number of sample batches analysed in the interval.

Interval	N	Domoic acid		Gymnodimine		Pectenotoxin-2		Okadaic acid		Yessotoxin	
		Rec. %	SD	Rec. %	SD	Rec. %	SD	Rec. %	SD	Rec. %	SD
Apr-June 2002	21	83.5	15.8	88.4	16.0	103.3	28.4	82.5	14.6	81.8	9.4
Jul -Sept 2002	34	90.1	20.4	95.1	14.8	105.8	28.2	95.5	11.5	87.3	13.7
Jan –Mar 2003	25	81.6	17.0	108.6	26.7	97.0	20.5	106.7	20.7	75.2	22.6
Apr-June 2003	25	76.4	12.4	91.1	12.3	103.2	13.3	97.3	18.8	76.9	16.8
Jul -Sept 2003	27	85.7	10.8	107.9	10.9	109.6	17.4	87.3	16.6	89.6	16.0
Oct–Dec 2003	38	82.2	12.4	99.1	13.2	102.5	15.0	85.2	10.2	82.2	15.2
Jan –Mar 2004	45	93.5	14.0	121.3	18.8	113.2	20.0	101.4	18.2	95.9	19.6
Apr-June 2004	44	78.4	14.3	99.3	19.7	92.5	18.6	85.0	14.0	87.4	14.8
Grand Mean	259	84.4	14.9	102.7	17.2	103.4	20.8	92.6	15.9	85.7	16.4

Conclusions

The thorough within-laboratory validation, inter-laboratory study and long-term QA/QC data for this LC-MS method for ASP and 'DSP' toxins have established that the technology can provide reliable, high-precision data for biotoxins in shellfish within routine monitoring programs. The method has provided a sound basis for enhanced management of HAB events through the faster sample turnaround, higher sensitivity and greater specificity. It has also made a major contribution towards elimination of small animal testing for shellfish monitoring.

Acknowledgments

Technical assistance was provided by Veronica Beuzenberg and Navreet Kaur. We are indebted to the following colleagues for provision of purified algal toxins and advice: Dr Philipp Hess, Lincoln MacKenzie, Dr Christopher Miles, Dr Michael Quilliam, Dr Masayuki Satake, Dr Toshiyuki Suzuki and Professor Takeshi Yasumoto. We are also grateful for the willing collaboration of the participants in the inter-laboratory study.

References

Draisci, R., Palleschi, L., Giannetti, L., Lucentini, L., James, K.J., Bishop, A.G., Satake, M., and Yasumoto, T., 1999. New approach to the direct detection of known and new DSP toxins in mussels and phytoplankton by LC-MS. *J. Chromatogr. A* 847: 213-221.

Draisci, R., Lucentini, L., and Mascioni, A., 2000. Pectenotoxins and yessotoxins: chemistry, toxicology, pharmacology, and analysis. In: *Seafood and Freshwater Toxins: Pharmacology, Physiology, and Detection,* Botana, L.M. (Ed.), Marcel Dekker, Inc, New York, pp. 289–324.

EU, 2002. Commission decision 2002/225/EC laying down detailed rules for the implementation of Council Directive 91/492/EEC as regards the maximum levels and the methods of analysis of certain marine biotoxins in bivalve molluscs. *Offic. J. Eur. Commun.* L75: 62–64.

Eurachem, 2000. *Quantifying Uncertainty in Analytical Measurement,*. Ellison, S.L.R., Rosslein, M. and Williams, A. (Eds), EURACHEM /CITAC guide, 2nd edition.

Goto, H., Igarishi, T., Yamamoto, M., Yasuda, M., Sekiguchi, R., Watai, M., Tanno, K., and Yasumoto, T., 2001. Quantitative determination of marine toxins associated with DSP by LC-MS. *J. Chromatogr. A* 907: 181-189.

Holland, P.T., and McNabb, P., 2002. *Within-laboratory validation of method 40.105: ASP and DSP toxins in shellfish by LC-MS.* Cawthron Institute, Nelson, New Zealand, Report No. 667, 75 pp.

Holland, P.T., and McNabb, P., 2003. *Inter-laboratory Study of an LC-MS Method for ASP and DSP Toxins in Shellfish.* Cawthron Institute, Nelson, New Zealand, Report No. 790, 25 pp.

Holland, P.T., McNabb, P., Selwood, A.I., and Neil, T., 2003. Amnesic shellfish poisoning toxins in shellfish – estimation of the uncertainty of measurement for an LC-MS/MS method. *J. AOAC. Int.* 86: 1095-1100.

Holland, P.T., McNabb, P., Selwood, A.I., Page, T, Bell, K., and Mackenzie, L., 2004. Marine biotoxin monitoring of New Zealand shellfish - a new management programme based on LC-MS. In: *Harmful Algae Management and Mitigation*, Hall, S., Etheridge, S., Anderson, D., Kleindinst, J., Zhu, M., and Zou, Y. (Eds.). Proc. 2nd HAMM Conference Qingdao, PRC, Nov. 2002. Publication #204-MR-04.2, APEC, Singapore, pp. 165-170.

ISO 5725-2, 1994. *Accuracy (trueness and precision) of measurement methods and results. Part 2. Basic methods for the determination of the repeatability and reproducibility of a standard test method.* International Standards Organisation, Geneva.

Mackenzie, L., McNabb, P., Holland, P.T., Beuzenberg, V., and Selwood, A., 2002. Complex toxin profiles in phytoplankton and Greenshell mussels (*Perna canaliculus*) revealed by LC-MS/MS analysis. *Toxicon* 40: 1321-1330.

McNabb, P., Selwood, A.I., and Holland, P.T., 2005. Multiresidue method for determination of algal toxins in shellfish: Single-laboratory validation and interlaboratory study. *J. AOAC. Int.* 88: 761-772.

Mountfort, D.O., Suzuki, T., and Truman, P., 2001. Protein phosphatase inhibition assay adapted for determination of total DSP in contaminated mussels. *Toxicon* 39: 383-390.

Quilliam, M.A., Hess, P., and Dell'Aversano, C., 2001. Recent developments in the analysis of phycotoxins by LC-MS. In: *Mycotoxins and Phycotoxins in Perspective at the Turn of the Century*, de Koe, W.J., Sampson, R.A., van Egmond, H.P., Gilbert, J., and Sabino, M.M., (Eds.), W.J. de Koe, Wageningen, The Netherlands, pp. 383-387.

Quilliam, M.A., 2003. The role of chromatography in the hunt for red tide toxins. *J. Chromatogr. A* 1000: 527-548.

Index

E

F

G

Q

R

S

X

Y

Z